DICTIONARY OF ENVIRONMENT
and
SUSTAINABLE DEVELOPMENT

D1559764

DICTIONARY OF ENVIRONMENT
and
SUSTAINABLE DEVELOPMENT

Alan Gilpin

JOHN WILEY & SONS
Chichester • New York • Brisbane • Toronto • Singapore

Copyright © 1996 by John Wiley & Sons Ltd,
Baffins Lane, Chichester,
West Sussex PO19 IUD, England

National   01243 779777
International (+44) 1243 779777

All rights reserved.

No part of this book may be reproduced by any means,
or transmitted, or translated into a machine language
without the written permission of the publisher.

Other Wiley Editorial Offices

John Wiley & Sons, Inc., 605 Third Avenue,
New York, NY 10158-0012, USA

Jacaranda Wiley Ltd, 33 Park Road, Milton,
Queensland 4064, Australia

John Wiley & Sons (Canada) Ltd, 22 Worcester Road,
Rexdale, Ontario M9W 1L1, Canada

John Wiley & Sons (Asia) Pte Ltd, 2 Clementi Loop #02-01,
Jin Xing Distripark, Singapore 0512

Library of Congress Cataloging-in-Publication Data

Gilpin, Alan.
  Dictionary of environment and sustainable development / by Alan Gilpin.
    p. cm.
  Includes bibliographical references.
  ISBN 0-471-96219-8 (cloth : acid-free paper). — ISBN
0-471-96220-1 (ppk. : acid-free paper)
  1. Sustainable development — Dictionaries.   2. Environmental economics — Dictionaries.
  I. Title.
  HC79.E5G518   1996
  363.7'03 — dc20   95-47707
                                                                                      CIP

British Library Cataloguing in Publication Data

A catalogue record for this book is available from the British Library

ISBN 0-471-96219-8 (Cl)
      0-471-96220-1 (Pr)

Typeset in 8/10pt Times by Laser Words, Madras
Printed and bound in Great Britain by Biddles Ltd, Guildford and King's Lynn

This book is printed on acid-free paper responsibly manufactured from sustained forestation,
for which at least two trees are planted for each one used for paper production.

To Carole, Leslie and Dorothy

Introduction

In early times, the world was large and people few. Today, the world is small and people many. Initially, the prevalent environmental influences and impacts were essentially natural: forest fires, volcanic eruptions, floods, droughts, earthquakes, with dramatic climatic, faunal, and flora variations. Human influences were relatively minor.

With the growth of towns and the increasing urbanization of an expanding population, the afflictions of humanity intensified turning on insanitary dwellings, water pollution, vector breeding, congestion, overcrowding, contaminated and adulterated food, poor hygiene and inadequate waste — disposal procedures. Humanity was wracked with the diseases of smallpox, cholera, typhoid, typhus, dysentery, malaria, yellow fever, tuberculosis, diphtheria, schistosomiasis, and a raft of other diseases. The Black Death of the 14th century, transmitted by the rat flea, killed between one-third and one-half of the European population, Over 1000 towns and villages in England were depopulated. Countermeasures, without germ theory or vaccines, were essentially crude; public health measures slow to evolve. Isolated steps included closing suspect wells, and separating animals from dwellings.

In 1273, the use of coal was prohibited in London as being 'prejudicial to health'. In 1306, a Royal Proclamation prohibited artificers from using sea coal (coal from Newcastle) in their furnaces; the execution of one offender is recorded. A further proclamation during the reign of Queen Elizabeth I made illegal the burning of coal while Parliament was sitting. However, the use of raw coal continued to expand for the great forests were dwindling rapidly and wood was becoming both scarce and dear. In 1648, Londoners presented a petition to Parliament in an attempt to prohibit the importation of coal from Newcastle because of its ill-effects. In 1661, John Evelyn submitted his famous *Fumifugium, or the Smoake of London Dissipated* to King Charles II. He wrote: 'London's inhabitants breathe nothing but an impure and thick mist accompanied with a fuliginous and filthy vapour which renders them obnoxious to a thousand inconveniences, corrupting the lungs so that catharrs, phthisicks, coughs and consumption are more in this city than the whole earth besides.' Eventually, a smoke abatement section was embodied in the Public Health Act 1875, known as the Charter of Public Health, and the first industrial inspectorate was appointed. However, not until the 1952 London smog disaster, claiming up to 4000 lives, were comprehensive national measures adopted.

In the meanwhile, vaccinations, anti-vector measures, better public health measures, the elimination of noxious 'accumulations and deposits', the appointment of environmental health officers, improved food hygiene, and a host of other measures gradually reduced the mortality and morbidity rates in many industrially

Introduction

advanced countries. However, many great cities today in the developing world remain largely unsewered.

Despite many successes and an increasing expectation of life in all countries, environmental problems have tended to intensify. This has led in more recent years to a whole series of United Nations conferences relating specifically to the condition of the human environment and the quality of life. The explosion in world population, the massive movement of people from country to town, increasing industrialization; the burgeoning of the automobile; the increasing widespread use of chemicals and fertilizers; noise problems; hazardous wastes (including radioactive wastes); the contamination of air, water, land, and food; major industrial and marine mishaps; soil erosion, the destruction of forests, overfishing, a general overexploitation of natural resources, and evidence of global climate change with progressive loss of biodiversity, has led to a worldwide recognition that development cannot be sustained in the longer term, in its present form, without a significant depletion of the resource base upon which all development ultimately depends. Industry itself, as well as government and its agencies, have become aware of new responsibilities that go beyond traditional concerns: a duty of care for the planet itself.

This dictionary attempts to identify the many strands of concern that are encompassed today within the term 'environment', with an eye to many progressive steps taken in North America, Europe, Asia and the Pacific, and elsewhere. To assist readers who might like a rapid coverage of the subject, a summary of the key mainstream terms and concepts follows.

Key Terms and Concepts: A User's Guide

Acoustics
Aesthetics
Agenda 21
Air pollution
Allocative efficiency, social
Alternative energy
Alternatives
Amenity
Aral Sea
Assimilative capacity
Atmosphere
Attitudinal changes
Bacteria
Balance of nature
Barriers
Beneficial use
Berlin Climate Conference
Best practicable means
Biodiversity
Biological remediation
Biosphere
Brandt Commission
British Standard 7750
Buffer zone
Capital
Carbon cycle
Carbon dioxide
Carbon tax
Carrying capacity
Chernobyl nuclear catastrophe
Cleaner production
Climate change
Coastal protection
Commission for Sustainable Development, UN
Common property resource
Conservation
Conserver society
Contaminated sites

Conventions (there are many of these)
Cost–benefit analysis
Council of Europe
Council on Environmental Quality, US
Cultural environment
Cumulative effects
Dam
DDT
Debt-for-nature swaps
Deep repositories
Demand management
Demographic transition
Development
Dioxins
Diseases and afflictions
Drinking-water quality guidelines
Drought
Earthwatch
Ecofeminism
Ecofundamentalism
Eco-labelling
Ecological environmental impact assessment (EEIA)
Ecology
Economic development
Economic impact assessment
Economic instruments
Economic system, functions of
Economic welfare
Ecopolitics
Ecosystem
Ecotaxes
Emission trading program
Endangered species
Energy conservation
Environment
Environmental agencies
Environmental audit
Environmental economics

Key Terms and Concepts: A User's Guide

Environmental health impact assessment (EHIA)
Environmental impact assessment (EIA)
Environmental impact statement (EIS)
Environmental management
Environmental planning
Environmental protection policies
Environmentally hazardous chemicals
Eutrophication
Evo-economics
Externality
Fauna impact statement
Fertilizer
Fisheries
Forestry
Geographic information system (GIS)
Global Environmental Facility
Greenhouse effect
Greens
Gross domestic product (GDP)
Habitat
Hazard and risk assessment
Hazardous wastes
Heavy metals
Heritage conservation
Human development index
Hydrological cycle
Hygiene
Ice age
Industrial-waste strategy
Industrialization
Integrated pollution control
Interaction matrix
Intergenerational equity
Intractable wastes
Intragenerational equity
Land capability
Landfill
Land-use planning
Love Canal
Mangroves
Marine park
Maximum sustainable yield
Mekong River Commission
Meterological influences
Methane
Mitigating measures
Motor vehicle pollution
Multiplier
National estate
National park
Natural environment
Natural forest
Natural resource accounting
Nature reserve
Noise
Non-renewable resources
Ocean dumping
Oil spill
Opportunity cost
Ozone layer
Packaging
Photochemical smog
Polluter-pays-principle (PPP)
Pollution-control strategy
Precautionary principle
Principles of sustainability
Programme impact statements
Public health
Quality of life
Radioactive waste
Rainforest
Regulatory impact statement (RIS)
Remote sensing
Renewable resource
Residual
Resource
Review of environmental factors (REF)
Sanitation
Social evolution
Social impact assessment (SIA)
Soil
Solid-waste management
State of environment reports
Strategic EIA (environmental impact assessment)
Sustainable development
Sustainable yield

Thermal pollution
Times Beach
Trade wastes
UN Conferences (on the Human Environment and related issues)
UN Environment Program (UNEP)
User pays
Value
Visual pollution

Waste management
Water pollution
World Bank
World Commission on environment and development
World conservation strategy
World Health Organization (WHO)
World Heritage List
World population

Abbreviations

CITES	Convention on international trade in endangered species of wild fauna and flora
DA	Development application
ECE	Economic Commission for Europe (UN)
EHIA	Environmental health impact assessment
EIA	Environmental impact assessment
EIS	Environmental impact statement
ESCAP	Economic and Social Commission for Asia and the Pacific (UN)
ESD	Ecologically sustainable development
EU	European Union
GATT	General Agreement on Tariffs and Trade
GDP	Gross domestic product
GEF	Global Environmental Facility
GEMS	Global Environmental Monitoring system
GIS	Geographic information system
km	Kilometre
m	Metre
mg	Milligram
NGO	Non-government organization
ppb	Parts per billion
ppm	Parts per million
SIA	Social impact assessment
SD	Sustainable development
t	tonne
UN	United Nations
UNEP	United Nations Environment Program
UNESCO	UN Educational, Scientific and Cultural Organization
UNICEF	UN Children's Fund
US/USA	United States of America
WHO	World Health Organization
WMO	World Meteorological Organization

A

abatement A reduction in amount, degree or intensity, for example, the abatement of a public or private nuisance either voluntarily or through legal action, or the abatement of pollution from a particular source either through the adoption of appropriate technical measures, change of technique, or cessation of activities.

Aberfan A mining village in Mid Glamorgan, Wales; in 1966, the liquefaction of coal waste overwhelmed a school and houses. Of the 144 dead, 116 were children.

abiotic Non-biological factors, such as light, temperature, humidity, and other parameters, not directly dependent on the activity of living organisms. (See BIOTIC.)

absorption The passing of a substance or force into the body of another substance. A liquid may be absorbed and held by cohesion or capillary action in the pores of a solid, or gaseous molecules may be held between the molecules of a liquid. The characteristic waves of heat or light radiation may be retained by a solid, liquid, or gas, being transformed into either kinetic energy or greater molecular vibrations when the temperature of the absorbing substance rises, or into excited atoms or molecules when the substance becomes fluorescent. (See ADSORPTION.)

absorption coefficient (1) The volume of gas, measured at normal temperature and pressure, dissolved by unit volume of a liquid under a pressure of one atmosphere. (2) The degree to which a substance will absorb radiant energy. (3) If a surface is exposed to a sound field, the ratio of the sound energy absorbed by the surface to the total sound energy which strikes it, if it absorbs 70% of the incident energy, the absorption coefficient is 0.70.

absorption tower (1) A structure in which gaseous or particulate pollutants are removed from an air stream, using a scrubbing liquid. The polluted air stream is passed slowly either over a surface flooded with scrubbing liquor or into an air space saturated with scrubbing liquor. Absorption towers, often known as scrubbers, have a low capital cost but generally high operating costs. (2) A structure commonly found in chemical factories for the manufacture, for example, of sulphuric acid in which sulphur dioxide/sulphur trioxide are absorbed by liquid.

abstraction The act or process of removing or separating; for example, the withdrawal of water from a source of supply either temporarily for, say, cooling purposes, or permanently for, say, irrigation or domestic purposes.

abyssal zone The cold, dark ocean depths at some 2000-6000 m below the EUPHOTIC ZONE; no PHOTOSYNTHESIS can occur at this level and animal life is sparse.

Acanthamoeba A genus of free-living amoebae common in soils and water. Some species can cause encephalitis, particularly in immuno-compromised people, or corneal infection among contact-lens wearers. They are resistant to normal disinfection.

acaricide A pesticide for killing mites, ticks, and spiders (order Acarina), as distinct from insects. Acarina are invertebrate animals with eight legs.

acceptable daily intake (ADI) The daily intake of a chemical which, during an entire lifetime, appears to be without appreciable risk to the health of the consumer on the basis of the known facts at the time. ADIs are expressed in milligrams of chemical per kilogram of body weight (mg/kg).

accessibility The ease by which the public, or a segment of it, may reach places of employment, shopping centres, community facilities, and services including higher-order services, expressed in terms of time, trouble, effort, or cost. One of the objectives of planning is to improve accessibility. However, in some circumstances it may reduce accessibility through, say, the establishment of a GREEN BELT, where the travelling times of those who live beyond it, but work within it, are increased in consequence.

acclimatization Physiological variations induced in an organism following exposure to changed environmental conditions; the process of adapting to changed ABIOTIC factors in environmental conditions by phenotypic variation, as distinct from genetic variation. Such changes may arise from differences in altitude or climate, or variations in sources of nutrition.

accommodation The location of a population within a specific geographical area or HABITAT.

accounting costs Those recorded costs which find their way into the accounts of an enterprise, company, or development. They do not include those costs of a social or environmental nature caused by the activity or by the subsequent depletion of a natural resource, unless remedial measures are specifically adopted and paid for by the enterprise. Accounting costs are subject to internal audit and appear in annual reports; an ENVIRONMENTAL AUDIT takes a broader view of a company's activities.

accretion Any increase or growth in size by gradual additions, for example, ice accretions on aircraft wings, or the clustering or adhering together of a number of small particles to form a larger single entity or agglomerate. Accretion or agglomeration has been associated with small particles in chimney stacks.

Achelous River dams, Greece In 1994–95, opposition by conservationists to the plans of a British, French, and Italian consortium to construct a hydroelectric and irrigation project on the Achelous River. It has been argued that the three dams and tunnel, located between the western Pindus Mountains and the Plains of Thessaly, would destroy a wetland protected under EU law and the Ramsar Convention, and would partly drain the Mesolongion Lagoon.

acid A substance which, in solution in water, forms hydrogen ions with a pH VALUE of less than 7.

acid chimney A chimney or stack in which the temperature of the flue or process waste gases is below the ACID DEW POINT of the gases, so that acidic condensates are formed. Such chimneys require special acid-resisting linings or insulation. The presence of acid tends to promote the formation of ACID SOOT or particles.

acid deposition The falling of acid substances onto land or water. When fossil fuels are burnt, the oxides of sulphur and nitrogen are released to the atmosphere. They may react with water vapour to form acids and if these acids adhere to particulates in the air, ACID SOOT may be deposited close to the source of pollution. This is known as acid deposition. On the other hand, the acid may be absorbed by rain, snow, or hail being carried far from the source of pollution. This is known as ACID RAIN.

acid dew point The temperature at which sulphuric acid appears as a condensate in a stack or duct as a flue gas containing sulphur trioxide cools.

acid mine drainage The discharge of mine water which has become contaminated with the sulphur compounds of ores or coal seams, creating acidity. Acid mine drainage can make ground water, streams, and lakes unsuitable for recreation, domestic water supply, industry, and agriculture. It often becomes necessary to neutralize the drainage water.

acid rain The acidification of rain associated with the combustion of fossil fuels: coal, oil, and natural gas. The constituents of flue gases which contribute to the acidity of rain are oxides of sulphur and nitrogen. Geographical areas significantly affected by acid rain include the northeast of the USA and Ontario, Canada, parts of Scandinavia, notably Sweden, and southern Norway. Other parts of Europe are affected by acid rain, possibly contributing to the dieback of the Black Forest in Germany. Rain water is normally mildly acidic due to the presence of carbon dioxide in the atmosphere. The oxides of sulphur come mainly from the combustion of coal and heavy fuel oil in industrial plant and power stations, while the oxides of nitrogen arise both from stationary sources and transportation, most notably the automobile. Negotiations under the auspices of the UN Economic Commission for Europe began in Geneva in 1983 with the aim of establishing a protocol for the reduction of sulphur dioxide emissions. The WORLD BANK has also promoted an international scientific network to map the impact of acid rain in Asia.

acid soot An agglomeration of carbon particles held together by acidic moisture. Acid soot emitted from chimneys leaves brown stains on materials and damages paintwork. The problem has been mainly associated with oil-fired installations equipped with metal chimneys.

acid sludge A black, viscous residue left after the treatment of petroleum oils with sulphuric acid for the removal of impurities.

acid sulphate soil Soil, clay, or sand with profiles or layers containing significant amounts of sulphide, yielding acid under moist or flooded conditions.

acoustic Having properties or characteristics affecting or connected with sound; for example, acoustic tiles.

acoustic enclosure The surrounding of a source of NOISE by an enclosure of brick or concrete, or lighter material acoustically lined, to achieve satisfactory noise absorption.

acoustic interferometer A device for measuring the velocity and absorption of sound waves in a gas or liquid.

acoustic reflex The mechanism by which the ear protects itself from extra-loud sounds by reducing them, just as the eye protects itself from extra-bright light by contracting the pupil. However, this mechanism operates successfully only over a certain range and its response is too slow to protect the ear from sudden noise as from a blast, detonation, or explosion. Pain occurs as the ear unsuccessfully attempts to protect itself.

acoustics The science concerned with the production, control, transmission, reception, and effects of noise and sound. The principal branches of the subject are architectural acoustics and environmental acoustics, though there are many other divisions. Environmental acoustics deals essentially with the problems of noise control: noise from aircraft, factories, heavy machinery, air-conditioning plant, trucks and cars, and residential nuisances. Noise control ranges from: the production of quieter machines and equipment and the use of absorbent mountings, to more careful design of moving parts to reduce noise and vibration; more attention to road surfaces and noise screening; better use of natural features and

acoustic trauma

BUNDS; improved insulation and partitioning; and ACOUSTIC SITE PLANNING.

acoustic site planning The careful arrangement of buildings on a site so as to minimize the effect of traffic or other noise. Acoustic site planning may need to be reinforced with architectural acoustic design techniques involving aspects such building height, room arrangement, placement of windows, and courtyard design. (See Figure 1.)

acoustic trauma Physiological changes in the body caused by sound waves. Excessive noise exposure can cause hearing loss and physically damage components of the ear.

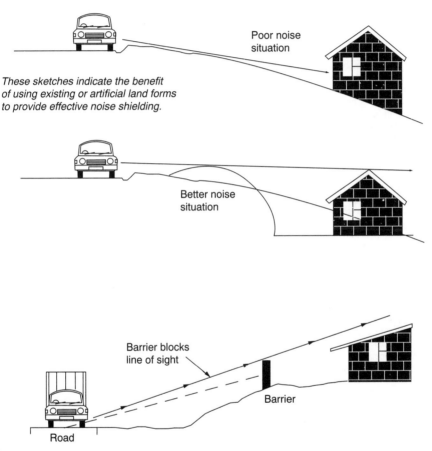

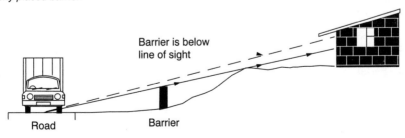

ACOUSTIC SITE PLANNING: Figure 1. Noise shielding as part of acoustic site planning.

Pain experienced in the ears suggests that the threshold for damage has been reached. More constant noise at moderate to high levels can cause stress, fatigue, and irritability.

actinides (Actinoides) A group of elements with atomic numbers ranging from 89 to 103, including thorium, uranium, and plutonium. Many are long-lived alpha-emitters. (See RADIOACTIVITY.)

activated carbon Carbon which has been treated to remove hydrocarbons thereby increasing its powers of ABSORPTION. It is used in many industrial processes for the recovery of valuable components from gaseous mixtures, or for the removal of unwanted components; it may also be used to treat liquid streams and in motor vehicles for the treatment of hydrocarbons. It is extensively used for odour control and in air-freshening applications. (See AIR POLLUTION.)

activated sludge process An aerobic process used in the secondary treatment of sewage and in the treatment of other substances. In sewage plants, organic wastes are brought into contact with biologically active microorganisms (in returned sludge) in the presence of mechanically produced air. The activated sludge process is in effect an artificially accelerated self-purification process.

activity patterns The distribution of centres of human activity throughout a city, town, or settlement, during daytime and night-time and over a period of time.

acute Immediate or short-term. In relation to exposure, conditions which may develop rapidly and cause harm within a short period. In relation to effects, physiological effects which appear promptly after exposure, reaching a crisis rapidly. (See CHRONIC.)

adaptation In biology, any change in the structure or function of an organism that enables it to survive and reproduce more effectively in its particular environment. In the theory of evolution, adaptation occurs as a consequence of random variation in the genetic make-up of organisms coupled with natural selection.

adaptive radiation In the theory of evolution, the production of new species adapted to different ways of life from a single parent stock. Adaptive radiation is most likely to occur whenever a species occupies a new habitat that contains few, if any, similar species.

ADB ASIAN DEVELOPMENT BANK.

added value In economic science, the difference between the cost of producing something and the price at which it is sold, measured in accountancy terms. Added value is the basis of value-added tax, that is, a percentage tax on the value added at each stage of the production process.

addiction A state of human dependence on alcohol, drugs, or other substances, or a habitual or compulsive commitment to some clearly harmful activity.

Addo Elephant National Park Established in 1931, a national park in the southern Cape Province, South Africa, north of Port Elizabeth. The park is largely covered with dense, impenetrable evergreen scrub; it provides a refuge for elephants, hippopotamuses, buffaloes, antelopes, black rhinoceroses, and many small birds, mammals, and reptiles.

adenovirus A VIRUS that causes pharyngitis, conjunctivitis, and gastroenteritis. It is spread by inhalation, ingestion, and direct contact. It may contaminate water through sewage.

ADI ACCEPTABLE DAILY INTAKE.

adiabatic Without loss or gain of heat to a system. Thus, an adiabatic change is a change in the volume and pressure of a parcel of gas without exchange of heat between the parcel and its surroundings.

adjustment A behaviour response of organisms and creatures as a result of experience, resulting in modifications to the social organization or in physical adaptation to new or changed surroundings.

administered pricing The pricing of a good or service which is kept constant by the supplier for a period of time and over a long series of transactions; it has emerged as the most common form of pricing, replacing haggling and bargaining.

administrative law Law regulating the powers, procedures, and actions of public administrators, departments, and agencies. Administrative law embraces: rule-making, or the power to make regulations having the force of law; the issuing of licences, with or without attached conditions (or the refusal or revocation of licences) affecting the operation of a range of businesses; powers of inspection and investigation; and powers of enforcement, including prosecution. All these features are common to pollution-control legislation in general and much environmental protection legislation. Such laws bestow considerable powers on ministers, public officials, departments, and agencies; they also restrict what can be done by these bodies, often naming avenues of appeal.

Adriatic Sea Since 1989, a sea known for the formation every summer of large evil-smelling mucilaginous mats. In 1993, scientists from the Italian Institute of Water Research and the University of Milan established that the mats were probably due to zeolites and polycarboxylic acids used in place of phosphates in 'green' detergents.

adsorption The taking up of one substance on the surface of another; adhesion. Adsorbents in industrial use include activated carbon, activated alumina, silica gel, and fuller's earth. (See ABSORPTION.)

adversarial procedure A court procedure observed primarily in countries in which the Anglo-American legal system of common law prevails. The system requires the opposing sides to bring out pertinent facts and information, and to present and cross-examine witnesses, each side hoping to gain an advantage to its side of the case. The adversarial approach is used also in UK public inquiry procedures, but rarely in Australian public inquiries, where cross-examination is only occasionally allowed.

'Aegean Sea' incident In December, 1992, the foundering of the Greek tanker the *Aegean Sea* in severe weather on rocks near the Tower of Hercules off La Coruna, Spain. The ship broke in two, rupturing seven of its tanks, and caught fire. About 97 km of the Galician coast was contaminated, the slick covering 52 km^2.

aeration The bringing about of intimate contact between liquid and air by one of several methods: agitating the liquid to promote surface absorption of air; injecting air through the liquid; or spraying the liquid into the air. (See ACTIVATED SLUDGE PROCESS.)

aeroallergens Vegetable dust and pollens suspended in the air which cause allergic diseases such as hay fever.

aerobic In BIOLOGY (1) descriptive of living organisms which use molecular oxygen for the release of energy; (2) Relating to or caused by aerobic organisms; (3) In the presence of free oxygen; (4) Describing a process which requires free oxygen to proceed. An aerobe is an organism needing free oxygen for growth. (See ANAEROBIC.)

aerodynamics A branch of fluid physics that studies the flow of gases, particularly as it applies to solid objects moving at speed through air (such as vehicles, rockets, and aircraft), or subjected to wind pressures (such as structures and chimneys). The design of the stack tops of modern power stations, with projecting individual flues, was the result of aerodynamic studies.

aero-electric generation See WIND ENERGY.

Aeromonas A genus of bacteria occasionally isolated in drinking water, but without a clear link to disease.

aerosol A particle of solid or liquid matter of such small size that it can remain suspended in the atmosphere for a long period of time. Aerosols diffuse light, the larger particles settling out on horizontal surfaces or clinging to vertical surfaces. All air contains aerosols, the larger particles above 5 μ being filtered out in the nose and bronchi. The smaller particles may enter the lungs. Aerosols are classified as smoke, fumes, dust, and mists.

aesthetic guideline value A concentration or measure of a water- or air-quality characteristic (a chemical or physical property) that is associated with good-quality water (clean, safe, and attractive) and a general atmosphere of acceptable quality.

aesthetics The study of beauty, form, and taste, including the manifestations of natural beauty and nature. As a concept it is applied in environmental planning, architecture, city form and structure, open space, parks and gardens, landscaping, and in preserving pleasant perspectives.

aestivation (1) The state of inactivity (dormancy) which certain animal species adopt during a dry period, usually in summer (as opposed to hibernation during a cold period). Many frog species aestivate in arid conditions, burrowing deep into the sand to conserve moisture until rains alleviate the situation. It is also characteristic of lung fish and snails. (2) In BOTANY, it is descriptive of the way in which flower petals and sepals are folded in the buds and is important in plant classification.

aetiology The science of the causes of disease; the study of a causal factor and its nature and relationship with the host.

affluent society A society in which all basic needs and requirements have been met with respect to food, clothing, housing, transport, health, education, and employment, and in which there is generally a surplus to be devoted to the 'better things of life'. In all societies it may be said that affluence is enjoyed by some, or even many; yet in all societies affluence is denied to many and in the world at large is comparatively rare. Poverty remains the lot of most of the world's population.

African Development Bank (ADB) Established in 1963, a bank dedicated to economic development and social progress among its members. Its membership includes 50 African states and 25 non-African countries. Its headquarters are in Abidjan, Ivory Coast. The bank is not formally associated with the UN.

after-burner A burner located in the exit for gases from a combustion process, providing sufficient heat to destroy smoke and odours. After-burners may be used also in other industrial and process plant to control emissions.

Agency for International Development, US (AID, US) The US government agency that manages foreign aid directed to diplomatic aims, trade and investment, humanitarian assistance and, increasingly, sustainable development, the environment and overpopulation. Total aid disbursements in recent years for the US have been about 0.2% of GDP, against about 0.3% for the OECD countries as a whole. On the other hand, some 80% of US aid is untied, that is, recipients are not obliged to purchase from the US. This is one of the lowest percentages in the world. (See FOREIGN AID.)

Agenda 21 A document adopted by the UN CONFERENCE ON ENVIRONMENT AND DEVELOPMENT in Rio de Janeiro, Brazil, in 1992, representing a programme for the 21st century, in particular a move towards the world's goal of SUSTAINABLE DEVELOPMENT. Agenda 21 is a massive document of over 800 pages embracing all the issues referred to the Conference by the UN General Assembly. Its objective is the alleviation of poverty, hunger, sickness, and illiteracy worldwide, while at the same time arresting the deterioration of the ecosystems on which humankind depends to sustain life. Agenda 21 is divided into four sections: social and economic dimensions; conservation and management of resources for development; strengthening the role of major groups; and the means of implementation through finance and the arrangements necessary for the funding of Agenda 21 programme in developing countries. The role of the GLOBAL ENVIRONMENT FACILITY in the funding of programmes was endorsed. A major recommendation was the creation of a UN COMMISSION FOR SUSTAINABLE DEVELOPMENT to monitor and continue to develop Agenda 21; the Commission reports to the UN General Assembly.

Agent Orange A defoliant widely deployed in the Vietnam War by US forces to eliminate ground cover for opposing troops. Otherwise, a selective weedkiller containing highly poisonous DIOXIN as a by-product.

agglomeration The clustering or adhering together of a number of small particles to form a larger single entity or 'agglomerate'. Agglomeration has been associated with the performance of particles in chimney stacks.

aggradation The building up of land by the deposition of material; for example, by the deposition of detritus by streams where they flow over a surface of reduced gradient.

Agreement on Cooperation for the Sustainable Development of the Mekong River Basin See MEKONG RIVER COMMISSION.

agricultural chemical Any substance or mixture of substances used or intended to be used solely for preventing, destroying, repelling, attracting, inhibiting, or controlling any insects, rodents, birds, nematodes, bacteria, fungi, weeds, or other forms of plant, animal, or microbial life regarded as pests; or which is used or intended solely for use as a plant regulator, defoliant, promoter, desiccant, or fertilizer. (See PESTICIDES.)

agricultural economics The study of the allocation, distribution, and utilization of the resources consumed by, and commodities produced by, farming.

agricultural sciences The sciences dealing with farm production, including soil cultivation and conservation, water control and management, crop growing and harvesting, animal husbandry, the processing of animal and plant products, machinery and engineering, and the use of fertilizers and pesticides.

agrochemicals See AGRICULTURAL CHEMICAL.

agroforestry The combination of timber production and agriculture on the same land.

AID, US AGENCY FOR INTERNATIONAL DEVELOPMENT, US.

airblast overpressure The shockwave transmitted through the atmosphere, caused by explosive blasting or accidental explosion; it presents a significant hazard in urban areas where residences are close to hazardous industries. Overpressure is the pressure developed above atmospheric pressure at any stage or location in a pressure pulse or blast wave. Airblast overpressure can cause discomfort to people and damage to structures; its severity depends on the magnitude of the blast, the position of the explosives, the topography, and atmospheric conditions.

air conditioning The control of temperature, humidity, quality, and movement of air within an enclosed space, independently of outside conditions. (See LEGIONNAIRE'S DISEASE.)

air mass A large body of air with relatively homogeneous characteristics, which can be plotted as it moves over a particular region of the earth's surface. Air masses are often comparable in size to continents, moving in one of the atmospheric currents of the general atmospheric circulation. Fronts are theoretical surfaces dividing one air mass from another. An air mass is designated by geographic origin as either polar (cold and dry), arctic (cool and moist), equatorial (hot and dry), or tropical (warm and moist) in international terminology.

air pollution The contamination of the atmosphere with undesirable particles, liquids, and gases. In a strict sense, air may be considered to be polluted when there is added to it any substance foreign or additional to its normal composition, but for practical purposes, the term is restricted to those conditions in which the general atmosphere contains substances in concentrations which are harmful, or likely to be harmful, to people or their environment. In 1993, a joint report by UNEP and WHO warned of health problems caused by urban air pollution, urging a reduction in pollution levels. International measures now extend to combating the GREENHOUSE EFFECT, and protecting the ozone layer.

air pollution control legislation Legislation and regulations now common in many countries to restrict emissions from POINT SOURCES such as industrial plant and vehicles, and to restrict pollution of the atmosphere generally.

Measures may extend to combating the GREENHOUSE EFFECT and protecting the ozone layer. (See ENVIRONMENTAL IMPACT ASSESSMENT.)

air pollution control strategy A national plan to direct resources to the abatement or prevention of air pollution problems. Such a strategy involves invariably a variety of measures such as setting standards for new equipment to restrict emissions, while restricting existing sources through licensing, and prosecutions. Before major enterprises can be undertaken, many countries require the project to go through an EIA process. Special measures have been taken in many countries against lead in petrol (gasoline). Research continues on the relationship between pollutants and general health. (See ENVIRONMENTAL IMPACT ASSESSMENT.)

air pollution index An index calculated hourly or daily in many cities to give an indication of the quality of the atmosphere and the possible need for special measures. The index may take account of the concentrations of various pollutants, giving a weight to each in the calculation. Thus, out of diverse readings, a single index may provide a satisfactory general guide.

air quality standards Standards set by various countries, states, and provinces in respect of desirable air quality, setting targets for individual pollutants. Box 1 displays the national ambient air quality standards for the USA; such standards have been adopted by several other countries.

airshed A delineated area within which emitted pollutants interact or tend to increase collectively in concentration; comparable with the concept of a WATERSHED.

Aitken nuclei Small particles a few hundredths of a micron in diameter, which normally exist in the atmosphere in concentrations varying from a few thousand to a few hundred thousand per millilitre. They are produced in large numbers by most combustion processes.

Box 1 National ambient air quality standards

Pollutant	Primary standards	Averaging time	Secondary standards
Carbon monoxide	9 ppm (10 mg/m^3)	8-hour^a	None
	35 ppm (40 mg/m^3)	1-hour^a	
Lead	1.5 µg/m^3	Quarterly average	Same as primary
Nitrogen dioxide	0.053 ppm (100 µg/m^3)	Annual (arithmetic mean)	Same as primary
Particulate matter (PM-10)	50 µg/m^3	Annual (arithmetic mean)^b	Same as primary
	150 µg/m^{3c}	24-hour	
Ozone	0.12 ppm (235 µg/m^3)	1-hour^d	Same as primary
Sulfur dioxide	0.03 ppm (80 µg/m^3)	Annual (arithmetic mean)	—
	0.14 ppm (365 µg/m^3)	24-hour^a	—
		3-hour^a	0.5 ppm (1300 µg/m^3)

^{a} Not to be exceeded more than once per year.
^{b} The standard is attained when the expected annual arithmetic mean concentration is less than or equal to 50 µg/m^3.
^{c} The standard is attained when the expected number of days per calendar year with a 24-hour average concentration above 150 µg/m^3 is equal to or less than 1.
^{d} The standard is attained when the expected number of days per calendar year with maximum hourly average concentrations above 0.12 ppm is equal to or less than 1.
Source: United States of America National Report 1992.

'Akatsuki Maru' incident In October 1992, protests by GREENPEACE at the port of Cherbourg, France, when the Japanese ship the *Akatsuki Maru* loaded about 1.5 tonnes of reactor-grade plutonium delivered from The Hague reprocessing plant. The ship reached Tokai, Japan, in January 1993, when there were further Greenpeace demonstrations.

Alaskan National Parks A range of national parks, wildlife refuges and wilderness areas established since the initial creation of the Denali National Park and Preserve in 1917. This was followed by the establishment of the Katmai National Park and Preserve in 1918 (including the Valley of Ten Thousand Smokes), Glacier Bay National Park and Preserve in 1925, and the Sitka National Historic Park in 1910. The Tongass and Chugach national forests in the southeast and south-central regions, respectively, are also public land reserves. The national parks and wildlife sanctuary system was expanded by the US Alaska National Interest Lands Conservation Act 1980.

Alaskan pipeline A 1277 km pipeline from Prudhoe Bay to the southern port of Valdez, to carry oil from the North Slope oilfield. Environmental and other objections imposed a five-year delay on the project; oil flowed in 1977. (See EXXON VALDEZ DISASTER.)

albedo That fraction of incoming solar energy which is reflected directly without being absorbed; a measure of the reflectivity of the Earth. Vegetation and ocean water have a low albedo, nearer to zero, as they absorb a large fraction of the incoming energy. On the other hand, snow and cloud surfaces have a high albedo, nearer to unity, as most of the incoming energy is reflected or scattered. The Earth's average albedo is about one-half. An albedo of one means complete reflection; zero means no reflection, or a completely dark body.

aldehydes A class of chemical compounds intermediate between alcohols and acids; most are colourless, volatile fluids while some have suffocating odours. One of the simpler aldehydes, namely, formaldehyde, when combined with phenol produces Bakelite and when combined with urea produces Formica. Formaldehyde is also used as a preservative. Many aldehydes are involved in the production of vitamins. Other aldehydes are useful as solvents, polymer compounds, perfume ingredients, and intermediates in the production of other compounds. As air pollutants in, say, diesel exhausts, aldehydes can be unpleasant and an irritant to the nose and eyes.

aldrin A white insecticide containing a chlorinated derivative of naphthalene. The insecticide DIELDRIN is made from aldrin by treatment with peroxyacetic acid. In October 1974, both aldrin and dieldrin were banned by the US Government because of clear evidence that they were carcinogenic and progressively restricted in other countries. Also the use of isodrin and endrin, along with other CHLORINATED HYDROCARBONS, are being discontinued as insecticides for similar reasons.

alert level That concentration of any pollutant or indicator at or above which most vulnerable members of the exposed population are likely to be adversely affected. Warnings of this may be publicly announced by the responsible environment protection authority.

alfisol soil A soil type of humid wooded regions, characterized by a well-developed clay horizon. Alfisols occupy between 10 and 15% of the world's land area, occurring on all continents, especially where broad-leaved forests are established. These soils are among the most productive. Wheat and maize are major crops in the three largest areas of alfisols, namely the USA, parts of Canada, and northern Europe.

algae Simple plants, containing chlorophyll or other photsynthetic pigments found widely in water. Often microscopic, these plants may be freely suspended or attached to surfaces. They do not have distinct stems, roots, or leaves. The pigments frequently include colours other than green. The most familiar algae are the large seaweeds attached to rocks in shallow water around the coasts of the world. There are several thousand species of algae, mainly small, the main colours being green, brown, rose, gold, and blue. (See BLUE-GREEN ALGAE; EUTROPHICATION.)

algicide A pesticide used to control ALGAE, especially in stored or industrial water supplies.

algology The study of ALGAE; a discipline of immediate interest as algae play an important role in ECOLOGY, certain algae being a vital segment in food chains or are used as a direct food source by people.

alienation A sense of estrangement from society coupled with feelings of frustration and powerlessness. The term has been used in a variety of ways and contexts by many writers, but the essence of isolation remains.

aliphatic hydrocarbons See HYDROCARBONS.

alkali A substance which in solution in water forms hydrogen ions with a pH VALUE of more than 7.

alkyl sulphonates Surface-active agents and basic components of synthetic detergents. Some, such as alkyl benzene sulphonate (ABS), are stable and resistant to biodegradation. The discharge of detergent residues containing ABS results in foaming of the receiving waters and interferences with sewage treatment processes. Those with a linear molecular structure such as linear alkyl sulphonates

(LAS) show much better biodegradability. In most countries commercial detergents sold for domestic use are usually of the LAS type and biodegradable.

allergy Hypersensitive bodily reaction to substances such as drugs, antibiotics, pollens, foods, dust, or microorganisms in amounts that do not adversely affect most people. The substances are known as allergens.

Alliance of Small Island States (AOSIS) Formed following the UN CONFERENCE ON ENVIRONMENT AND DEVELOPMENT and the adoption of the CONVENTION ON CLIMATE CHANGE, an alliance of 36 nations, many of them very small, who fear that global warming could swamp them. They want the industrialized nations to reduce CARBON DIOXIDE emission by 20% by the year 2005.

allocative efficiency, social The flow of goods, services, resources, and funds which are likely to maximize the QUALITY OF LIFE of society over time, as distinct from the interests of special groups or classes. Allocative inefficiencies may occur due to the maldistribution of income distorting effective demand, inefficient markets, the influences of monopoly and oligopoly, restrictive practices, outdated industrial relations policies, antiquated regulations, or distortions in government policies.

allotment A piece, parcel, or subdivision of land, the boundaries of which are separately defined on a plan deposited with a survey office or registered with an official registrar of land titles.

alluvium Material deposited by rivers, often forming floodplains and deltas. Alluvium often consists of silt, sand, clay, gravel, and probably a good deal of organic matter. The result is often very fertile soils such as those of the deltas of the Ganges, the Nile, the Brahmaputra, the Huang, and the Mississippi rivers.

alpha diversity (niche diversification) Diversity occurring as a result of competition between species in favourable environments. As a consequence, the variation in individual species becomes more limited. (See BETA DIVERSITY.)

alpha particle A heavy particle produced by a radioactive decay process and in various nuclear reactions. It consists of two protons and two neutrons and thus carries two positive charges. It is identical with the nucleus of a helium atom. It was discovered and named by Ernest Rutherford in 1899. It is damaging when in contact with living tissue following, say, inhalation or ingestion.

alternative dispute resolution The settling of disputes between contending parties by mediation or arbitration, thus avoiding the need for expensive legal actions before the courts, while minimizing delays. Alternative means of dispute resolution in respect of environmental issues has been developing rapidly in several countries.

alternative energy Energy that is renewable and ecologically safe. The most important source of alternative energy is hydroelectric power, harnessing the energy of flowing water. Other sources are windmills, wind turbines, tidal power, solar power, wave power, and geothermal energy.

alternatives In ENVIRONMENTAL IMPACT ASSESSMENT of projects, an examination of alternative locations for the project, methods, and techniques, including the alternative of not proceeding. It may be demonstrated that a project is not actually needed if demand-management approaches (for example, curbing the demand at peaks for electricity) are adopted or strengthened. Other locations may be feasible and more suitable. Alternative methods may be employed which reduce the generation of wastes, promote clean production, encourage recycling, and reduce the cumulative effects and the use of resources. An assessment of a single project may suggest a wider review at regional, national, and international levels with a view to alternative policies and plans minimizing environmental adverse effects.

alum See ALUMINIUM SULPHATE.

aluminium sulphate (alum) Dissolved in water, it may be used as a flocculant in water treatment plants to bind together fine sediments, to be removed in mechanical filters; and as a coagulant to remove particles that discolour water. Guidelines for public water supplies normally limit aluminium to less that 0.1 mg/l. Aluminium has been very tentatively linked to Alzheimer's disease, although scientific evidence on this appears to be inconclusive. Some water treatment authorities have switched to the use of an alternative filtering agent such as ferric chloride.

alveoli Innumerable, minute air-filled sacs in human and animal lungs; they are thin-walled and surrounded by blood vessels. Through their surfaces the respiratory exchange of oxygen and CARBON DIOXIDE occurs. Oxygen is absorbed and carbon dioxide released to the atmosphere, contributing to greenhouse gases.

Amazon rainforest An extensive tropical rainforest, occupying the drainage basin of the Amazon River and its tributaries in Brazil, covering about 7 000 000 km^2; it comprises about 40% of the total area of Brazil. With more than 1000 tributaries, the Amazon is the largest basin area in the world. Characterized by luxuriant vegetation, the rainforest has a wide range of trees including acacia, myrtle, rosewood, rubber tree, Brazil nut, mahogany, cedar, and palm. The associated wildlife is rich and diverse. Until 1979, the

Amazonia National Park

aim of the Brazilian government was to open up the Amazon forest for development. However, neither the cattle ranches nor many small farms established prospered. The failures were due in part to the belief that the tropical soil was prodigiously fertile, but the topsoil was in fact very thin and ecologically fragile. In 1979, a forest policy committee reviewed the situation, recommended a reversal of policy, and urged the establishment of several national parks and ecological reserves. The committee affirmed that the Amazon forest was a valuable source of drugs, fibres, fuel, crops, and resins, as well as providing gene pools of rare and valuable species of plants and animals. This reversal of policy should slow down the rate of clearing of the Amazon forest. (See AMAZONIA NATIONAL PARK.)

Amazonia National Park Established in 1974, a park comprising areas of submountainous forest with rock outcroppings along the sedimentary Amazon river basin, with moderately undulating plains and meandering rivers. The park is characterized by a wide variety of flora. Trees include mangrove, rubber, and palm. There is also great diversity of fauna including armadillos, deer, tapirs, weasels, dolphins, and a great variety of monkeys.

ambience The distinct or special atmosphere of an environment or setting.

ambient Surrounding, encompassing, prevailing, or encircling.

ambient air quality standards See AIR QUALITY STANDARDS.

ambient noise The background noise or prevailing general noise in an area, perhaps in the absence of a noise under investigation.

ambient quality standards Also known as environmental quality standards, maximum permissible limits, maximum allowable concentrations, maximum acceptable levels of pollutants. (See AIR QUALITY STANDARDS; WATER QUALITY STANDARDS.)

amenity A word that frequently appears in environmental legislation usually referring to non-marketable environmental benefits such as beauty and tranquillity. Public agencies are increasingly required to weigh economic efficiency, narrowly construed, against potential losses in amenity.

ammonia A colourless pungent gas composed of nitrogen and hydrogen; it is the basis for the production of many commercially significant nitrogen compounds. However, the major use of ammonia is as a fertilizer apppplied either directly or as AMMONIUM NITRATE, ammonium phosphate, and other salts. Ammonia is highly toxic, though readily identifiable.

ammonifying bacteria Nitrifying bacteria that break down the proteins in the tissues of dead plants and animals, releasing AMMONIA.

ammonium nitrate A salt of AMMONIA and nitric acid, used widely in fertilizers and explosives. It is the most common nitrogenous component of artificial fertilizers and is also used in herbicides and insecticides. Ammonium nitrate is the main source of nitrate runoff from farm land, a result of the excessive use of fertilizers. In water, it may contribute to EUTROPHICATION.

Amoco Cadiz disaster A major oil spill, occurring in 1978 when the supertanker the *Amoco Cadiz* went aground near Portsall, France; some 220 000 tonnes of oil were poured into the sea, the spill destroying oyster and shellfish beds and polluting beaches quite extensively. Almost all the oil drifted onto the coast of Brittany.

amoeba One of the simplest living creatures, consisting of a single cell and belonging to the class Protozoa; it feeds by flowing round and engulfing organic debris.

amortization (1) The gradual repayment of a debt, both capital and interest, over a prescribed period as in the case of mortgage repayments. (2) The annual writing down of the value of an asset by depreciation. Amortization is synonymous with depreciation, but is usually applied to intangible assets, while the word depreciation is confined to tangible assets.

amplitude The maximum value or peak; the maximum movement of an oscillation from the normal equilibrium position. In respect of waves, it is the height of a wave or the depth of a trough. With sound waves, amplitude corresponds to the loudness or intensity of the sound. A maximum value taken on by a quantity whose value fluctuates.

anabatic wind (mountain wind) A breeze warmed by contact with the ground, flowing uphill; the reverse of a KATABATIC WIND. This kind of wind is found most especially in mountain regions, where one side of a valley may be warmed much more than the other.

anabolism A facet of METABOLISM relating to the synthesis of simpler molecules into more complex compounds thus building up structures, metabolites and storage substances in cells. These reactions require energy from adenosine triphosphate (ATP) produced by CATABOLISM.

anadromous fish Fish, such as salmon, which spend most of their growing years in the ocean and, after attaining sexual maturity, ascend freshwater streams in order to spawn. The erection of power and irrigation dams, or the presence of thermal pollution, may isolate considerable numbers of fish from their traditional spawning grounds. One solution has been to provide fish ladders to enable them to leap by steps to the higher water levels.

anaerobic Living in the absence of air or free oxygen; the opposite of AEROBIC. An anaerobe is an organism able to grow without free oxygen. In anaerobic processes, HYDROGEN SULPHIDE may be formed giving rise to an objectionable odour.

anaerobic digestion A digestion process that permanently removes the offensive odour of many organic wastes, so they can be utilized on agricultural land without causing nuisance. A high proportion of the CHEMICAL OXYGEN DEMAND is removed, with the recovery of the organic carbon as METHANE, while most of the lipids and other constituents which otherwise might attract flies and vermin are degraded. A wide variety of bacteria are involved in the process.

ando soil A major soil type, rich in amorphous clays. It has a low apparent density and is very permeable to air and water. Decay of organic matter is slow while the HUMUS content is unusually high. Ando soils are common in volcanic regions, being usually quite fertile. As a result, settlements spring up around volcanoes, despite the dangers.

anechoic chamber An acoustic test chamber, almost totally sound absorbent at a very wide range of frequencies. An anechoic chamber provides almost free field conditions, in which no significant reflections of sound occur.

anemometer An instrument for measuring the speed of airflow in the atmosphere, in wind tunnels, and in other gas-flow situations. That most widely used for wind speed measurements is the revolving-cup electric anemometer. In this, the revolving cups drive an electric generator operating a meter that is calibrated in wind speed from, say, 5 to 100 knots.

Angkor National Park Established in 1925 in Cambodia, a relatively small national park intended mainly to protect many temple ruins and other archaeological remains of Khmer civilization.

angle of repose The maximum acute angle that the inclined surface of a pile of loosely divided material can make with the horizontal, while remaining stable. An increase in angle would result in material running down the outside of the pile. It is also the angle which material tends to assume when discharged from an overhead source.

animal welfare A concept of welfare that goes beyond the conservation and management of animals for the immediate needs of humanity. It involves respect for animals for their own sake condemning ill-treatment, neglect, and wanton or systematic cruelty. The Royal Society for the Prevention of Cruelty to Animals (RSPCA) was founded in Britain in 1824, though with much indifference elsewhere. Concern for the ENVIRONMENT from the 1970s has extended somewhat to the questions of misuse and cruelty to animals. In the 1980s, the European Parliament introduced a ban on the import of baby seal products from Canada, brought about by the International Fund for Animal Welfare. In 1986 came the European Union Directive on Animal Experimentation, a commitment to minimizing the suffering and misuse of animals wherever possible. Later the European Union created the European Centre for the Validation of Alternative Methods in Ispra, near Milan, Italy. It promotes the development of non-animal testing methods. In 1991, the European Union issued a Directive on the Protection of Animals during Transport. The MAASTRICHT TREATY (1993) included an appendix, a Declaration on the Protection of Animals. Such restrictions have not yet touched blood-sports. (See RIGHTS OF NATURE.)

anion A negatively charged ION; in an electrolytic process, a potential gradient causes anions to migrate towards the anode (the positive electrode), away from the cathode (the negative electrode).

anion exchange resins Resins which are used to exchange with anions in a fractionation test or treatment.

Annual International Conferences on Environmentally Sustainable Development Initiated in 1994, annual conferences planned by the WORLD BANK on the furtherance of SUSTAINABLE DEVELOPMENT as envisaged by the UN CONFERENCE ON ENVIRONMENT AND DEVELOPMENT held in 1992. The Bank's Agenda is: (1) to assist borrowers to promote environmental stewardship through lending and policy advice; (2) to assess and mitigate any adverse impacts associated with World Bank projects; (3) to build on the positive working relationships between development and the ENVIRONMENT; and (4) to address global environmental challenges through the GLOBAL ENVIRONMENTAL FACILITY.

anomaly A deviation or departure from what is normal or common; something that is unusual, irregular, or abnormal. A geological feature considered capable of being associated with commercially valuable minerals or hydrocarbons.

anoxia Depleted or deficient in oxygen; for example, in the tissues or in the blood or in a body of water. In the Baltic, anoxic bottom-water conditions tend to persist in the deeper parts.

antagonistic effect The tendency of some chemicals and processes to react together to form combinations which may have a less powerful effect than the substances or processes taken separately; the opposite of a synergistic effect. The term is also applied where the growth of one organism is inhibited by another through the creation of unfavourable

Antarctic Treaty

circumstances such as the exhaustion of the food supply. (See SYNERGISM.)

Antarctic Treaty Signed in 1959, a treaty which allows freedom of scientific research and movement on and around the Antarctic continent, setting aside questions of sovereignty. Twenty-five nations are parties to the Treaty. Important international agreements initiated by the Treaty nations include Agreed Measures for the Conservation of Antarctic Fauna and Flora; Antarctic Protocol on Environmental Protection; the Convention for the Conservation of Antarctic Seals; the Convention on the Conservation of Antarctic Marine Living Resources; the Convention for the Regulation of Antarctic Mineral Resources Activity. All these conventions and agreements have been reinforced by the imposition by the treaty nations of a general ban on mining and exploration in Antarctica. (See Figure 2.)

anthelmintics Chemical substances used to destroy parasitic worms.

anthrax A disease of cattle and sheep occasionally transmitted to humans, usually through infected hides and fleeces. It may appear as a skin lesion or take the form of pneumonia. Anthrax may form highly resistant spores capable of retaining their virulence in contaminated soil or other material for many years. Anthrax is one of the oldest recorded diseases of animals.

anthropocentric Regarding human beings as the central feature of the world; interpreting environmental and resource issues solely in terms of human values and standards. Hence, the rights of animals, for example, can only be derived from the interests of humans. The prevention of cruelty to animals and the practice of human slaughtering is seen therefore as providing comfort and peace of mind to humans, rather than a recognition of independent animal rights.

anthropology The scientific study of human beings; it deals with the human species biologically, physically, socially, and culturally. As a subject, it overlaps with sociology, linguistics, psychology, archaeology, zoology, and medicine.

antibiotic A drug which inhibits or destroys the growth of bacteria or fungi. Since the introduction of penicillin in 1941, antibiotics have revolutionized the treatment of bacterial infections in humans and other animals. However, some bacteria can develop resistance to antibiotics, and resistant strains often proliferate.

antibody A protein molecule produced by lymphocytes in response to the presence of invading substances, offering a measure of immunity in humans. An invading foreign body or microbe is known as an antigen. Antibody production may continue for several days until the antigen (say, a common cold virus) is removed.

anticyclone A region of high atmospheric pressure caused by descending air, which becomes warm and dry. Winds radiate from a relatively calm centre, taking a clockwise direction in the northern hemisphere, and an anticlockwise direction in the southern hemisphere. In summer, an anticyclone generally means fine, warm sunny weather; in winter, however, dense clouds may form giving a condition known as anticyclonic gloom; a deep INVERSION layer may be formed.

anti-urbanism A sense of disaffection with urban development; a genuine belief that urban life is in many ways less desirable than rural life. This may be linked with a view that if cities are inevitable, they have grown too large. Urban growth should be accepted, therefore, only by way of new, relatively small, communities in new locations.

anti-vibration mounting A noise insulation measure achieved by isolating a noisy machine from its mounting or base by means of springs, rubber pads, or other resilient mountings. Other measures may be necessary also. (See ACOUSTIC ENCLOSURE; ACOUSTIC SITE PLANNING.)

AOSIS ALLIANCE OF SMALL ISLAND STATES.

APEC forum ASIAN–PACIFIC ECONOMIC COOPERATION FORUM.

aphicide A pesticide used to control aphids, which are small insects that suck the juices of plants. Some carry plant virus diseases.

Apia convention CONVENTION ON THE CONSERVATION OF NATURE IN THE SOUTH PACIFIC.

apiculture Commercial beekeeping.

application factor The ratio between the concentration of some substance producing a specified chronic response in an organism, and that causing a 50% mortality in the population in a specified period. From this may be derived the concept of a 'safe application factor', that is, that fraction of the lethal level of substance that would be environmentally safe for the organisms concerned.

aquaculture Also called fish farming, fish culture, or mariculture; the rearing of fish, shellfish, and some aquatic plants under controlled and managed conditions. The fish may be confined in ponds, pools, cages suspended in the open sea, or in barricaded coastal waters.

aquatic ecosystem Any watery environment, from small to large, from pond to ocean, in which plants and animals interact with the chemical and physical features of the environment.

aquatic weeds Certain aquatic plants which have the ability to propagate rapidly in wetland environments. Most spread vegetatively in the presence of water, and so can colonize wetlands. These plants may disturb the wetland

aquifer

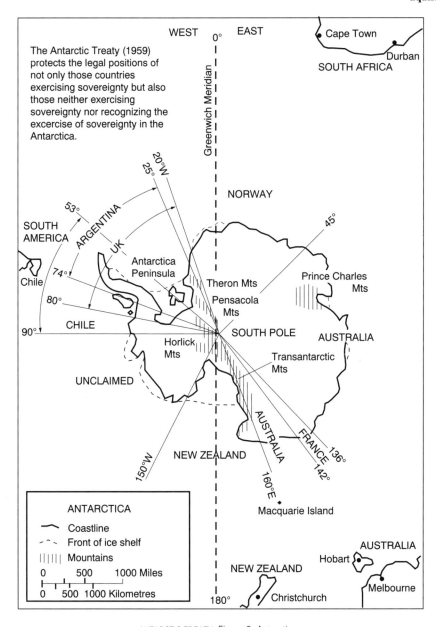

ANTARCTIC TREATY: Figure 2. Antarctica.

ecosystem, and may also adversely affect the use of water by interfering with irrigation systems and clogging channels, streams, and drains. Control measures are often expensive and may cause widespread ecological damage.

aquiclude A geological formation of rock or soil which, although porous and capable of absorbing water slowly, will not transmit it fast enough to furnish an appreciable supply for a spring or well.

aquifer A porous water-bearing bed or stratum of permeable sand, gravel, or rock, capable of yielding a continuous supply of water. The protection of aquifers from pollutants requires the constant vigilance of environmental agencies.

aquifer recharge area An area of land through which surface water percolates and assists in replenishing ground water in an AQUIFER.

aquitard A geological formation of rock or soil which retards the movement of GROUND WATER.

Aral Sea A large saltwater lake in south-central Russia, formerly the world's fourth largest body of inland water. There has been a remarkable shrinkage of its area and volume in the second half of the 20th century, due primarily to the diversion for irrigation purposes of water from two of its tributary rivers. By the late 1980s, the lake had lost more than half the volume of its water and had been reduced to about half its former depth. The remaining water was unfit for drinking purposes, and the fishing industry was virtually destroyed along with all the fish. The two ports on the lake were now many kilometres from the lake's shores. A partial depopulation of the areas around had occurred. The local climate became harsher. By the end of the decade and the century, the Russian government was attempting to use less water for agriculture to assist the lake.

arbitration A method of ALTERNATIVE DISPUTE RESOLUTION; the settling of disputes between parties outside ordinary court procedures by deferring to a mutually agreed-upon third party who is given the authority to make a legally binding decision. Arbitration is most commonly resorted to in cases involving commercial disputes, but it has also found a place in industrial disputes and in environmental controversies. The third party who will make a decision binding on the parties to the dispute may be a single individual or a tribunal chosen by the parties.

arboreal Relating to trees; hence an arborist is a person who studies trees, and ARBORICULTURE is concerned with forestry and the culture of trees.

arboriculture The cultivation of trees, shrubs, and woody plants for shade and ornamentation. It involves propagation, transplanting, pruning, fertilizing, spraying, identifying, diagnosing, treating damage and ailments, and other related matters. Arboriculture is concerned with the welfare of individual plants in contrast to silviculture and agriculture, in which the major concern is the welfare of a large number of plants collectively.

architecture The art and science of building structures for living, working, and recreational purposes. Architecture embraces utilitarian, aesthetic, social, and environmental purposes. Its aim is to enhance a neighbourhood as well as to serve its occupants and investors to their satisfaction. Not all purposes may be successfully achieved.

architectural acoustics The production and reception of sound in enclosed spaces such as concert halls and theatres, churches, school classrooms and lecture halls, and residences. (See ACOUSTIC SITE PLANNING.)

Arctic The region north of the Arctic Circle. Land areas in the Arctic have mainly stunted tundra vegetation but Arctic wildlife is rich and varied, especially in the warmer months after the break-up of pack ice. Animals include reindeer, caribou, musk ox, fox, hare, lemming, wolf, polar bear, seal, and walrus. The Aboriginal people are the Inuit.

Arctic haze A winter haze occurring as an aerosol of fine liquid and minute solid particles, adversely affecting visibility and possibly modifying the climate of the polar region. The haze contains sulphate compounds, together with soot and hydrocarbons. The prevailing winds pick up pollution particles from European and Russian sources and carry them north into the Arctic air mass.

area-wide EIA The extending of the principle of ENVIRONMENTAL IMPACT ASSESSMENT to entire areas or regions to take account of the cumulative effects of a range of similar or contrasting developments. This procedure recognizes the limitations of the EIA of individual development projects.

arid areas Those areas in which rainfall is too low or erratic for the economic cropping of sown pastures, although perhaps permitting grazing. The arid areas embrace most of the world's deserts. Arid zones are to be found in the USA, Australia, Pakistan, Morocco, and elsewhere.

aridisol soil The principal soil type of deserts and other relatively arid regions, covering about one-sixth of the total area of the world's soils. Aridisols have a low organic matter content, supporting far less vegetation than the soils of humid regions. Indeed, in their natural state aridisols are unsuitable for cultivation. However, when irrigated they become highly productive.

arsenic A chemical element in the nitrogen family, existing in both grey and yellow crystalline forms; it is widely distributed in nature, being present in minute quantities in soil, the sea, and the human body. Among the most important commercial compounds are arsenious oxide (known as white arsenic) and arsenic pentoxide. Arsenious oxide is obtained as a by-product from the roasting of the ores of copper, lead, and other metals, as well as by the roasting of arsenopyrite and arsenic sulphide ores. Arsenious oxide is the starting material for most other arsenic compounds; arsenic pentoxide is a major ingredient of insecticides and herbicides. Arsenic compounds of importance in agriculture include arsenic acid, lead arsenate, and calcium arsenate, which are used for sterilizing soil and controlling pests.

arsenic poisoning Most often, the results from the ingestion or inhalation of insecticides containing arsenious oxide or other arsenical compounds. It is an occupational hazard among agricultural workers handling insecticidal sprays and dusts. Sprayed fruit and vegetables, if not washed, may also carry enough arsenic to be potentially toxic to the consumer. In 1987, a district in Nakhon Si Thammarat, Thailand, was found to be heavily affected by chronic arsenic poisoning. It was found that the well water was heavily contaminated, with people having few alternative sources of clean and safe water. Tin mining was blamed, with arsenopyrite contaminating groundwater sources.

artesian bore An artificial means of bringing water to the surface from an artesian basin by mechanical pumping. An artesian well achieves the same thing without pumping, through water pressure alone. The Great Artesian Basin, a huge reserve of underground water, covers 1.8 million km^2 of the Australian continent.

Arthropoda A phylum of invertebrate animals with jointed limbs and a body divided into segments; the body is usually covered with a chitinous shell. It is the largest group in the animal kingdom in terms of numbers of species and includes crustaceans, insects, spiders, centipedes, and related forms.

artificial reef A method of solid-waste disposal, which consists of the tipping into deep water of junk and debris, rusting car bodies, used car tyres, and other material. The reef provides cover and breeding sites for various marine fish.

Arusha National Park Established in 1960 in Tanzania, a relatively small national park, yet rich in flora and fauna. The area embraces mountain rainforest, acacia groves, grassland, swamps, and moorland. Fauna include the hippopotamus, elephant, black rhinoceros, giraffe, lion, leopard, buffalo, antelope, and a variety of monkeys. Water birds and flamingo abound.

asbestos A name given to a group of fibrous minerals which provide great heat resistance because of their poor conductivity and non-inflammability. The principal asbestos mineral is chrysotile or white asbestos, obtained by mining or quarrying. Canada is the major source of asbestos. (See ASBESTOSIS.)

asbestosis A type of pneumoconiosis, caused by the inhalation of ASBESTOS fibres. The disease is not limited to asbestos workers, but has also occurred among people living near mines, factories, and construction sites. A proportion of those exposed to dust may acquire a rare, but fatal, form of cancer known as mesothelioma. For this reason, substitutes for asbestos have been vigorously sought. Blue asbestos or crocidolite has been extensively banned, mesothelioma being closely associated with it, while all forms of asbestos are strictly controlled.

ASEAN ASSOCIATION OF SOUTH-EAST ASIAN NATIONS.

Asian Development Bank (ADB) Established in 1966, a development finance institution for the Asia-Pacific region. It is owned by 51 member countries, of which 35 are regional members. The ADB promotes economic, social, and environmental progress in its developing member countries, providing both concessional and non-concessional finance. The bank has its headquarters in Manila, Philippines.

Asian-Pacific Economic Cooperation Forum (APEC forum) An international forum established on the initiative of Australia in 1989 to encourage free trade in the Pacific area and to consider development and security issues. APEC comprises 17 nations: Australia, Brunei, Canada, Chile, China, Indonesia, Japan, South Korea, Malaysia, Mexico, New Zealand, the Philippines, Papua New Guinea, Singapore, Taiwan, Thailand, and the United States. In 1994, APEC decided to abolish all trade barriers, tariff and non-tariff, between members by the year 2020, if not sooner. This decision, following upon that of the GENERAL AGREEMENT ON TARIFFS AND TRADE (GATT), has attracted similar reactions from environmentalists that boosting trade in all goods and services promotes the rate of attrition of the region's natural resources and limits the scope for effective environmental policies, both domestic and international.

Asian-Pacific Seminar on Climate Change A conference held in Nagoya, Japan, in January 1991, involving about 70 participants from 19 Asian-Pacific countries and 10 international agencies. The seminar aimed to assist the exchange of information and to generate awareness on the climate change issue. Participants agreed that there was considerable scope for future regional cooperation.

asset Any valuable item, tangible or intangible, owned by an individual, corporation, agency, government, or society, not necessarily measurable in monetary terms. For example, the skills, health, and education of the community, its social and political stability, and its natural and environmental resources are all assets to society at large.

assimilative capacity The capacity of natural assets such as the atmosphere, bodies of water, oceans, and forests to absorb pollutants within certain limits without detrimental effects. Thus, a body of water may recover naturally from contamination by organic wastes.

The process of self-purification depends very largely upon biochemical reactions in which bacteria and other microorganisms in the presence of sufficient dissolved oxygen use the organic matter as food. Other factors such as dilution, sedimentation, and sunlight also play an important part in the process. If the dissolved oxygen is used up, self-purification will cease and a septic, offensive condition will prevail.

Association of South-east Asian Nations (ASEAN) Formed in 1967, an association comprising Brunei, Indonesia, Malaysia, the Philippines, Singapore, and Thailand with a combined population of over 300 million. The association promotes regional security and trade, facilitates the settling of disputes between members and promotes joint ventures in the region. Environmental issues are collectively discussed, all members contributing to the UN CONFERENCE ON ENVIRONMENT AND DEVELOPMENT 1992.

asthma A condition in which the air passages in the lungs (the bronchi) become narrowed because of a combination of muscular spasm, inflammation of the lining, and production of excessive mucus. Some of the recognized trigger mechanisms for the condition include: spring pollens; air-borne spores; dust and dust mite; animal feathers and fur; some organic chemicals; smog components; cigarette smoke; cold, dry air; certain food preservatives and colourings; certain medications; bronchial infections; and excessive exercise and emotional stress. Asthma is a common ailment, shows a familial incidence, affects all races, and occurs equally in males and females.

astrogeology A science concerned with the geology of the solid bodies in the solar system such as the asteroids and the planets and their moons. Research in this area offers a better understanding of the evolution of the Earth in comparison with that of its neighbours.

Aswan High Dam A rock-fill dam across the Nile River at Aswan, Egypt, completed in 1970; the dam impounds a reservoir, Lake Nasser. Lake Nasser backs up the Nile for some 320 km in Egypt, and a further 160 km further upstream in the Sudan, which shares the waters of the Nile. Creation of the reservoir necessitated the costly relocation of the ancient Egyptian temple complex at Abu Simbel. The flooding necessitated the relocation of 90 000 Egyptian fellahin and Sudanese nomads. The Aswan High Dam has yielded enormous benefits to the Egyptian economy, controlling the annual Nile flood and for the first time releasing the waters when needed onto irrigated land. The dam generates 2100 MW of electric power, while the reservoir supports what was a new fishing industry. On the other hand, the dam has produced some marked negative effects. There has been a gradual reduction in the productivity of riverside agriculture downstream, for the silt that was formerly deposited on the farmlands by the flood waters is now retained by the dam. The application of artificial fertilizers has been an inadequate substitute. Also, it promoted the spread of the disease SCHISTOSOMIASIS, through snails that live in the irrigation channels.

Athens Charter 1933 A statement which emerged from a series of discussions conducted during the International Congress for Modern Architecture held in Athens, Greece in 1933. The charter addressed a range of contemporary problems relating to building standards, open space, slum clearance, industrial location, traffic, and heritage matters. It has had considerable influence on urban design. The Athens Charter was largely the work of Le Corbusier (1887–1965) whose visionary city planning schemes had a profound influence on modern architecture and town design.

Athens Treaty An international treaty relating to land-based sources of pollution around the Mediterranean, drawn up in 1980 and ratified in 1983. Sixteen governments were parties to the Treaty, which was part of the Regional Seas Program promoted by UNEP to help protect particularly vulnerable seas and oceans. Out of the treaty came the Mediterranean Action Plan.

atmosphere The gaseous envelope of air surrounding the Earth, the principal constituents of which are nitrogen and oxygen, together with CARBON DIOXIDE and very small amounts of inert gases such as argon, krypton, xenon, neon, and helium. Also present are water vapour, traces of ammonia, organic matter, ozone, salts, suspended solid particles, and other GREENHOUSE GASES. The atmosphere is commonly divided into six layers on the basis of the vertical distribution of temperature and other parameters. (See HOMOSPHERE; IONOSPHERE; MAGNETOSPHERE; MESOSPHERE; STRATOSPHERE; THERMOSPHERE; TROPOSPHERE.)

atmospheric turbulence Irregular air movements in which the wind constantly varies in speed, direction, and agitation. Turbulence is an important additional factor in the dispersal of AIR POLLUTION as it helps to dilute and mix pollutants, dispersing them vertically and horizontally. Turbulence near the Earth's surface has a marked diurnal variation, reaching a maximum value around midday. Towards evening, the surface of the ground cools rapidly, chilling the air near the ground. When the air near the ground is colder than the air immediately above it, a TEMPERATURE INVERSION exists, with wind speed and gustiness decreasing sharply. In these conditions, the concentration of pollutants tends to intensify and may remain so all night until dawn and sunlight breaks up the inversion.

atmospheric sciences Sciences which focus on the structure and dynamics of the ATMOSPHERE of the Earth, divided traditionally into three topical areas: meteorology, climatology, and aeronomy. In meteorology, the primary concern is the day-to-day, hour-to-hour changes in weather within the lower STRATOSPHERE and TROPOSHERE. Climatology concentrates on weather in the same region, but over much longer periods ranging up to millions of years. Aeronomy is concerned with atmospheric regions above the lower stratosphere, dealing with such matters as the photochemical processes of the upper atmosphere, ionospheric physics, magnetospheric storms, and auroral phenomena.

attemperation The process of cooling combustion and other waste gases. Gases may be cooled by mixing air with them, by passing them through an indirect cooling system, or by introducing water in the form of a fine spray. Attemperation may often be necessary before passing gases through a cleaning system.

attenuation (1) The natural purification of polluted water by passage through the ground or prepared material; the gradual removal of suspended materials from a liquid by passage through porous material. (2) The reduction in magnitude of some variable in a transmission system; for example, the loss of power suffered by radiation of any form as it passes through matter.

attitudinal changes Fundamental shifts in the attitudes of a society or large part of a society over time to various issues and practices. For example, attitudes to slavery, women working in coal mines, child employment, physical handicaps, mental derangement, homosexuality, *de facto* relationships, illegitimacy, race and colour, alcohol and drugs, crime and punishment, suicide, cigarette smoking, pollution, environment protection, and SUSTAINABLE DEVELOPMENT.

audio frequency Any frequency in the range of normal human hearing. There is no general agreement on the range; some authorities quote 16 Hz to 16 kHz, others 20 Hz to 20 kHz.

audiogram A graph, usually plotted by an audiometer, showing a person's hearing response as a function of frequency.

audiometer An instrument for measuring a person's hearing with respect to tones of normally audible frequencies.

audit (See ENVIRONMENTAL AUDIT.)

Audubon Society A US society named after John James Audubon (1785-1851), ornithologist, artist, and naturalist, who became famous for his drawings and paintings of North American birds. The Audubon Society (more correctly the National Audubon Society) founded around the beginning of the 20th century is concerned with wildlife conservation, particularly with the protection of birds.

Australia, evolution of environmental policy See Box 2.

autotrophs Organisms which use inorganic materials such as CARBON DIOXIDE, water, and inorganic ions to manufacture complex organic substances. Most derive energy from light to do this and are known as the photosynthetic

---

Box 2 Australia: evolution of environmental policy

1879	Royal National Park established
1958–64	Australian States introduce air pollution control legislation
1964–79	Australian States introduce solid- and liquid-waste management and anti-litter legislation
1970	New South Wales (NSW) establishes the State Pollution Control Commission; Victoria (Vic.) establishes the Environment Protection Authority and Land Conservation Council
1970–71	Australian States introduce water-pollution control legislation
1971	Australian Environment Council established (later to become the Australian and New Zealand Environment and Conservation Council). Western Australia (WA) establishes Environment Protection Authority
1972	South Australia (SA) establishes an Environment Protection Council
1972–78	Australian States introduce noise control legislation
1973	Tasmania establishes a Department of the Environment
1974	Commonwealth (federal) Government introduces EIA legislation

continued overleaf

availability

Box 2	(continued)
1975	Commonwealth Government creates the Great Barrier Reef Marine Park Authority and the Australian Heritage Commission
1978	Victoria introduces EIA legislation
1979	NSW introduces EIA legislation and creates a Department of Environment and Planning
1980	Northern Territory establishes Conservation Commission
1981	SA establishes a Department of Environment and Planning
1983	National Conservation Strategy adopted by the Commonwealth and most states
1985	NSW introduces environmentally hazardous chemicals legislation; national Unleaded Petrol (gasoline) Program inaugurated
1987	State conservation strategies announced for Vic. and WA; Murray–Darling Basin Commission established; Ningaloo Marine Park created off WA coast
1989	Commonwealth Government creates Resource Assessment Commission (was abolished in 1994); national ozone protection legislation introduced; national industrial chemicals notification and assessment scheme operational
1992	Intergovernmental agreement on the environment between the Commonwealth, States, Territories and local government; national strategy adopted for ecologically sustainable development (ESD), following the reports of several ESD working parties; national water-quality management strategy; Commonwealth Environment Protection Agency created; NSW creates Environment Protection Authority (replacing the State Pollution Control Commission)
1994	National strategy for the conservation of Australia's biological diversity

autotrophs. A few derive energy from the oxidation of substances such as sulphur and iron and are known as chemosynthetic autotrophs. All other organisms depend on the autotrophs, directly or indirectly, for the complex organic compounds they need. Autotrophic organisms are described as the producer organisms of a community. Green plants are autotrophic. Those depending on autotrophs for food are known as heterotrophs and are also termed consumer organisms.

availability Plant capacity actually available for use, usually expressed as a percentage of the potential maximum capacity of installation. Availability may refer to a particular time, such as a time of peak demand for steam or electricity, or it may be the average taken over a period of, say, one year.

average individual risk The average chance of any individual in a defined population sustaining a given level of harm from incidents likely to affect that population. The peak individual risk is the highest individual risk for any person in the exposed population.

avicide A PESTICIDE used to kill birds.

Awash National Park Established in Ethiopia in 1969, a national park embracing the semi-arid open plains of the Awash Valley at the foot of the Shewa escarpment and includes Mount Fantale. The flora includes grassland, acacia, savannah, and thorny thicket. The fauna include hippopotamus, leopard, lion, cheetah, zebra, gazelle, antelope, and crocodile. The park stands adjacent to the Awash West Game Reserve.

B

Bacharach smoke scale A scale of ten shades from white to black used for the assessment of smoke concentrations in flue gases. A smoke stain is obtained by drawing a sample of flue gases through a filter paper in a prescribed manner, the resultant stain being compared with the scale. (See RINGELMANN CHART.)

backfill The process of refilling an excavation, or the material used for this purpose.

backflushing The cleaning of a water filter or ion-exchange column by reversing the fluid flow through the filter or bed, so that dirt and debris are removed from the bed. After a prescribed time, normal flow is resumed.

background The total landscape that is perceived by an observer from a particular viewpoint, ranging from 4 to 6 km to infinity. Within this range, the observer is unaware of individual details and discerns broader landscape units as patterns of light and dark.

background noise The level of NOISE experienced when all more immediate and intrusive sounds are temporarily eliminated; a constant level of noise from more faraway sources.

background radiation level The naturally occurring level of RADIATION to which all people are exposed over a wide area; for example, from rocks and soil.

backing wind The anti-clockwise change of direction of a wind, for example, from N through NW; an opposite change of direction to veering. The same definition applies whether in the northern or southern hemisphere.

backwater area A flood-prone area that does not experience significant velocities in water flow. As such, flood water may fill the area without adverse effects of flood levels or velocities.

BACT BEST AVAILABLE CONTROL TECHNOLOGY.

bacteria Microscopically small elementary forms of life, which can live anywhere organic matter is to be found. They do not require light for growth, but live by attacking and breaking down organic matter. In soils and compost heaps, they bring about the decay of animal and plant tissue, reducing complex materials to simple chemicals. Some cause diseases in humans, such as cholera, typhoid, and dysentery. The bacteria may be spherical (coccus type), rod-shaped (bacillus type), or spiral and cylindrical (spirillum type). (See ESCHERICHIA COLI.)

bactericide A chemical that kills vegetative forms of BACTERIA.

bacteriostatic Describing the inhibition of bacterial cell growth and propagation due to the effect of environmental factors.

BACTNEEC BEST AVAILABLE CONTROL TECHNOLOGY NOT ENTAILING EXCESSIVE COSTS.

bagasse The fibrous portion of sugar cane remaining after the juice has been extracted. Mills crushing sugar cane commonly use bagasse as fuel for steam production. There has been little incentive to burn bagasse efficiently, steam boilers serving also as incinerators. Low-cost scrubbers have been developed to restrict the emission of carbon particles to the general atmosphere.

bag filter A device for removing particulate matter from the waste gases of industrial processes. The filter medium is woven of felted fabric usually in the form of a tube. The bags may be up to 10 m in length and up to 1 m in diameter. The collecting efficiencies of bag filters are high, between 99 and 99.9%, but low gas velocities are required. Bag filters are used in US, UK, and Australian power stations, as well as a whole range of other industrial plant.

balance of nature In ECOLOGY, the concept of an inherent stability in most ECOSYSTEMS. Thus, organisms in an ecosystem are adapted to each other. The resources used by some organisms are replenished by others, and the waste products of one species are used by another. Human activity of one sort or another can interfere, therefore, with this stability. (See also ECOBALANCE.)

ballast water Usually, water taken on board by ships travelling empty to maintain stability at sea. Such water, perhaps contaminated, may then be discharged into national waters prior to taking on cargo. Sea water carried in oil tanks is discharged along with the residual oil during ballast blowing thus contributing to marine pollution. Some oil terminals are equipped with facilities to recover oil and minimize the contamination of water bodies. Ballast waste may be treated by gravity separation, passing then through interceptors and pressure sand filters. After three or four days, the cleansed effluent is discharged offshore via a diffuser on the sea bed. Recovered oil is injected into a crude-oil loading line.

Baltic Marine Environment Commission See CONVENTION ON THE PROTECTION OF THE MARINE ENVIRONMENT OF THE BALTIC SEA.

BANANA An acronym for a certain attitude to development: 'Build absolutely nothing anywhere near anybody'. It falls short of opposing any development anywhere and does not reflect a belief in zero economic growth shared by some.

band A segment of the frequency spectrum, for example, an octave, half-octave, one-third octave. The spectrum is a group of continuous frequencies rising from low to high.

Banff National Park Established in 1885, Canada's first national park located in Alberta on the eastern slopes of the Rocky Mountains embracing several large ice fields and glacial lakes. It has been enlarged over the years. The park's vegetation includes alpine meadows, while the fauna include bears, elk, deer, moose, and wild sheep and goats. A special feature are the hot springs at Sulphur Mountain. Banff National Park is contiguous with Jasper National Park in Alberta and Yoho National Park in British Columbia, Canada,

Barcelona convention A convention drawn up in 1976 by nations bordering the Mediterranean Sea, forbidding the discharge of a wide range of noxious substances into the sea, including organosilicon compounds, petroleum hydrocarbons, radioactive wastes, and acids and alkalis. (See ATHENS TREATY ON LAND-BASED SOURCES OF POLLUTION; CONVENTION FOR THE PROTECTION OF THE MEDITERRANEAN SEA AGAINST POLLUTION.)

barriers Defensive mechanisms and arrangements to protect the public against adverse effects. In the supply of clean and safe drinking water, for example, barriers should include most of the following measures: water sources should be protected from animal and human faeces with an active catchment-management programme in place; water should be pre-treated in reservoirs to allow bacteria to die off; water storages should be protected with coagulation, settling, and filtration carried out; water must be disinfected before it enters the distribution system; an adequate residual concentration of disinfectant should be maintained in the water throughout the distribution system; the distribution system should be secured against possible recontamination; and regular monitoring should be carried out. (See BACTERIA.)

basal energy requirement The energy required by a person when resting quietly, producing just enough energy to maintain basic body processes such as heartbeat and breathing.

Basel convention See CONVENTION ON THE CONTROL OF TRANSBOUNDARY MOVEMENTS OF HAZARDOUS WASTES AND THEIR DISPOSAL.

baseline information Information relating to a defined segment of the environment in terms of air, land, or water, from which trends or changes, short-term or long-term, can be identified and assessed. Such information may be an average representation of conditions (for example, water quality over five years) intended to take climatic variability into account. Such information is highly relevant to ENVIRONMENTAL IMPACT ASSESSMENT.

base-load power station Power stations within an electricity supply system that are in continuous use, day and night, in comparison with mid-load and peak-load stations, which are in use less frequently. Base-load stations have low generating costs and therefore stand high in the operating 'order-of-merit'.

bathyal zone A marine ecological region extending down from the edge of the continental shelf to the depth at which the water temperature is 4°C. The zone may vary therefore between 200 and 2000 m below the surface. PHOTOSYNTHESIS is rare in the bathyal zone. In many areas bathyal waters are essentially stagnant, resulting in low oxygen concentration and impoverished faunal levels. The fauna occurring reflect the generally narrow ranges of temperature and salinity that occur. The bathyal zone lies above the ABYSSAL ZONE, which is deeper than 2000 m. No photosynthesis occurs at that level and faunal life is sparse. Above the bathyal zone lies the EUPHOTIC ZONE.

batter The excavated or constructed face of a dam wall, bank, or cutting, resulting from earth-moving operations involving cutting and filling. In describing batter grade, one in three means a fall (or rise) of one vertical metre in a horizontal distance of 3 m. For SOIL CONSERVATION purposes, bank and dam batters are identified by numbering them 1, 2, and 3, commencing upslope.

Bavarian Forest National Park Established by Germany in 1970 as a contribution to the European Conservation Year, a national park located in southeastern Germany along the frontier with the Czech Republic. Tree-covered and mountainous, with the peaks Rachel and Lusen, the park is a refuge for the rarer kinds of wildlife.

Bayer Australia public inquiry In 1987, a public inquiry conducted under New South Wales (NSW), Australia, planning laws for a controversial development application by Bayer Australia to construct facilities for the formulation and storage of agricultural and veterinary products at Kurnell, on the south side of Botany Bay. The commission of inquiry found that the Bayer project could not operate without harmful effect on the local environment of Kurnell, in particular, the nature reserves, oyster leases, fishing and prawn breeding grounds, and the wetland generally of Botany Bay. The NSW Minister for Planning endorsed the commissioners' recommendations and the development application was refused.

beach The shore of a body of water, usually sandy or pebbly. It is usually defined by high- and low-water marks. In Britain, since 1988, beaches free of industrial pollution, litter and sewage have been entitled to display a blue flag.

beach and dunal system Typically, a sculptured sand mass with parts under water (offshore bar and surf zone), parts adjacent to the water with no vegetation (the BEACH) and parts further from the water with a vegetative cover (frontal, secondary, and back dunes). The vegetative cover stabilizes the dunes and enables them to function as a natural barrier between ocean forces and the land behind. Access to the beaches is often across the vegetated dunes that are readily disturbed. Damage is readily extended by the use of recreational vehicles. Exposure of bare sand often results in blowouts and movement of sand inland. The adoption of remedial measures often requires careful management.

beach barrier A prominent and wide ridge of sand rising above normal high-tide level, running parallel to the coast, and perhaps separated from it by a lagoon or marsh.

beach grass Also called marram grass, psamma, or sand reed, coarse perennial grasses about 1 m tall which grow on the sandy coasts of temperate Europe, North America and northern Africa. American beach grass grows along the Atlantic coast and round the Great Lakes. European beach grass has been introduced on the northern Pacific coast of the US as a dune stabilizer

Beaufort Sea hydrocarbons project An oil and gas development in the Beaufort Sea, Canada. A review of the proposal was carried out between 1980 and 1984 under the Canadian federal environmental assessment procedures to embrace the economic, developmental, technological, social, cultural, as well as the environmental aspects. The assessment team consulted 29 potentially affected communities (including the indigenous people). A number of recommendations related to phased development, with careful management and monitoring. The panel concluded that the local peoples should be able to manage the effects of the changes while deriving long-term benefit from the development, and the degree of risk to renewable resources should prove acceptable to the northerners.

Beaufort wind-force scale A scale devised in 1805 by Francis Beaufort (1774–1857) of the British navy for observing and classifying wind force at sea. It was adapted in 1874 by the International Meteorological Committee for international use in weather telegraphy. It has a scale of 0 (calm) to 12 (hurricane). Beaufort force numbers 13 to 17 were added by the US Weather Bureau in 1955.

Beaver Report The report of the Committee on Air Pollution presented to the British Government in 1954. The committee was set up to review air pollution problems in Britain following the disastrous London smog of 1952. It was chaired by Sir Hugh Beaver. The outcome was the Clean Air Act of 1956, strengthening measures against industrial pollution while inaugurating a programme of smoke-control areas to combat domestic pollution throughout Britain.

bel A unit of sound volume equal to 10 decibels, rarely used. (See dB(A).)

Belgrade Charter A declaration with recommendations arising out of the Belgrade International Workshop on Environmental Education convened in 1975 as part of a UNESCO-UNEP programme. The purpose was to formulate guidelines which could be applied internationally.

beneficial use In the context of environmental planning, a use of the ENVIRONMENT, or any element or segment of the environment, that is conducive to public benefit, welfare, safety or health, and which requires protection from the effects of waste discharges, emissions, deposits, and despoilation. Beneficial uses include: potable water supplies for drinking and domestic purposes; agricultural and industrial water supplies; wildlife habitat; recreational activities; and navigation.

beneficiation (1) The removal of valueless material from pulverized metal ore. (2) The upgrading of a resource that was once too uneconomical to extract and process. Upgrading often depends upon technological improvements such as those which make possible the concentration of minerals to permit easier handling, transportation, and processing. In relation to ores, a concentration process may include drying to reduce water content, roasting to reduce the sulphur content, and washing to remove some of the gangue or undesirable impurities.

benefit-cost ratio See COST-BENEFIT ANALYSIS.

benthic region The bottom of a body of water, often occupied by BENTHOS.

benthos Plants and animals living near or on the sea bed, attached or unattached.

bentonite A clay material used as an impermeable lining material in waterlogged ground, for example, in the construction of refuse storage bunkers and as a lining for excavations or quarries to be used for the tipping of solid refuse, with the purpose of preventing polluting LEACHATE from reaching GROUND WATER.

benzene An aromatic liquid hydrocarbon obtained from coal and petroleum. It is used as a solvent, as a fuel, in the manufacture of plastics and is a minor constituent of petrol (gasoline). It is highly toxic and long exposure to it may cause leukaemia.

benzpyrene A hydrocarbon present in coal and cigarette smoke, strongly suspected of being carcinogenic to humans.

Bergen Ministerial Declaration on Sustainable Development The outcome of a regional ministerial meeting of the UN ECE nations (Europe and North America) held in Bergen, Norway, as a follow-up of the 1987 report of the WORLD COMMISSION ON ENVIRONMENT AND DEVELOPMENT (the Brundtland Commission) from 8 to 16 May 1990. The object of the conference was to evaluate progress in the carrying out of the recommendations of the Brundtland Commission, and to adopt a politically binding final declaration on new joint measures to promote SUSTAINABLE DEVELOPMENT in the region. There was also discussion on how the ECE nations could improve the opportunities of developing nations to achieve sustainable development. The outcomes of other follow-up conferences, such as the African conference in Uganda in June 1989, were also taken into account. The Bergen ministerial declaration on sustainable development recommended the integration of environmental considerations into long-term economic and social planning, nationally and internationally, supported the objective that energy consumption should be stabilized and the use of renewable sources of energy encouraged, urged industry to integrate environmental considerations in planning and investment decisions and supported public participation in decision-making and environmental education initiatives. This declaration became an important part of the preparations for the UN CONFERENCE ON ENVIRONMENT AND DEVELOPMENT, held in 1992. (See OSLO CONFERENCE ON SUSTAINABLE DEVELOPMENT.)

Berlin Climate Conference A conference of the world's environment ministers held in March 1995, to review progress towards achieving the objects of the CONVENTION ON CLIMATE CHANGE that emerged from the UN CONFERENCE ON ENVIRONMENT AND DEVELOPMENT 1992. The aim was to reduce emissions of CARBON DIOXIDE to 1990 levels by the year 2000. Representatives from 150 countries attended, but there was little to report in the way of solid achievement in respect of carbon taxes, promotion of renewable energy sources, or curtailing of actual carbon dioxide emission from transport or industry. The conference received a report from the INTERGOVERNMENTAL PANEL ON CLIMATE CHANGE. (See CARBON TAX.)

berm A narrow path, bank, or barrier; for example, the term is currently used to describe a barrier adjacent to a motorway to intercept and deflect sound (to create, in effect, an acoustic shadow), such a barrier being either an earth mound or a concrete shield. The term is also used extensively in coastal and riverine flood prevention works.

Berne convention CONVENTION ON THE CONSERVATION OF EUROPEAN WILDLIFE AND NATURAL HABITATS.

beryllium A chemical element used in metallurgy as a hardening agent and in many space and nuclear applications. Berrylliosis is an industrial disease caused by poisoning with beryllium, usually involving the lungs. It can affect workers extracting beryllium metal from ore or manufacturing beryllium alloys, usually in an acute form. It can also affect scientific and industrial workers who are exposed to beryllium-containing fumes and dusts, usually in a chronic form. In both categories, the results can be serious.

best available control technology (BACT) Emission controls or production methods, techniques, processes, or practices which are capable of achieving a very high degree of reduction in the emission of wastes from a particular source. Financial and economic considerations are excluded from this concept.

best available control technology not entailing excessive costs (BACTNEEC) Essentially a refinement of BEST AVAILABLE CONTROL TECHNOLOGY (BACT), allowing the level of costs to be taken into consideration. A concept incorporated in the British Environmental Protection Act in relation to INTEGRATED POLLUTION CONTROL (IPC).

best practicable environmental option (BPEO) A term introduced by the ROYAL COMMISSION ON ENVIRONMENTAL POLLUTION in its Fifth Report (1976) in order to take account of the total pollution from an enterprise or activity and the technical possibilities of dealing with it; possibly a successor to BEST PRACTICABLE MEANS, which was primarily concerned with emissions to air. Apart from dealing with all pollutants, BPEO was to take into account the risk of pollutants transferring from one medium to another. (See BEST AVAILABLE CONTROL TECHNOLOGY; BEST AVAILABLE CONTROL TECHNOLOGY NOT ENTAILING EXCESSIVE COSTS; BEST PRACTICABLE MEANS; GOOD CONTROL PRACTICE; MAXIMUM ACHIEVABLE CONTROL TECHNOLOGY.)

best practicable means (BPM) A commonly used approach to pollution control requirements from industrial and other premises. The word 'practicable' is taken to mean 'reasonably practicable' having regard, among other things, to the state of technology, to local conditions and circumstances and to the financial implications. The concept is much easier to apply than the AIR QUALITY STANDARDS approach. (See BEST AVAILABLE CONTROL TECHNOLOGY; GOOD CONTROL PRACTICE; MAXIMUM ACHIEVABLE CONTROL TECHNOLOGY.)

best professional judgement A degree of judgement often required in the finalization of an EIS or EIA, which often lies beyond the realms of methodology in the narrower sense. The matrix, the checklist, together with advice of outside consultants and the results of exhaustive analyses may still lead to some doubt about the final outcome, that is whether the document is now ready to withstand challenge by the public and the makers. One of the basic problems is that in most assessments there are winners and losers. A new or extended airport may guarantee increased noise levels for some, or a hydroelectric dam may involve the relocation of residents. In other words, many issues have to balance, not simply development in relation to environment protection as an optimization exercise, but a balance in respect of interests and human welfare.

beta decay Any of three processes of radioactive disintegration. The three processes are called electron emission, positron (positive electron) emission, and electron capture. Beta decay was named by Ernest Rutherford in 1899 when he observed that radioactivity was not a simple phenomenon. He called the less penetrating rays 'alpha' and the more penetrating rays 'beta'. Beta decay is a relatively slow process.

beta diversity (habitat diversification) Diversity occurring as a result of competition between species, thus revealing an increasingly narrow range of tolerance of environmental factors.

beta particle A light particle produced in many nuclear reactions and in radioactive decay processes. It may carry a negative or positive charge, but in common usage the term refers to the negatively charged particles which are identical to electrons.

beta radiation Radiation consisting of high-velocity negative electrons. Their ionizing power is far less than that of alpha particles, but their penetrating power is greater.

betterment Also called unearned increment, a profit likely to accrue to the owner of land as a result of advantageous rezoning of the land, for example, from agricultural activities to housing. Some form of betterment tax is sometimes suggested as a way of capturing for the public some of the increased value resulting from government decisions about land use. Indeed, in Britain, the Town and Country Planning Act 1947 laid down the basic principle that simply to own land in an area scheduled for a profitable change of use did not entitle the owner to anything more than the return that would have been obtained from its sale for the original use. If sale occurred, the profit or betterment should go to the public authority. The experiment ended in the early 1950s because of difficulties in administration, rather than a change of view regarding the principle itself.

Bhopal disaster, India A catastrophic gas leak at a pesticide plant in Bhopal, India in December 1984 as a result of which over 2000 people, mostly children and older people, died and some 50 000 suffered from various degrees of blindness, temporary and permanent. Methyl isocyanate, a gas used in the manufacture of the pesticide Sevin, had leaked from the Union Carbide plant and flowed into congested neighbouring slum areas. (See MEXICO CITY INDUSTRIAL DISASTER.)

Bialowieski National Park Established in Narododowy, Poland, in 1947, a relatively small low-plains area, being the best preserved remnant of primeval European lowland forest with a range of vegetation. Some trees are hundreds of years old. The park is a breeding centre for the rare European bison.

bilharziasis See SCHISTOSOMIASIS.

bioaccumulation (bioconcentration) The process whereby a PESTICIDE becomes concentrated in living organisms; the build-up of chemicals in organisms at concentrations greater than the levels in the environment of the organisms; the uptake of substances from the environment, their concentration and retention by organisms.

bioassay The quantitative estimation of the potency or concentration of biologically active substances by the extent of their effects, under standardized conditions, on specific living organisms; the laboratory determination of the effects of substances upon specific living organisms.

biochemical oxidation The process by which microorganisms within an aerobic treatment process transform organic pollutants into settleable organic or inert mineral substances.

biochemical oxygen demand (BOD) An index of WATER POLLUTION, which represents the content of biochemically degradable substances in a sample of water. A test sample is stored in darkness for five days at 20°C; the amount of oxygen taken up by the microorganisms present is measured in grams per cubic metre. However, when samples contain substances such as sulphides, which are oxidized by a purely chemical process, the oxygen absorbed may form part of the BOD result. For this reason, the BOD test is no longer considered an adequate criterion by itself for judging the presence or absence of organic pollution. Further, it cannot be used to assess the presence of many more recent pollutants such as pesticides, industrial organic compounds, fertilizing nutrients, dissolved salts, and soluble iron. However, the test remains widely used, particularly in sewage treatment.

biochemistry The field of science concerned with the chemical substances and processes that occur in animals, microorganisms, and plants. Biochemistry involves the study of all the complex interrelated chemical changes that occur within the cell, for example, those relating to the conversion of food to energy, the transmission of hereditary characteristics, the analysis of the organic compounds that comprise the basic constituents of cells (proteins, carbohydrates, and lipids), and of those substances that play a key role (nucleic acids, vitamins, and hormones).

biocide Any agent that kills organisms.

bioclimatology The scientific study of the relationships between living organisms and climate.

biocoenology The study of complex associations, with the object of understanding the numerical determination of species.

bioconcentration See BIOACCUMULATION.

bioconversion (biological conversion) The conversion of the energy stored in plant materials such as trees, crops, algae, and water plants to produce ethanol, methanol, and methane. Crop wastes, urban wastes, sewage, and animal excreta can also be converted to liquid and gaseous fuels such as methane and hydrogen.

biodegradable Readily decomposed by bacterial activity. Soaps are readily decomposed shortly after discharge into a sewerage system. Earlier synthetic detergents of the alkylbenzene-sulphonate (ABS) type were non-biodegradable, possessed high foam stability, and remained intact for years. Subsequently, the detergent industry changed to linear alkylate sulphonate (LAS) which were amenable to the decomposing action of bacteria.

biodiversity (biological diversity) An umbrella term to describe collectively the variety and variability of nature. It encompasses three basic levels of organization in living systems: the genetic, species, and ecosystem levels. Plant and animal species are the most commonly recognized units of biological diversity, thus public concern has been mainly devoted to conserving species diversity. This has led to efforts to conserve endangered species and to establish specifically protected areas. However, sustainable human economic activity depends upon understanding, protecting, and maintaining the world's many interactive diverse ecosystems with their complex networks of species and their vast storehouses of genetic information.

bioengineering The application of engineering science to biology and medicine. Common applications include heart–lung machines, kidney machines, life-support systems, instruments for monitoring biological processes, and artificial organs and limbs. Of relatively recent origin, bioengineering has provided some of the most remarkable breakthroughs in medical science.

bioethics The discipline dealing with the ethical implications of both biological research and the application of that research, particularly in medicine. Bioethics involves such issues as the definition of death, the withdrawal of life-sustaining medical treatment, the storage of frozen human embryos, the use of animals and humans for scientific research, the expansion of genetic engineering, and the disposal of toxic wastes.

biofilter In the treatment of a gaseous pollutant, the passing of the contaminated gas through a porously packed bed in which bacteria feed on the chemical constituents of the gas. For example, the odours that are to be removed from a polluted gas may be the very substances on which bacteria thrive. Capital costs are relatively high, but biological treatment of gas in biofilters can be carried out with negligible operating costs.

biogas The gaseous product of anaerobic digestion of organic materials such as sewage sludge or whey. Mainly METHANE, such gases can be used for heat and power purposes.

biogenic Produced by the action of living organisms.

biogeochemical cycle The paths followed by the essential chemical elements in living organisms, from the ENVIRONMENT to the organisms and back to the environent. (See CARBON CYCLE; NITROGEN CYCLE.)

biogeography The study of the geographic distribution of fauna and flora and the factors responsible for variations in that distribution.

biological benchmark A concept in which fauna and flora are used to measure pollution, perhaps to augment the physicochemical studies traditionally employed. Benchmarks of population levels and fitness are required against which future changes can be evaluated.

biological conversion See BIOCONVERSION.

biological control The control of pests by using predators, and disease-producing organisms instead of using pesticides; the use of living things to control pests. For example, the use of enemy insects (wasps versus caterpillars) and the release of sterilized males to produce infertile matings.

biological diversity See BIODIVERSITY.

biological filter See TRICKLING FILTER.

biological monitoring The direct measurement of the changes in the biological status of a HABITAT, for example, to determine variations in the composition and abundance of BIOTA above and below an effluent outfall, or before and after the commencement of a potentially detrimental waste discharge.

biological pesticides A large number of organisms such as viruses, bacteria, protozoa, nematodes, and fungi, which are pathogenic to pests, or which in some way interfere with the normal biological pattern of pests so as to modify their damaging activities.

biological remediation The use of microorganisms and nutrients to deal with hazardous-waste problems, for example, contaminated land and manufacturing sites. Bioremediation has significant advantages over conventional methods of waste disposal; it can be carried out on site and avoids the need to transport and store hazardous and toxic waste. Different microorganisms are capable of degrading specific industrial contaminants including straight-chain hydrocarbons, polyaromatic hydrocarbons, chlorinated phenols, and other complex organic molecules.

biological shield A thick wall or shield, usually consisting of a 3 to 4 m thickness of concrete, surrounding the core of a nuclear reactor to absorb neutrons and gamma radiation for the protection of operating personnel and the public.

biology The study of living organisms and their vital functions. The two main divisions of biology are zoology (the study of animals) and botany (the study of plants). Other biological disciplines include: physiology (the functioning of organs); cytology (the study of cells); ecology (the study of organisms and their interactions with their environments); anatomy (the study of structures); morphology (the study of the form and structure of living organisms); genetics (the study of inheritance); biochemistry (the study of chemical processes in living organisms); and molecular biology (the study of biological processes at the molecular level).

bioluminescence The emission of light by an organism. Examples include glow-worms, phosphorescence of protozoans in tropical seas, signals of fireflies, many deep-sea fishes, bacteria and fungi.

biomagnification The increase in concentration of certain stable chemicals, such as HEAVY METALS or fat-soluble pesticides, that occurs in successively higher trophic levels of a food chain, each organism at the higher level consuming several at the lower level. Biomagnification does not occur with biodegradable substances. Also known as biological magnification, bioconcentration, or BIOACCUMULATION.

biomass (1) The total weight of all living matter in a particular HABITAT. Biomass is often expressed as grams of organic matter per square metre. Biomass differs from productivity, which is the rate at which organic matter is created by PHOTOSYNTHESIS. (2) Animal or plant matter used as fuel.

biomass energy The concept of using vegetation as a continuous source of future energy, relying on crops which offer a high yield in energy terms. It has been argued that biomass could contribute a significant proportion of society's energy requirements during the 21st century.

biome A community of plants or animals extending over a large natural area; a major regional ecological community such as a tropical rainforest.

biometeorology Studies involving pollutants and infectious agents, and their interaction with weather factors.

biophysics A discipline concerned with the application of the principles and methods of the physical sciences to biological problems. Much work in biophysics has been concerned with the role of electric pulses in the conduction of information by nerves and in muscular contraction. The ways in which birds fly and fish swim have involved complex methodologies and instrumentation. Studies have ranged from the use of ultrasonics by bats, low-frequency radar used by certain fish, the use of magnetic fields by pigeons, and polarization of sunlight by bees.

biosphere The sphere of living organisms; it comprises parts of the atmosphere, the hydrosphere (oceans, inland waters, and subterranean water), and the lithosphere. The biosphere includes the human HABITAT or ENVIRONMENT, in the widest sense of these terms. A contemporary view is that all these layers, both BIOTIC and ABIOTIC, as components of an integral complex of interdependent systems, should be known as the ECOSPHERE.

biosphere reserves An international network of reserves forming part of UNESCO's biosphere programme. The purpose of the reserves is to ensure the conservation of representative ecological areas and the genetic resources they contain, so as to implement the WORLD CONSERVATION STRATEGY and to strengthen international cooperation in the field of ecological research and monitoring.

biostimulants Substances which stimulate the growth of aquatic plants, for example, the addition of large amounts of nitrogen and phosphorus compounds to lakes may stimulate massive growth of microscopic plants such as blue-green algae or the larger waterweeds. The process is called cultural EUTROPHICATION. Sewage is a major source of biostimulants, particularly nitrogen and phosphorus.

biosystematics The study of the BIOLOGY of populations, particularly in relation to their breeding systems, reproductive behaviour, variation, and evolution.

biota FLORA and FAUNA; all organisms.

biotechnology The application to industry and commerce of developments in the techniques, methodologies, and instruments of research in the biological sciences.

biotic

biotic Relating to life and living systems, rather than the solely physical and chemical characteristics of the ENVIRONMENT.

biotic element The organisms which form the populations and communities within an ECOSYSTEM.

biotic index A rating system used in assessing the ecological quality of an ENVIRONMENT; the index ranges from 0 to 10. Very clean water with a wide variety of species, including pollution-sensitive creatures such as stonefly and mayfly nymphs, has a high biotic-index score. With increasing pollution, the oxygen levels decrease and the more sensitive species disappear. Very polluted water with a low or zero level of oxygen, possessing only a few tolerant species such as annelid worms and red midge larvae, has a very poor biotic score.

biotic interaction The relationships between living organisms in a biological or ecological community.

biotic potential The maximum reproductive capacity of an organism, assuming optimal environmental conditions; biotic potential is restricted in practice, however, by environmental resistance, that is, any condition that inhibits an increase in the number of the population. Resistance includes: unfavourable climatic conditions; lack of space, surface, or light; an inadequate supply of minerals; the presence of predators, parasites, or disease organisms; or unfavourable genetic variations.

biotic pyramid A pyramid presented by a stable FOOD CHAIN with primary producers at the base (mainly vegetation), then primary consumers (herbivores), secondary consumers (omnivores such as humans and some carnivores), and then tertiary consumers (large carnivores). The number of individuals at each level decreases upwards. (See AUTOTROPHS.)

biotope A region of relatively uniform environmental conditions occupied by a given plant community and its associated animal community. The interdependent biological and physical components are in equilibrium if their relative numbers remain more or less the same, forming a stable ecological community or system.

bioturbation A disturbance of sediments by the activities of living organisms.

biotype A group of individuals exhibiting the same genetic characteristics.

Birds of North America An outstanding work of the US naturalist John James Audubon (1785–1851), who produced a similar work on American quadrupeds. (See AUDUBON SOCIETY.)

bituminous coal The best-known of solid fuels. The description 'bituminous' was originally applied because of the tendency to burn with a smoky flame and melt when heated in the absence of air. The volatile content varies between 20 and 35%, and the fixed carbon content between 45 and 65%. Most bituminous coals have a banded or laminated structure, and a shiny black appearance. The introduction of automatic stokers and stack monitoring not only ensured efficient combustion, but also steam raising without the emission of dark or black smoke, under most conditions. This was the first major achievement in the battle against AIR POLLUTION.

Black Country A central area of England around and to the north of Birmingham. Heavily industrialized, it gained its name during the 19th century for its belching chimneys and dirty environment. The progressive implementation of clean air legislation in respect of both industrial and residential pollution has done much to clean up the Black Country.

Black Death Name given to the great epidemic of plague, probably both bubonic and pneumonic, originating in China and Inner Asia, which ravaged Europe between 1347 and 1351, killing between one-third and one-half of the population. The cause of the Black Death was a bacillus transmitted by rat fleas. A rough estimate is that about 25 million people in Europe died from plague. In England, about 1000 villages were completely depopulated. The Great Plague of London in 1664–65 resulted in more than 70 000 deaths. An outbreak in Canton and Hong Kong in 1894 spread throughout the world with more than 10 million deaths. In 1994, bubonic and pneumonic plague broke out in a region of central India that had been recently devastated by an earthquake. It was thought that tremors had driven swarms of rats carrying fleas out of the forests into the villages. The epidemic was fought with pesticides and tetracycline.

Black Sea An inland sea of southeast Europe, linked via the Dardenelles with the Mediterranean. It receives the inflows of several major rivers including the Danube, the Don, the Dnieper, and the Dniester. It is bordered by the Ukraine and Russia to the north, Turkey to the south, and Bulgaria and Romania to the west. Oxygen is dissolved only in the upper water levels; below a certain depth there is no oxygen because the sea is permeated by a high concentration of dissolved hydrogen sulphide, forming a dead zone inhabitable only by specially adapted bacteria. In addition, increasing pollution loads from the rivers, especially the nutrients nitrogen and phosphorus, have led to algal blooms and the destruction of important nursery areas for fish. The damming of the major rivers has considerably altered the seasonal flow patterns of these rivers resulting in an increase in SALINITY in critical coastal and estuarine areas, especially in the Sea of Azov which creates further problems for fish breeding, which has seriously declined. With assistance from the GLOBAL ENVIRONMENT FACILITY, the six Black Sea countries (Bulgaria, Georgia, Romania, Russia, Turkey, and the Ukraine) have begun a regional programme to analyse the causes of the environmental degradation and propose solutions.

blasting The use of explosives to break up rocks; a common feature of surface coal mining and quarrying operations. Overburden, for example, is often removed or loosened with the aid of explosives. In some mines explosives are used to remove or loosen up to 50% of the overburden. Blasting is also used in many mines to break the coal seams. This activity can be a significant source of noise, shock, vibration, and dust for nearby residents. Awareness of wind direction, careful consideration of charge size, and the restriction of blasting to daylight hours can assist in minimizing NOISE POLLUTION and surface damage.

blast wave A pressure pulse formed by an explosion; a blast wave consists of an initial positive-pressure phase followed by a negative-pressure phase.

BLEVE BOILING LIQUID EXPANDING VAPOUR EXPLOSION.

BLM, US BUREAU OF LAND MANAGEMENT, US.

bloom A readily visible proliferation of PHYTOPLANKTON, macrophytes, or zooplankton in a body of water. (See ALGAE; BLUE-GREEN ALGAE.)

blue-green algae (cyanophytes) Single-celled, primitive organisms that resemble bacteria in their internal cell organization, sometimes joined together in colonies or filaments. Blue-green algae are widely distributed being extremely common in fresh water, tide pools, coral reefs, and tidal spray zones. On land blue-green algae are common in soil down to a depth of 1 m. They live in some of the most inhospitable environments known, in hot springs, cold lakes, beneath ice, and in deserts. Blue-green algae can reproduce at times at explosive rates, blooms colouring an entire body of water. They can consume so much of the dissolved oxygen that fish and other aquatic life die. In 1991, a startling bloom occurred along a 1000 km stretch of the Darling River in New South Wales, Australia. Many blooms are capable of producing toxins that have caused death to birds, fish, and animals in many countries and have been linked to human illness. Blue-green algae are not necessarily blue or green; some may be brown, yellow, black, or red.

BOD BIOCHEMICAL OXYGEN DEMAND.

body burden The total amount of a substance present in the body tissues and fluids of an organism.

boiling liquid expanding vapour explosion (BLEVE) A term introduced originally in the US to describe the sudden rupture, as a result of fire impingement, of a vessel or system containing liquefied flammable gas under pressure. The release of energy from the rupture (the pressure burst) and the flashing of the liquid to vapour creates a localized blast wave and potential missile damage, while the immediate ignition of the expanding fuel-air mixture and intense combustion creates a fireball. BLEVE accidents have proved quite devastating, with tanks and equipment being projected a considerable distance. The most stringent measures are required by way of separation and protection of storage tanks to minimize the possibilities of fire and explosion. (See Figure 3.)

bone conduction The means by which sound can reach the middle ear and be heard without travelling via the air in the ear canal. Sounds vibrating the ear drum (or tympanic membrane) are transferred to the membrane of the inner ear by three small bones, the auditory ossicles.

Bonn conventions See CONVENTION ON THE CONSERVATION OF MIGRATORY SPECIES OF WILD ANIMALS; CONVENTIONS FOR THE PROTECTION OF THE RHINE AGAINST POLLUTION.

borrow pit or area An excavation from which material such as sand, gravel, clay, or soil is removed to assist an industrial development. The material may be 'borrowed' but is not usually returned.

botany The branch of BIOLOGY that deals with plants; it involves the study of the structure, properties, and biochemical processes of all forms of flora, including trees. The principal branches of botany are morphology, physiology, systematics (the identification and ranking of all plants), and ecology.

bottom sediment and water Description of contaminants of fuel oil; the greatest amounts of sediment and water are found in residual fuel oils.

bottoms Liquid which collects in the bottom of a vessel during a distillation process or while in storage, hence, tank bottoms, tower bottoms.

botulism Poisoning by a toxin known as botulinus toxin or botulin, produced by *Clostridium botulinum* bacteria, often linked with the eating of improperly sterilized canned foods containing the toxin. However, it is very rare in foods that are canned commercially.

BPEO BEST PRACTICABLE ENVIRONMENTAL OPTION.

BPM BEST PRACTICABLE MEANS.

'Braer' incident On 5 January 1993, the foundering of the oil tanker *Braer*, being driven onto rocks in Quendale Bay in the Shetland Islands off Scotland. The cargo of

Brandt Commission

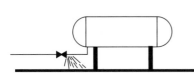

1. Pipework associated with a liquefied petroleum gas (LPG) pressurized storage tank leaks.

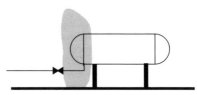

2. The gas leak is ignited by a flame or spark.

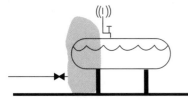

3. The flame heats the tank. LPG boils, raising the internal pressure; the relief valve operates and the escaping vapour ignites.

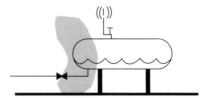

4. The level of liquid steadily falls, until the flame impinges on the metal which is above the liquid level.

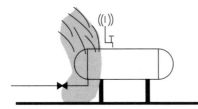

5. The metal heats to a point where it fails.

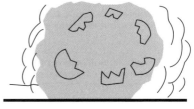

6. The tank tears apart almost instantaneously. The depressurized liquid erupts into vapour spray, which is ignited by the existing fire. The fireball spreads to its full diameter within perhaps one second. Large fragments of the tank may travel 300 m.

BOILING LIQUID EXPANDING VAPOUR EXPLOSION: Figure 3. Sequence of events leading to a boiling liquid expanding vapour explosion (BLEVE).

85 000 tonnes of light-crude and 5000 tonnes of heavy-fuel oil were lost, with extensive contamination of both the east and west coasts of the Shetlands and damage to farmed salmon.

Brandt Commission A commission set up in 1977 at the instigation, among others, of the WORLD BANK; the twenty members of the Commission were drawn from rich northern industrial countries and poorer southern countries. The purpose of the commission was to recommend an emergency programme in respect of international trade imbalances and longer-term reforms, The 1980 report of the commission *North-South: A Program for Survival* envisaged a large transfer of resources to the less developed countries to reach 0.7 of the GDP of the rich countries by 1985, and 1.0% by the year 2000, together with other measures by way of increased lending and major reforms in the international economic system. The report proposed a new institution, a World Development Fund, with universal taxes flowing into its coffers. Since that time, few countries have achieved that target, or even approached it. In 1992, at the UN CONFERENCE ON ENVIRONMENT AND DEVELOPMENT, in endorsing AGENDA 21, the developed countries reaffirmed their commitment to reach the target of 0.7% of GDP for foreign aid as soon as possible. It was also decided that the COMMISSION FOR SUSTAINABLE DEVELOPMENT would regularly review and monitor progress towards this target. (See SUSTAINABLE DEVELOPMENT.)

Box 3 Brazil: evolution of environmental policy

1939 Iguaca, Parque Nacional do established
1966 Amazon Basin opened up for ranches
1973 Special Environment Agency (SEMA) established to identify environmental problems and provide solutions, recommend standards in respect of pollution, and protect fauna and flora
1974 Amazonia, Parque Nacional do established
1975 Limits set for mercury in water
1976 Limits set for bathing waters; classification of waters; water quality criteria; ban on non-biodegradable detergents; air quality standards proclaimed
1977 Programme of ecological stations introduced
1978 Policy on rational use of water resources; treatment of effluents
1979 Policy on treatment and disposal of solid wastes; Forest Policy Committee recommends reversal of policy on opening-up the Amazon with the establishment of national parks and ecological reserves
1984 Cubato disaster; deaths of some 500 people following leaking gasoline pipeline
1988 Creation of National Environment Guard to deter further deforestation by ranchers and miners and lumber and rubber companies; murder of internationally known ecologist Francisco Mendes Filho
1992 Brazil hosts the UN Conference on Environment Development, Rio-de-Janeiro, 3–14 June
1993 Two Brazilian conservationists murdered, one opposed to the extraction of sand from beaches, dunes, and salt marshes, the other opposed to the logging of mahogany in tribal lands and ecological reserves
1994 The World Bank funds the Water and Sanitation Programme for low-income urban populations, assisting some 800 000 people in eleven Brazilian towns and cities

Brazil, evolution of environmental policy See Box 3.

Bretton Woods Agreement An agreement on the aims of post-Second World War monetary policy reached at an international conference held at Bretton Woods, New Hampshire, USA in 1944. A direct result of the agreement was the creation of the INTERNATIONAL MONETARY FUND and the WORLD BANK.

Britain, evolution of environmental policy See Box 4.

British Standard 7750 A product of the British Standards Institute on environmental management systems. BS 7750 is similar to the EUROPEAN ECO-MANAGEMENT AND AUDIT REGULATION (EMAR), both including elements such as the ENVIRONMENTAL AUDIT. Yet, there are also some significant differences. The British standard offers a framework for responsible environmental management in industry, a code which many companies will find it advantageous to adhere to. A system of independent certifiers will assess individual companies and organizations as to their compliance both with BS 7750 and the EU EMAR Scheme.

bronchitis Inflammation of the mucous membrane of the bronchial tubes existing in either an acute or chronic form. Epidemiological studies have indicated that there is a correlation between high concentrations of air pollution and morbidity and mortality among chronic bronchitics, although the ultimate progress of this disease is sensitive, particularly in winter, to many stimuli of which AIR POLLUTION is but one. Cigarette smoking also contributes to this disease.

Brundtland Commission See WORLD COMMISSION ON ENVIRONMENT AND DEVELOPMENT.

brunisolic soil A major soil type, associated with deciduous temperate forests; a brown forest soil. The most important brunisolic regions are in western Europe, the eastern United States, Washington State, and the adjacent Canadian coast.

bubble concept In the US a basis for emission trading; firms are encouraged to combine the emissions from their numerous outlets for the purpose of defining the emission limitations for the entire industrial facility, and then to develop their own strategies for different levels of control at different sources. The entire

bubonic plague

Box 4 Britain: Evolution of environmental policy

1661	John Evelyn's thesis on London smog
1848	Public Health Act
1875	Public Health Act
1906	Alkali, etc. Works Regulation Act
1932	Town and Country Planning Act
1936	Public Health Act; Housing Act
1937	Food and Drugs Act
1945	Water Act
1947	Town and Country Planning Act
1948	Rivers Boards Act
1949	National Parks Act
1951	Rivers (Prevention of Pollution) Act
1952	Town Development Act
1954	Report of the committee on air pollution (Beaver report)
1956	Clean Air Act
1960	Clean Rivers (Estuaries and Tidal Waters) Act; Noise Abatement Act; Radioactive Substances Act
1961	Pippard report on the pollution of the tidal Thames
1964	Harbours Act
1968	Town and Country Planning Act
1970	Department of the Environment established; Royal Commission on environmental pollution created
1971	Town and Country Planning Act; Merchant Shipping (Oil Pollution) Act
1972	Poisonous Waste Act
1974	Control of Pollution Act; Dumping at Sea Act; Health and Safety at Work Act
1978	Commission on energy and the environment
1980	Highways Act
1988	Town and Country Planning (Assessment of Environmental Effects) Regulations
1989	Electricity Act
1990	Town and Country Planning Act
1991	Environmental Protection Act; Planning and Compensation Act
1992	Transport and Works Act; Town and Country Planning (Assessment of Environmental Effects) Regulations; Harbour Works (Assessment of Environmental Effects Regulations); EIA guides published by Kent and Essex County Councils, the Passenger Transport Executive Group and the Department of the Environment
1993	Clean Air Act; Radioactive Substances Act
1994	Environmental Agency for England and Wales established; National Environmental Technology Centre created to bring best environmental research under one roof; Conservation (Natural Habitats) Regulations; Strategy on Sustainable Development

---

plant must stay, however, within the overall emission ceiling set by the US Environmental Protection Agency. The bubble concept enable managers to impose more severe limitations on emissions which can be restricted relatively inexpensively, in exchange for reduced controls on outlets more expensive to control. (See EMISSION TRADING PROGRAM.)

bubonic plague See BLACK DEATH.

buffer zone An area of land separating land uses which are incompatible with each other, which should be of sufficient width to prevent any conflict between them. Buffer zones may be established to separate industrial from residential areas or to separate airports from all other developments. Buffer zones may be planted with various forms of suitable vegetation, shrubs and trees, and may be contoured to form noise BUNDS. Buffer zones may form also a part of an open-space programme. Certain innocuous forms of activity may be permitted within a buffer zone, such as recreational facilities.

building codes   Rules imposed by planning authorities on the actual construction details of an approved development, relating to such matters as: the position of the building in relation to other buildings and the boundary of the site; the proportion of the site to be covered by the building; means of access and egress; landscaping; the height of the building; design and materials; parking; evidence of a sanitary site; stability; and the application of specific codes such health and sanitary codes, fire codes, plumbing codes, electrical codes, mechanical codes, and traffic regulations.

building standards   See DEVELOPMENT STANDARDS.

built environment   A reference to buildings, structures, utilities, and services that enable people to live, work and play, circulate and communicate, and fulfill a wide range of functions. The quality may range from grandeur to blight, but all is a product of human activity, as distinct from the natural ENVIRONMENT, which is nearer to its original state.

bund   An earthwork or wall surrounding a tank or tanks to retain the contents in the event of the fracture of the tank; or an earthwork or screen separating a source of noise from residents to minimize impact.

Bureau of Land Management, US (BLM, US)   An agency created by the US Federal Land Policy and Management Act 1976. Initially, the BLM focused on the traditional uses of land such as grazing, timber harvesting and the production of minerals and energy. More recently, the BLM has concentrated on a multiple-use approach which balances the economic uses of natural resources with conservation and outdoor recreation, including the management of fish and wildlife, wilderness, wetlands, and riparian ecosystems.

Burley Griffin Plan   The plan adopted for Australia's capital city, Canberra, following a competition in 1911. Walter Burley Griffin (1876-1937), the winner, was an American architect from Chicago. The main features of his design were a system of grand avenues radiating from a central point at which some imposing structures of administration would be built and a lake (later Lake Burley Griffin) formed from damming the Molonglo River. There have been modifications through the years, but the broad principles remain unchanged in a city with a population of over 300 000.

Business Charter for Sustainable Development   A charter developed by the International Chamber of Commerce during the early 1990s. It is composed of 16 principles for environmental management that have been endorsed by over 1500 companies worldwide. These principles serve as a template for companies to create environmental standards and pursue the objectives of SUSTAINABLE DEVELOPMENT.

business park   An area zoned for business premises, usually located out of town, serviced by a developed transport system, preferably on a motorway and close to an airport.

C

cadmium A white metallic element; used as an alloy, it gives strength and ability to withstand high temperatures. Alloys include cadmium-nickel alloy, cadmium-silver alloy and cadmium-copper alloy. It is used in bearing metals, in electroplating, and for some types of nuclear-reactor control rods. Cadmium has toxic qualities. It occurs in minerals used for zinc production and is a potential pollutant around zinc smelters. Human intake of cadmium can be from food and drink as well as air. A painful malady, known as the itai-itai or ouch-ouch disease afflicted the residents of the Japanese town of Fuchu in the Toyama Prefecture north-west of Tokyo for many years. Deaths were attributed to the affliction. The blame was eventually attributed to water pollution, responsibility being placed on a mining company whose waste from a cadmium mine polluted the Jintsh River, which flows through Fuchu. Crops were also damaged.

Cairo Guidelines Developed by UNEP and adopted in 1987, a set of guidelines and principles for the environmentally sound management of HAZARDOUS WASTES. (See CONVENTION ON THE CONTROL OF TRANSBOUNDARY MOVEMENT OF HAZARDOUS WASTES AND THEIR DISPOSAL.)

calcium cycle The circulation of calcium atoms brought about mainly by living things. Calcium is taken up from the soil by trees and other plants and deposited in roots, trunks, stems, and leaves. Insects and cattle and other herbivores obtain their share of calcium from the plants and leaves; birds acquire it by eating the insects. Animals and birds die, leaves and branches fall and decay, and thus the calcium component returns to the soil. Runoff may carry some calcium to bodies of water where it is recycled through phytoplankton, zooplankton, and fish. In a balanced ECOSYSTEM gains equal losses.

Campylobacter A microorganism, transmitted in water and food, that may cause gastroenteritis.

Canada: evolution of environmental policy See Box 5.

Canada-Ontario Agreement respecting the Great Lakes Basin Ecosystem An agreement reached in 1994, between the Canadian federal government and the provincial government of Ontario in respect of an ecomanagement strategy for the improvement, protection, and preservation of the fragile ecosystem of the Great Lakes. Remedial action plans have been introduced.

Canadian Council of Ministers of the Environment (CCME) An intergovernmental forum for discussion and joint action on environmental issues of provincial, national, international, and global concern. The CCME comprises all environmental ministers at the federal, provincial, and territorial levels. It has issued 'a national commitment to pollution prevention', which stresses the need for anticipating and preventing pollution instead of simply reacting to it after the fact. A guiding principle is that prevention should be considered at the earliest possible point in the development of any concepts, plans, policies, products, projects, or processes.

Canadian Green Plan A programme initiated by the Canadian federal government in 1990 with the purpose of improving the environmental wellbeing of Canadians. The Green Plan sets out programmes to achieve the sustainable use of renewable resources, including forests, agriculture, and fisheries, and to promote biological diversity. From this evolved the Canadian Biodiversity Strategy of 1994.

Canadian pollution release inventory See POLLUTION RELEASE INVENTORY, CANADIAN.

Canaima National Park A Venezuelan national park, established in 1962. A mountainous region, it has primary tropical rainforest, thickly wooded riverine areas, grassland savannah, and open meadow. Fauna include the jaguar, tiger, tapir, armadillo, opossum, numerous monkeys, and many birds. A special feature is Salto Angel, the highest cataract in the world.

Canberra convention See CONVENTION ON THE CONSERVATION OF ANTARCTIC MARINE LIVING RESOURCES.

canopy The total cover provided by the upper layers of tree foliage. Canopy cover is often expressed as the percentage of the ground that is covered when the extremities of the canopy are projected to the ground.

Canyonlands National Park A national park established in 1964 in Utah, USA. Essentially a wilderness area of colourful rock formations at the confluence of the Colorado and Green Rivers. Both rivers have cut deep gorges through the sandstone in the area. The area has many mammals and birds, including a herd of endemic desert bighorn sheep.

capacity The estimated maximum level of production from an industrial plant or enterprise on a sustained basis, allowing for all necessary shut-downs such as maintenance periods and holidays. Variations in demand due to market conditions usually ensure that few plants run continuously at full capacity. In times of recession, excess capacity becomes the norm. (See CARRYING CAPACITY.)

> **Box 5 Canada: evolution of environmental policy**
>
> 1868 Fisheries Act
> 1885 Banff National Park, Alberta, established
> 1895 Waterton Lakes National Park, Alberta, established
> 1909 Boundary Waters Treaty (USA–Canada)
> 1946 Atomic Energy Control Act
> 1957 Fertilizers Act
> 1970 Canada Water Act; Canada Shipping Act; Fisheries Act; Northern Inland Waters Act; National Parks Act
> 1971 Clean Air Act; Department of Environment (Environment Canada) established
> 1972 Great Lakes Water Quality Agreement (USA–Canada); Ministry of the Environment (Environment Ontario) established; Pest Control Products Act
> 1973 Federal Environmental Assessment Review Office (FEARO) established
> 1975 Environmental Contaminants Act; Ocean Dumping Control Act; Environmental Assessment Act (Ontario)
> 1980 Memorandum of Intent on Transboundary Air Pollution (USA–Canada)
> 1985 Amendments to the Canada Water Act, Clean Air Act, Canada Shipping Act, Fisheries Act, Northern Inland Waters Act, Pest Control Products Act, Environmental Contaminants Act
> 1989 Environmental Protection Act
> 1992 Canadian Environmental Assessment Act; creation of Canadian Environmental Assessment Agency
> 1993 Canadian round tables on sustainable development
> 1994 Kitlope Valley temperate rainforest wilderness preserved, with 800-year-old trees; Wild Animal and Plant Protection and Regulation of International and Interprovincial Trade Act; Canada–Ontario Agreement on the Great Lakes Basin; Canadian Biodiversity Strategy

capacity costs Fixed or overhead costs or charges, costs and derived charges that do not vary with the total amount of goods and services produced, but only with additions to the total CAPACITY of the system. In respect of electricity generation, capacity costs are those incurred in the initial construction of power stations (including the cost of land and, in the case of nuclear power stations, the initial charges of fuel and heavy water), transmission and distribution networks, and control centres.

capacity building See UN DEVELOPMENT PROGRAM.

capital One of the FACTORS OF PRODUCTION that may be defined as wealth used for the production of further wealth, or simply a commodity used in the production of other goods and services. The term is also commonly used for the money subscribed by shareholders or stockholders, or lent by banks or financial institutions, for use in a business. It may also be used with reference to an individual's savings. Reference is sometimes made to human capital, being the sum of all the skills and energy embodied in the community at large or, more narrowly, to the workforce. Sometimes natural resources such as soil, water, fossil fuels, minerals, and the natural environment are referred to as natural capital.

capital-intensive Describing those forms of production in which there is a considerable use of capital equipment per person employed. In a capital-intensive industry, the capital charges (the interest on capital employed and provision for depreciation and repayment of the principal) may account for 50% or more of production costs, excluding labour, raw materials, and energy. In chemical processes, the value of capital equipment per worker is very high.

capitalization of gains, socialization of losses A tendency in both the private and public sectors of the economy to keep all gains during the good times, and to plead for taxpayers' support during adverse market conditions. The arguments are usually that the enterprise is vital to the economy, that many jobs are at stake, and that the help sought at public expense either through direct financial assistance, or measures of protection, would be needed for a short period only. Such assistance may be sought by all types of activity such as car manufacturing, steel-making, exporting, developing, or farming.

capitalized value The capital equivalent of an annual payment; capital values are determined by the current rate of interest. For example, if the prevailing rate of interest for investments is 10% per annum and a parcel of land has a net rental of $50 000, then the capitalized value will be $500 000. If the rate of interest falls, then the capital value of the land increases, the rent still being regarded as corresponding to the current rate of interest, all other factors remaining the same. However, if the site has some prospect of development or rezoning, the market value may greatly exceed, at least temporarily, the capitalized rental value. Opposition to future development or an application for rezoning may, on the other hand, depress the market value to below its capitalized rental value.

Capitol Reef National Park A national park in Utah, USA, created as a national monument in 1937 and as a national park in 1971. It comprises a great buttressed cliff of coloured sandstone extending for 32 km along the western edge of the Water Pocket Fold. The Fremont River and its tributaries cross the park through deep canyons. The vegetation is desertlike and sparse.

carbamates Synthetic organic pesticides which contain carbon, hydrogen, and nitrogen; carbamates may be fungicides, herbicides, or insecticides. They have a low toxicity for mammals and persist in the soil for only a short period. They include carbamates, thiocarbamates, and dithiocarbamates.

carbohydrate An organic compound consisting of a chain of carbon atoms to which hydrogen and oxygen are attached in a 2:1 ratio. Examples include sugars, starch, cellulose, and glycogen. Green plants utilize the energy of sunlight to convert CARBON DIOXIDE and water into carbohydrates. This process, called PHOTOSYNTHESIS, releases oxygen into the atmosphere.

carbon cycle The cycling of the element carbon from non-living surroundings through organisms and back again. All communities in which there are AUTOTROPHS show some cycling of carbon. CARBON DIOXIDE is absorbed from the ATMOSPHERE through the stomata of leaves, and is converted into complex compounds such as sugar, starch, and other carbohydrates, and proteins and fats. These form parts of the structure of the plant. Plants are consumed by HETEROTROPHS, and these may be eaten in turn. Thus, carbon atoms have been taken from the air by producer organisms and have been included in organic compounds, which are then passed from one consumer organism to another. Carbon atoms are also progressively released into the ENVIRONMENT by every organism in the community through the decomposition of organic compounds in respiration and through the decay of dead material. (See GREENHOUSE EFFECT.)

carbon dioxide A normal constituent of the ATMOSPHERE, relatively innocuous in itself but playing an important role in the GREENHOUSE EFFECT. It is produced during the combustion of fossil fuels when the carbon content of the fuels reacts with the oxygen during combustion. It is also produced when living organisms respire. It is essential for plant nutrition and in the oceans phytoplankton is capable of absorbing and releasing large quantities of the gas. (See CARBON CYCLE.)

carbon monoxide A colourless odourless gas that readily combines with the haemoglobin of the blood to produce carboxyhaemoglobin, which inhibits the absorption of oxygen. The principal source of carbon monoxide in the general ATMOSPHERE is from motor vehicle exhaust gases.

carbon tax A tax on the consumption of fossil carbon-containing fuels in order to discourage consumption, to reduce CARBON DIOXIDE emissions, and to provide funds to promote other measures to reduce the GREENHOUSE EFFECT. Coal would attract the highest tax, followed by oil, and then natural gas. Renewable non-carbon-emitting energy sources such as solar energy, wind, and geothermal power would not be taxed. In Sweden, during 1990-91, environmental charges were imposed on fossil fuels and a value-added tax on energy. The taxes embrace carbon dioxide and also sulphur and nitrogen oxides. The present carbon dioxide tax for industry may be adopted by the EU for general application.

carboxyhaemoglobin (COHb) A combination of CARBON MONOXIDE and haemoglobin, which is the respiratory pigment occurring in blood. The haemoglobin is deprived of its oxygen-exchanging properties, with resultant poisoning and suffocation. Haemoglobin, while having a marked affinity for oxygen to form oxyhaemoglobin, has an even greater affinity for carbon monoxide. Individuals absorb carbon monoxide from smoking and smoke produced by other smokers, from the emissions of petrol-driven vehicles, and sometimes from occupational exposure.

carcinogenic compounds Complex chemical compounds producing cancer in humans. Carcinogenic compounds have been identified in soot, carbon black, processed rubber, exhaust gases, and cigarette smoke.

Caribbean Action Plan See CONVENTION FOR THE PROTECTION AND DEVELOPMENT OF THE MARINE ENVIRONMENT OF THE WIDER CARIBBEAN REGION.

'Caring for the Earth' See WORLD CONSERVATION STRATEGY.

carnivore An organism, animal or plant feeding on and digesting animal substances as a source of energy and nutrients.

car pool system A system aiming at the more efficient use of the car by increasing occupancy from typically one person per vehicle to several; it has been urged that higher-occupancy vehicles should be permitted to use preferential bus/car-pool lanes and many traffic authorities allow this.

carrying capacity The maximum number of individuals of a species that can be supported in an area. The carrying capacity of an area is usually limited by components of the ENVIRONMENT, such as food, nesting sites, and other habitats. The carrying capacity represents the point of balance between reproduction potential and environmental resistance and is the maximum population that can be supported on a sustainable basis.

car scrapping scheme A Swedish measure to regulate the disposal of worn-out cars. The owner is paid a sum of money when the car is deleted from the register; deregistration requires, however, a scrapping certificate made out by an authorized car scrapping firm.

Carson, Rachel (1907-1964) US biologist, well known for her writings on environmental pollution and the natural history of the sea. In 1951 she published the *Sea Around Us*, which won the National Book Award. In 1962 came her prophetic *Silent Spring*, warning against the excessive use of chemicals that not only kill insects but the birds in the air, the fish in the sea, the earth that supplies our food, and, inevitably, human beings themselves.

Cartagena convention See CONVENTION FOR THE PROTECTION AND DEVELOPMENT OF THE MARINE ENVIRONMENT OF THE WIDER CARIBBEAN REGION.

catabolism A facet of METABOLISM in which complex molecules are broken down into simpler compounds. Such reactions produce energy in the form of ATP for work, the maintenance of functional efficiency in the cell, and the synthesis of new structures. (See ANABOLISM.)

catadromous fish A fish which spends most of its growing years in fresh water and, after attaining sexual maturity, descends to the ocean in order to spawn. An example is the American eel which descends the New England streams when sexually mature, swims hundreds of kilometres to a site in the Sargasso Sea east of Bermuda, spawns in deep water, and dies. The young immature fish eventually ascend the New England freshwater streams and the cycle begins again. (See ANADROMOUS FISH.)

catalyst A substance that accelerates or retards the rate of a chemical reaction without itself undergoing any permanent change in its composition, and which is normally recoverable when the reaction is completed.

catalytic converter A device that uses a CATALYST to convert unwanted substances into less harmful residues. In cars, a catalytic converter in the exhaust gas system uses platinum or other compounds to convert CARBON MONOXIDE and HYDROCARBONS into CARBON DIOXIDE and WATER VAPOUR. As lead in petrol causes such catalysts to deteriorate, only unleaded fuel may be used.

catch crop A crop that is inserted between two principal crops in a rotation to provide livestock grazing at a time when the land would otherwise be lying idle.

catchment A drainage basin that collects all the rainwater that falls on it, apart from that removed by evaporation, directing it into a river, stream, or reservoir. The boundary of a catchment basin is defined by the ridge beyond which the water flows in the opposite direction, that is, away from the basin. Effective catchment management is central to maintaining water quality. It involves the control of any activities that contribute to specific or diffuse sources of pollution, including human habitation, agriculture, industry and mining. The clearing of vegetation should be carefully controlled as this can result in soil erosion and increased water salinity.

catch pit A chamber or pit provided in a drainage system to catch grit and other heavy matter. In a similar way, grease traps intercept grease.

cay A low island formed by an accumulation of sand or dead coral.

CBA COST-BENEFIT ANALYSIS.

CBD CENTRAL BUSINESS DISTRICT.

CCME CANADIAN COUNCIL OF MINISTERS OF THE ENVIRONMENT.

CEA COST-EFFECTIVENESS ANALYSIS.

cell In BIOLOGY, the basic unit of life and the smallest unit capable of an independent existence. All cells are similar in form, composition, and function. Humans are made up of billions of cells. Bacteria, protozoa, and many other microorganisms consist only of single cells.

cellular respiration The process by which organisms combine oxygen with food molecules, utilizing the chemical energy in these substances for life-sustaining processes. CARBON DIOXIDE and water, which are waste products, are returned to the atmosphere. Organisms that do not utilize oxygen degrade food substances by a process called fermentation. (See also CARBON CYCLE.)

cellulose An insoluble complex CARBOHYDRATE formed from glucose residues; it is the major constituent in the cell walls of all green plants. Cellulose comprises about 33% of all vegetable matter, 50% of wood, and 90% of cotton. It is the most abundant of all naturally occurring organic compounds. Cellulose is a food for herbivorous animals such as cows and horses. However, it is non-digestible by humans. Cellulose is processed to produce paper and certain fibres. Chemically modified, it yields substances used in the manufacture of plastics, photographic film, rayon, adhesives, explosives, thickening agents, and moisture-proof coatings.

CEMP   CENTRE FOR ENVIRONMENTAL MANAGEMENT AND PLANNING.

census of population   An enumeration of inhabitants carried out at regular intervals by government. The modern census, going far beyond numbers, is the most prolific source of information about a nation. It reveals information about the basic demographic structure and trends such as growth, internal distribution, and alterations in the age and sex structure of the population. It also reveals information about the occupational structure of society, its standard of living, housing accommodation, and education. The census makes possible the estimation of future trends, which is essential to all kinds of planning, both public and private, local and national.

central business district (CBD)   The centre of a metropolitan area, city, or large town. It is often the preferred location for: the head offices of major public- and private-sector organizations; a communications centre; a cultural, educational, and recreational centre; a commercial and financial centre; and an entertainment, tourist, and retail centre. There is also a residential component. The importance of CBDs has become somewhat modified by the growth of major suburban retail centres and by the promotion of decentralization policies by government agencies. Some CBDs have deteriorated badly and have required rehabilitation and urban renewal programmes to overcome social blight. In general, CBDs around the world appear to have survived and prospered, all offering major traffic congestion and acute parking limitations.

Centre for Environmental Management and Planning (CEMP)   Established in 1972, a training and research centre within the University of Aberdeen, Scotland, CEMP focuses on: the management, preparation, and review of environmental assessments; health, social, economic, water, air, ecological, visual, landscape, and risk assessments; environmental monitoring and baseline studies; environmental audits; resource and land-use planning; and SUSTAINABLE DEVELOPMENT. Participants include the WORLD BANK, UNIDO, WHO, UNEP, ILO, the British Council, the UK Department of the Environment, planning authorities, corporations, and international aid agencies. Conferences and workshops are conducted worldwide.

CEQ   See COUNCIL ON ENVIRONMENTAL QUALITY, US.

CERCLA   COMPREHENSIVE ENVIRONMENTAL RESPONSE, COMPENSATION, AND LIABILITY ACT 1980.

certificates of financial responsibility (COFR)   A requirement introduced by the USA in 1994 that COFR must be carried by all oil tankers entering US waters. Their purpose is to provide insurance against the possibilities of environmental damage. The financial liability certificates spring from the US Oil Pollution Act of 1990, legislation introduced following the EXXON VALDEZ DISASTER.

cesspool   A storage tank for sewage. It should be watertight and not allowed to overflow unless connected to a satisfactory treatment unit, otherwise, when full, it should be emptied in a nuisance-free manner with the safe disposal of the contents elsewhere.

ceteris paribus   Latin for 'other things being equal', or the holding of other factors constant. A standard practice in economic analysis so as to isolate the impact of a particular variable on another variable. It has relevance in other sciences when examining and testing hypotheses.

Cevennes National Park   Created in 1970, a nature reserve located in the Cevennes and Causses regions of southern France. It is dominated by limestone plateaus. Forests cover more than half of the park while the remainder abounds in flowering plants. Birds include golden eagles, peregrine falcons, and hen harriers; animals include otters, badgers, foxes, and wild boars. The park has also a very substantial buffer zone.

CFCs   CHLOROFLUOROCARBONS.

Chaco Culture National Historical Park   Established in 1907, a national historical park located in northwestern New Mexico, USA, 72 km south of Bloomfield. It consists essentially of a canyon containing 13 pre-Columbian Indian ruins and more than 300 other archaeological sites representing high points in Pueblo culture. The excavations indicate that the people excelled in weaving, toolmaking, pottery, farming, and masonry.

charcoal   The residue when carbonaceous material is partially burned or is heated with limited access to air, for example, the solid black residue resulting from the destructive distillation of wood. Charcoal burns without smoke and is used mainly for heating and cooking. It is also used in the manufacture of gunpowder, as an absorbent and a decolourizing agent, in hop drying and in work where a very pure fuel is required. Activated charcoal (produced by blowing steam through charcoal) absorbs organic compounds and is used in sugar refining, in solvent recovery, for filtration and purification of drinking water, and in many industrial processes for the recovery of valuable components from gaseous mixtures or for the removal of unwanted components. It may also be used to treat liquid streams and in motor vehicles for the recovery of hydrocarbons.

checklists A technique in ENVIRONMENTAL IMPACT ASSESSMENT to construct, develop, and apply lists of potential effects or impacts to proposed policies, plans, programmes, and projects to ensure exhaustive examination. SCOPING meetings are a valuable aid to the preparation of checklists.

chemical engineering The design and operation of plants and the development and application of new processes in which materials undergo marked changes in their physical and chemical state. Much of the work of the chemical engineer involves the engineering science and technology of fluid flow, heat transfer, evaporation, distillation, absorption and adsorption, filtration, extraction, crystallization, pollution control, and ENVIRONMENTAL MANAGEMENT.

chemical mutagens Agents that may give rise to mutations resulting in an increase in the incidence of congenital defects in future generations. Humanity is exposed to many thousands of synthetic chemicals whose mutagenic potential is unknown. (See also RADIOACTIVITY.)

chemical oxygen demand (COD) The weight of oxygen taken up by the total amount of organic matter in a sample of water without distinguishing between biodegradable and non-biodegradable organic matter. The result is expressed as the number of parts per million (or milligrams per litre, or grams per cubic metre) of oxygen taken up from a solution of boiling potassium dichromate in two hours. The test has been used for assessing the strength of sewage effluent and trade wastes. The BIOCHEMICAL OXYGEN DEMAND and the chemical oxygen demand tests are of equal importance in waste-water treatment.

chemical pollutants Pollutants presenting a hazard to HEALTH and the ENVIRONMENT of natural or synthetic origin, essentially chemical in nature. The most serious chemical pollutants include: the CHLORINATED HYDROCARBON pesticides such as DDT, aldrin, and dieldrin; the POLYCHLORINATED BIPHENYLS (PCBs) that have been used in a variety of industrial processes; and such metals as MERCURY, LEAD, CADMIUM, ARSENIC, and BERYLLIUM. All these substances persist in the environment, being very slowly, if at all, degraded by natural processes. Further, all are toxic to life if they accumulate in any appreciable quantity. The persistent insecticides have been responsible for serious ecological problems. (See PESTICIDES.)

chemisorption In chemistry, the attachment of a single layer of molecules, atoms, or ions of a gas to the surface of a solid or liquid. It is the basis of catalysis and of great industrial significance. (See CATALYST.)

chemotroph An organism that can obtain its energy directly from chemical reactions that do not involve light. Many bacteria are chemotrophs.

Chernobyl nuclear catastrophe Occurring in April 1986, the worst accident in nuclear power station history. Located 104 km north of Kiev in the Ukraine, the Chernobyl nuclear power station consisted of four reactors, each of 1000 MW capacity. The station was commissioned during 1977-83. During April 25 to 26, 1986, a test was being run on the No. 4 reactor involving apparently a breach of several safety precautions. These errors were compounded by others and a chain reaction in the reactor core went out of control. Several explosions and a large fireball followed, blowing the steel and concrete lid off the reactor. The result of these events and the subsequent fire in the graphite core released large amounts of radioactive material into the atmosphere, where it was carried great distances over Europe by air currents. On 27 April, the 30 000 inhabitants of Pripyat, the nearest town, began to be evacuated. Initially, Chernobyl caused the deaths of more than 30 people, mostly from radiation and many more were afflicted by radiation sickness. The environmental effects of the incident were severe, the soil and ground-water supplies being severely contaminated. Eventually 135 000 people had to be evacuated. In addition, several thousand extra cancer deaths were expected in the long term from radioactivity. The event strengthened opposition throughout the world to nuclear power stations.

chernozemic soil Also called 'black earth', a grassland soil with a dark thick humic horizon. First identified in the grasslands of the Russian steppes, it is a characteristic of the American Great Plains, the Canadian prairies, and the Pampas of Argentina.

Chesapeake Bay Program A programme for the restoration of the largest estuary in the USA. Agricultural runoff and urban waste discharges have resulted in a serious decline in the quality and productivity of the bay. An overabundance of phosphorus and nitrogen has contributed to explosive algal growth that blocks out sunlight vital to seagrasses and has depleted dissolved oxygen levels essential to all life in the bay. The Chesapeake Bay Agreement of 1987 between Maryland, Pennsylvania, and Virginia, the District of Columbia, and the Chesapeake Bay Commission has resulted in a coordinated programme on many fronts. These include improved sewage treatment, phosphate detergent bans, better compliance with environmental requirements, reduced fertilizer use, better animal-waste management, better erosion control, a moratorium on striped-bass fishing, construction of a fish-passage facility, restrictions on shoreline development, controls on urban and farm runoff, protection of the wetlands, and reduction of pollutant discharges from boats. One objective of the agreement is to reduce phosphorus and nitrogen entering the Bay by 40% by the year 2000.

China, evolution of environmental policy See Box 6.

China clay See KAOLIN.

China syndrome The concept that a nuclear core meltdown could burn a hole through the Earth 'to China'; the basis of a film distributed by Columbia Pictures around the time of the Harrisburg nuclear power plant incident.

Chipko Andolan Movement A 'Save-the-Trees' campaign that has spread among the Indian public. The Chipko Andolan (tree-hugging) movement began in Uttar Pradesh in 1973, when tree fellers were prevented from cutting down trees by villagers.

chlordane The principal component of a mixture of chlorine-containing organic compounds used as a contact insecticide. Although chlordane is highly toxic to many insects, its toxicity towards mammals is lower than that of other CHLORINATED HYDROCARBONS, such as aldrin. However, adverse results in laboratory animals has led to chlordane being banned in some countries.

chlorinated hydrocarbons Hydrocarbon compounds containing CHLORINE, also known as organochlorines. They include pesticides such as DDT (dichloro-diphenyl-trichloroethane); CHLORDANE; aldrin, endrin, heptachlor, benzene hexachloride, endosulphan, dieldrin, 2,4,5-T, and substances which have been used in electrical transformers such as PCBs (POLYCHLORINATED BIPHENYLS). These compounds are highly toxic to most organisms and they are not biodegradable. Through indiscriminate use and BIOMAGNIFICATION they can reach toxic levels in many species, especially those at higher trophic levels. The use of these insecticides has been restricted in most countries.

chlorination The use of CHLORINE for the treatment of water and effluents, a practice which has spread throughout all parts of the world. In waterworks, chlorination is used to kill bacteria and to remove algae. In swimming pools it is used to control bacterial contamination, inhibit algae, and improve water quality. In power generation, the chlorination of condenser cooling water helps to control slime-forming bacteria and to eliminate fouling by mussels. In sewage treatment works, it may be used to improve final effluent quality. In the food-processing industries, the chlorination of retort cooling water helps to prevent spoilage and blown cans. Chlorinated water helps also to reduce carcass contamination in meat-and poultry-processing factories. Chlorination also ensures a wholesome water supply in dairies and ice cream factories, and for bottle-washing plants in breweries and soft-drink plants. The effective sterilization or disinfection of water requires the constant presence of a 'chlorine residual', that is, free chlorine remaining in the water after a necessary period of contact. Frequent and reliable estimations of the chlorine residual are necessary.

chlorine A greenish poisonous strongly oxidizing gas manufactured by the electrolysis of brine. Industrially it is used in the production of organochlorine compounds and as a bleaching agent and bactericide in the food industry. Its derivatives are used as solvents, polymers (vinyl chloride), refrigerants, and aerosol propellants. Chlorine is frequently used as a disinfectant in the treatment of public water supplies and swimming-pool water. Compounds containing chlorine have been under criticism for some time for their detrimental effects. In consequence, POLYCHLORINATED BIPHENYLS (PCBs) and CHLOROFLUOROCARBONS (CFCs) have been largely phased out. Chlorine has also been largely eliminated from pulp-bleaching processes, together with the phase out of chlorinated solvents. In Sweden, chemical manufacturers, in close cooperation with pulp producers, have successively developed new bleaching systems where chlorine is completely replaced by chlorine dioxide, hydrogen peroxide, and oxygen, applied in special sequences. The use of PVC products poses a problem as incineration emits chlorine into the atmosphere. PVC uses about 40% of Europe's chlorine production. (See AGENT ORANGE; DIOXIN.)

chlorofluorocarbons (CFCs) Aliphatic carbon compounds containing open chains of carbon atoms (in contrast to the closed rings of the aromatic compounds), possessing both chlorine and fluorine atoms. The CFCs most commonly used as aerosol propellants, as refrigerants, in fire-extinguishing foams, and in the manufacture of plastic articles have been CFC-11 (trichlorofluoromethane) and CFC-12 (dichlorofluoromethane). They are known as freons in the USA and arctons in Britain. They are considered to be harmful in the upper atmosphere as ozone-depleting

Box 6 China: evolution of environmental policy

1956	Drinking-water quality standards; hygiene standards for industrial enterprises, promulgated
1958	Yujiang of Jiangxi Province becomes first county in China to have eliminated schistosomiasis (bilharzia)
1972	China represented at the UN Conference on the Human Environment, Stockholm, Sweden
1973	Office of the Environmental Protection Leading Group of the State Council established, issuing a wide range of environmental standards relating to water quality, industrial emissions, and agricultural herbicides and pesticides
1972–78	Preparatory stage for the introduction of EIA
1978	EIA regulations introduced; amendment of the Constitution of the People's Republic of China by the Fifth National People's Congress: 'The State protects the environment and natural resources and prevents and eliminates pollution and other hazards to the public.' (Article 11)
1979	Environmental Protection Law passed and promulgated by the National People's Congress based on Article 11 of the Constitution; Law provides for rational use of the natural environment, the control of pollution, the prevention of damage to ecological systems, and establishes a legal basis for EIA; Forestry Law introduced, permitting the planting of a 'Green Great Wall' across Northern China to reduce wind and sand
1981	EIA applied to the petrochemical industry, large water conservancy projects, coal mining, light industry, steel industry, and the construction industry
1986	EIA administrative and technical regulations issued, including management measures for construction projects
1988	The Three-Gorges hydroelectric and irrigation project endorsed by a feasibility study prepared by the CIPM Yangtze Joint Venture, a consortium led by Canadian International Project Managers Ltd and sponsored by the Canadian International Development Agency
1989	National Environmental Protection Agency (NEPA) created, reporting directly to the State Council; EIA requirements made more specific
1990	Environmental Protection Law adopted and promulgated by the National People's Congress, authorizing NEPA to supervise and implement environmental protection management measures throughout the country; technical guidance issued by NEPA for construction projects, industry, railways, highways, ports and airports, natural resources and agricultural developments
1991	A 10-year programme for economic and social development of China 1991–2000 approved; includes the controversial Three-Gorges water conservation and hydro-electric project
1992	The Three-Gorges project reaffirmed at the Fifth Plenary Session of the Seventh National People's Congress
1994	Demographic survey reveals a steady decline in fertility and population growth (1.65 children per woman in 1991)

agents. If the OZONE LAYER becomes depleted, more solar ultraviolet radiation (in the UV-B range), penetrates to the Earth's surface with adverse health and biological effects, such as an increasing incidence of skin cancer. See CONVENTION FOR THE PROTECTION OF THE OZONE LAYER; HYDROCHLOROFLUOROCARBONS; MONTREAL PROTOCOL ON SUBSTANCES THAT DEPLETE THE OZONE LAYER.

chlorophyll A green pigment present in algae and higher plants that is involved in PHOTOSYNTHESIS, the process by which light energy is converted to chemical energy through the synthesis of organic compounds. Chlorophyll is found in virtually all photosynthetic organisms, including green plants, blue-green algae, and certain bacteria. Chlorophyll absorbs energy from light, this energy then being used to convert CARBON DIOXIDE to carbohydrates.

chlorosis Evidence of a plant disease in which normally green tissue appears pale, yellow, or bleached. The failure of the CHLOROPHYLL to develop may be due to infection by a virus, lack of an essential mineral, damage from a fertilizer, exposure to excessive air pollution, or some other factor. Chlorotic plants may be stunted and shoots may die back to the roots.

Chobe National Park A national park established in Botswana in 1961. It comprises grassland, swamp, marsh, and forest. Fauna consist of the hippopotamus, white rhinoceros, elephant, lion, leopard cheetah, hyena, buffalo, zebra, giraffe, and many species of antelope. A special feature is a fossil lake bed.

cholera An acute bacterial infection of the intestines characterized by severe diarrhoea and rapid depletion of body fluids and salts. Without therapy the mortality rate can be high. The disease is usually transmitted through contaminated water or foods. The prevention of cholera outbreaks depends upon better sanitation, particularly the availability of clean and safe drinking water.

chromatography A method for separating chemical substances, taking advantage of the relative rates at which different substances are absorbed from a moving stream of liquid or gas. Separations are usually made in columns of selected sorptive powders or liquids, or in strips of fibrous media. Gas-liquid and gas-solid systems are used. Gas chromatography is used principally as an analytical technique for the determination of volatile compounds.

chronic In medicine, describing a condition which may have a slow onset, but which runs a prolonged course, such as chronic bronchitis.

cinnamonic soil A reddish soil, rich in dehydrated iron oxides. In cinnamonic soils organic matter decays rapidly and does not accumulate on the soil surface. Many cinnamonic regions have a Mediterranean climate, often associated with a mountainous topography and a high incidence of limestones and volcanic materials.

city core See CENTRAL BUSINESS DISTRICT.

class action A legal action initiated by a single person or a few people on behalf of a group of people with a similar claim or claims. A class action often provides improved access to the courts for many people who otherwise have only a theoretical right to justice and no practical means of achieving it. The costs are shared by those with limited financial resources, and the ultimate judgement applies to them all. The right to bring a class action, which may vary somewhat in form, differs with jurisdictions throughout the world and with the circumstances and character of the case. It seems to be most common in the USA.

clean air legislation National, state, provincial, or city legislation to control air-pollution emissions and to protect the general ENVIRONMENT. Most industrial countries have now introduced legislation. Britain, the oldest of the industrial nations, under limited industrial legislation in the 1860s created an alkali inspectorate, while more general problems were dealt with under the public health acts. The Clean Air Acts of 1956 and 1968 considerably broadened control activities in the industrial, commercial, and residential sectors. In the USA, the Environmental Protection Agency (EPA) sets national ambient air quality standards for a range of potentially hazardous gases and substances, bringing pressure to bear on individual states to achieve the desired targets. The boxes in this work indicate how clean air legislation has evolved in many countries, embracing pollution from industrial sources and motor vehicles.

cleaner production The use of environmentally friendly processes to produce environmentally friendly products, a route to SUSTAINABLE DEVELOPMENT, reducing the risk to the ENVIRONMENT of industrial activities in the most cost-effective way. The concept includes, *inter alia*, waste minimization, pollution prevention, process modification, and energy efficiency.

clear-felling The removal of an entire stand of trees in one felling, leaving a clear area in the forest, resulting in the complete destruction of the ECOSYSTEM in that locality. In contrast, selective logging involves the cutting of individual trees, or small stands of trees, for timber. This removes trees of the greatest economic value, but leads to changes in the forest structure due to the shift in abundance of various species. However, selective logging is preferable to clear-felling as it encourages natural regeneration while maintaining many of the natural characteristics of a mixed-age forest. It may not be so popular with forest workers, involving greater risks.

climate The average weather conditions of a place or region throughout the seasons. It is governed by latitude, position in relation to continents or oceans and local geographical conditions. It is described in terms of atmospheric pressure, temperature, solar radiation, wind speeds and direction, cloudiness, humidity, rainfall, evaporation, incidence of fog, temperature inversions, lightning, thunderstorms, typhoons, and monsoons. Climates have varied considerably over geological time. Climate must be distinguished from weather which is the condition of the atmosphere at a certain time or over a short period.

climate change   See CONVENTION FOR THE PROTECTION OF THE OZONE LAYER; CONVENTION ON CLIMATE CHANGE; GREENHOUSE EFFECT; HAGUE CONFERENCE ON THE ENVIRONMENT; INTERGOVERNMENTAL PANEL ON CLIMATE CHANGE; LONDON CONFERENCE ON CLIMATIC CHANGE; MONTREAL PROTOCOL ON SUBSTANCES THAT DEPLETE OZONE LAYER; TORONTO CONFERENCE ON THE CHANGING ATMOSPHERE; VILLACH CONFERENCE ON CLIMATIC CHANGE; WORLD COMMISSION ON ENVIRONMENT AND DEVELOPMENT; WORLD CONGRESSES ON CLIMATE AND DEVELOPMENT; WORLD METEOROLOGICAL ORGANIZATION.

climax community   The final or most stable community in a succession series, that is, the final outcome of a slow, orderly progression of changes in communities of animal and plant species in a particular area over time. Usually, the series becomes increasingly more complex until it reaches a climax community. A climax community is capable of maintaining itself indefinitely as long as the environment is not disturbed by, say, the introduction of other species.

Club du Sahel   Created within the ORGANIZATION FOR ECONOMIC COOPERATION AND DEVELOPMENT (OECD), an agency to assist the poorest countries of Black Africa. Membership comprises both FOREIGN AID recipients and OECD member countries. It has access to international finance and influences direct investment by the OECD members. EIA principles are now applied to developments.

Club of Rome   A voluntary association formed in 1968 by a group of thirty individuals from 10 countries: scientists, educators, economists, humanists, industrialists, and civil servants, following a meeting in the Accademia dei Lincei in Rome. The meeting was at the instigation of Aurelio Peccei. The purpose of the Club of Rome was to foster understanding of the varied but interdependent components that make up the global system and to promote new policy initiatives. The Club initiated a Project on the Predicament of Mankind, a research project, which resulted in the publication in 1972 of *The Limits to Growth*. The report concluded that, even under the most optimistic assumptions, the world cannot support present rates of economic and population growth for more than a few decades, but that a state of equilibrium might be achieved. The Club's second report, *Mankind at the Turning Point*, was published in 1975. Further reports followed with *Goals for Mankind* appearing in 1977.

coal washery reject material   The material discarded after the recovery of good-quality coal from run-of-mine coal. Coarse coal reject material, or chitter, consists of rock fragments, carbonaceous shales and poor-quality coal. This waste may contain up to 30% carbon. Most of the reject is in this category. The remainder is fine coal reject material, also known as slimes and tailings. It is usually discharged as a slurry of fine carbonaceous shales, clays and fine coal, containing perhaps 55% carbon. Many methods for the treatment and disposal of coal reject material have been developed. A number of these are indicated in Figure 4.

coastal erosion   The erosion of cliffs, beaches, and land by the battering of waves, with two effects: (1) a hydraulic effect, compressing air into pockets in the rocks and cliffs; and (2) an abrasive effect, flinging debris against rocks and cliff faces, wearing them away. Where there are beaches the waves can cause longshore drift in which sand and pebbles are carried in a direction parallel to the shore. A river mouth may shift as longshore drift builds a sand spit across it. Longshore sand and shingle may also pile up behind groynes, structures built to inhibit the movement of beaches. Waves will also enlarge and destroy caves, create blowholes, and undermine headlands.

coastal protection   Measures by way of planning, prior approval of works, prohibition of some activities, physical structures, and restoration efforts to protect the coastline against the ravages of nature and haphazard and unplanned developments. Coastal environments include tidal wetlands, estuaries, bays, shallow near-shore waters, mangrove swamps, and in-shore reef systems. It is important to protect critical habitats such as: feeding, breeding, nursery, and resting areas; economically, culturally, or ecologically important species of fauna or flora; threatened or unique species; and genetically significant areas such as coastal wetlands and coral reefs. Some areas will be protected by international convention or agreement, by World Heritage listing, and by conservation legislation. Often, however, special coastal protection legislation will be needed to provide a coordinated programme of coastal protection.

cochlea   A tiny snail-shaped structure forming the inner ear. The ossicles of the middle ear transmit the sound vibrations to a fluid contained in the cochlea within which there are microscopic hair cells that move back and forth in response to sound waves. It is the energy impulses created by the movement of these crucial hair cells that go to the brain, where they are interpreted as sound.

cocktail party effect   The faculty of 'locking-on' to one voice amid a hubbub of other voices, or to one particular sound in a sea of other sounds and noises.

COD

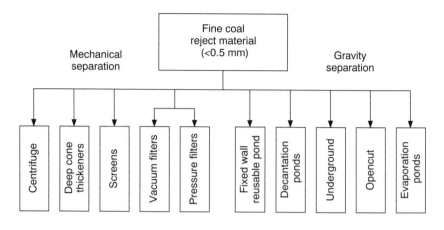

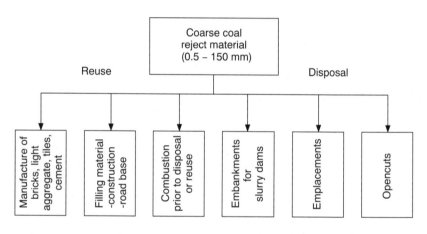

COAL WASHERY REJECT MATERIAL: Figure 4. Some methods of treatment and disposal for coal washery reject material.

COD See CHEMICAL OXYGEN DEMAND.

Codex Alimentarius Committee (Codex) A body set up jointly by the UN FOOD AND AGRICULTURE ORGANIZATION (FAO) and the WORLD HEALTH ORGANIZATION (WHO) to reach agreement between member countries on standards for foods, which provide adequate protection for the health of consumers. Codex publishes the Codex Maximum Residue Limits (MRL), being the maximum concentrations of various substances that should be legally permitted in or on a food or animal foodstuff, including pesticide residues. The role of Codex has been strengthened by the GENERAL AGREEMENT ON TARIFFS AND TRADE (GATT) as a result of the 1993 Uruguay Round agreement. Under the revised rules, Codex standards have become the effective international norm; indeed any country applying standards higher than Codex standards is liable to challenge under the GATT disputes procedures for applying those standards as a non-tariff barrier to trade. Codex standards thus emerge as minimum–maximum standards. The Codex international standards for food from animals are known as sanitary standards, and from plants as phytosanitary (PSS) standards.

COFR CERTIFICATES OF FINANCIAL RESPONSIBILITY.

cogeneration (combined heat and power generation) Usually the generation of heat in the form of steam, and the generation of power in the form of electricity. Combined heat and power plants are able to convert a much higher proportion of the energy in fuel into final output. The steam produced may be used through heat exchangers in a district heating scheme, while the electricity provides lighting and power. Such plants are associated with medium to large industry with a large immediate demand for both forms of energy.

coliform bacteria Bacteria, including the bacterium ESCHERICHIA COLI, commonly found in the human large intestine and the presence of which in the ENVIRONMENT usually indicates contamination by human wastes. Laboratory results are expressed as the number of organisms per 100 ml of sample. While the presence of *E. coli* indicates faecal pollution, the organism is generally considered to be non-pathogenic. To confirm that pathogenic organisms are present, further testing will be necessary.

colloidal particles Material that is in a very fine state of division. The particles will not settle out or pass through a semipermeable membrane.

combined cycle generation A type of electricity generating plant, in which power is generated from a gas turbine fuelled with oil or gas, and then from a steam turbine using the hot waste gases from the gas turbine. The thermal efficiency of about 50% is higher than that normally obtained from a conventional thermal power station. It can be built in two years compared with six years for a conventional power station.

combined heat and power generation See COGENERATION.

combined sewer A single conduit or pipe intended for the removal of sullage, sewage, and storm water, as distinct from a separate system in which the storm water and rain water are removed through a separate sewer.

combustion A state of chemical activity in which the reactive elements of a fuel burn or unite with oxygen, accompanied by the evolution of heat and often light. Combustion means, therefore, oxidation with the formation of the products of oxidation, CARBON DIOXIDE and WATER VAPOUR. Both are GREENHOUSE GASES contributing to the GREENHOUSE EFFECT.

command and control A US term for a regulatory scheme based on rules that apply specific technical performance standards, generally based on known feasible control technology, to every discharge point within a regulated process. A new era of environmental policy is emerging in which traditional command and control of 'end of pipe' emissions is being augmented by an emphasis on pollution prevention, minimization, recycling, and process review. This trend tends to emphasize reduced utilization of toxic chemicals, more efficient use of materials and energy, better process management, and innovations offering competitive advantage.

commensals Two or more kinds of organisms that live together, with one or several benefiting and without injury to any.

comminutor A primary treatment process for sewage combining screening with the grinding of large solids, the shredded material being returned to the sewage.

Commission for sustainable development A body created by the UN in 1992 following the UN CONFERENCE ON ENVIRONMENT AND DEVELOPMENT held in Rio de Janeiro, Brazil, earlier that year. A key function of the Commission is to promote international cooperation in the implementation of AGENDA 21, with particular emphasis on policies to promote ecologically SUSTAINABLE DEVELOPMENT.

Commission for the prevention of marine pollution by dumping from ships and aircraft A supervisory body created by the Oslo CONVENTION FOR THE PREVENTION OF MARINE POLLUTION BY DUMPING FROM SHIPS AND AIRCRAFT.

Commission for the prevention of marine pollution from land-based sources A supervisory body created by the Paris CONVENTION ON THE PREVENTION OF MARINE POLLUTION FROM LAND-BASED SOURCES.

common-cause failure The failure of more than one component, item or system in an industrial plant, due to the same cause.

common law Customary law based upon judicial decisions and embodied in reports of decided cases, which has been administered by the common-law courts of England since the Norman Conquest, with minimal reference to statutes or enactments, or judges' rules on individual cases in the light of custom and precedent. Common law continues to prevail, though undergoing considerable modification. This body of customary law is to be found in the USA and most of the member states of the Commonwealth of Nations. Common law stands in contrast to statute law (being the acts of legislative bodies) and to the rules developed by the courts of equity. While common law prevails today in appropriate cases, the trend has been more and more towards a system of statute law, developed by legislators not judges. Most environmental matters tend to fall within the scope of statute law, the task of judges being to interpret, not make, that law.

common-mode failure The failure of a component, item, or system in an industrial plant, in the same manner.

common-property resource Those attributes of the natural world which are valued by society but are not in individual ownership and do not enter into the processes of market exchange and the price system. Notable

commons, tragedy of

among those resources are the ATMOSPHERE, watercourses, many ecological systems, fisheries, wilderness, and the visual properties of the landscape. The key aspect of a common-property resource is that it can be used free of charge by any member of society wishing to use it.

commons, tragedy of A tragedy of overgrazing and lack of care which resulted in erosion and falling productivity of the English commons, prior to the enclosure movement when grazing rights became restricted to the few. At one time each community had its commons set aside for public use, essentially for sheep and cattle grazing. Eventually the number of animals became more than the commons could support, and no one had any interest in ensuring the future productivity of this resource. Exploitation continued until productivity collapsed, and this social system was superseded. Today, on a much larger scale, the natural resources of air and water may be regarded as the 'commons of the world'. In 1968, the theme was elaborated by Garret Hardin of the University of California and the term is associated with his name.

community In BIOLOGY, all organisms occupying a common ENVIRONMENT and interacting with each other; also called a biological community. The major biological communities of the world are called BIOMES, examples being tropical rainforest, savannah, marsh, Arctic tundra, and the marine pelagic zone. These communities change through a process known as SUCCESSION; the highest point of the slow process of succession is known as a CLIMAX COMMUNITY. The various species in a community each occupy their own ecological NICHE.

Comoe National Park An Ivory Coast national park established in 1926, embracing a series of ranges. The fauna include hippopotamus, elephant, lion, leopard, panther, hyena, many varieties of antelope, baboons, and monkeys. The flora comprises wooded Guinean savannah, sedge, bulrush, and bombax.

compacted soil Soil that has been compressed by the impact of hard-hooved animals, people, or vehicles, and which has lost many of its air-holding and water-holding pores. Few plants can germinate or thrive in such soils and coupled with a lack of ground cover, erosion becomes much more likely.

compactor Equipment for the compression of waste materials, to facilitate transportation to landfills, scrap yards, or recycling centres.

comparative advantage, theory of In international trade, a statement that countries tend to specialize in the production and export of those goods and services which they can make or provide most profitably.

compensation depth The depth in water at which, because of reduced light penetration, the rate of production of organic material by PHOTOSYNTHESIS exactly balances the rate of breakdown of organic material by plant respiration. There is no net production below the compensation depth.

competition (1) In BIOLOGY the interaction which occurs when two or more individuals of a single species (intraspecific competition) or two or more individuals of different species (interspecific competition) attempt to acquire and use the same resources such as food, nesting material, or living space, in the same ECOSYSTEM; (2) In ECONOMICS, a free-trading process in which individuals or businesses exchange goods and services through markets, with a view to mutually beneficial exchanges, buying and selling being characterized by rivalry or struggle to achieve an advantage in terms of market share and profit.

complex The meeting of several communities. (See COMMUNITY.)

composting A biological process in which the organic material in rubbish and refuse is converted to a useable stable material by the action of microorganisms present in the refuse. A composting plant can range from a simple windrow (that is, a piling up in the open) of raw refuse until it is broken down or a simple domestic bin or box to a sophisticated fully mechanical operation. Compost is a soil conditioner with some manurial value. It adds humus to the soil and helps break up the structure of the soil, improving the moisture-retaining properties.

Comprehensive Environmental Response, Compensation, and Liability Act 1980 (CERCLA) Known also as Superfund, a measure introduced by the US Congress to meet the costs of the clean-up of abandoned hazardous waste sites, scattered throughout the USA. The act imposed environmental taxes on the petroleum and chemical industries and set up a Hazardous Substance Response Trust Fund. The US Environmental Protection Agency identifies, ranks and supervises the clean-up or neutralization of the most hazardous waste sites. (See CONTAMINATED SITES.)

compulsory purchase The acquisition of private property and land without the owner's consent for an approved public purpose such as public housing, redevelopment, or new town construction. The power of compulsory purchase is embodied in British planning and housing legislation. The procedure for arriving at the purchase price is also laid down. In the USA, through the excercise of eminent domain, private property may be acquired for the public advantage, compensation being made to the owner. For example, a city may acquire land for a needed highway.

concentration The amount of a particular substance or gas contained in a given amount of a solution, gas, or mixture; for example, the concentration of salt in sea water. In environmental science and engineering, many concentrations are expressed as parts per million (ppm), parts per hundred million (pphm), or parts per billion (ppb), being proportional parts by volume. The International System of Units (SI) uses micrograms per cubic metre ($\mu g/m$) for air pollution, being weight per unit volume and milligrams per litre (mg/l) or grams per cubic metre (g/m^3) for water pollution. For water, parts per million by volume is identical with grams per cubic metre, as one gram (1 g) of water occupies one cubic centimetre (1 cm^3). One billion in this text is one thousand million.

conciliation See ALTERNATIVE DISPUTE RESOLUTION.

conditioning The preparation or modification of anything for further treatment, for example, the addition of water or steam to dusty coals or fines to improve the distribution of air through the fuel bed and reduce the quantity of dust carried forward towards the stack.

configuration dredging The dredging of a river or harbour bed in accordance with a plan to, say, avoid damage to seagrasses, influence wave patterns, or facilitate future developments.

confluence The point at which one stream or river flows into another or where two streams converge and unite.

coning Description applied to a chimney plume when it is expanding roughly along a cone. Chimney plumes do this when either efflux momentum or buoyancy or both are dominant in determining the path of the plume.

conservation Defined by the WORLD CONSERVATION STRATEGY as the 'management of human use of the BIOSPHERE so that it may yield the greatest sustainable benefit to present generations while maintaining its potential to meet the needs and aspirations of future generations'. Conservation is thus, something positive, embracing preservation, maintenance, sustainable utilization, restoration, and enhancement of the natural environment. Living-resource conservation is specifically concerned with plants, animals, and microorganisms, and with those non-living elements on which they depend. Living resources are renewable if conserved, and they are destructible if not conserved. Living-resource conservation involves the maintenance of essential ecological processes and life-support systems, the preservation of genetic diversity, and the sustained utilization of species and ecosystems. (See SUSTAINABLE DEVELOPMENT; SUSTAINABLE YIELD.)

conservation tillage (minimum tillage) An agricultural practice aiming to reduce erosion by minimizing the amount of ploughing (tillage) of the ground. The remains of the previous crop are left on the ground as a protective covering until just before the new crop is sown. This minimizes the time the ground is left bare and exposed to erosion.

conservation zones Statutory zoning controls which have the purpose of securing the preservation of land or buildings for the benefit of the community at large. Land may be set aside by the planning authority for open space or as a place of historical, or geological importance. Buildings may be marked for preservation as a group because of their architectural or historical significance.

conserver society A society consciously managed to minimize the consumption of natural resources yet aiming to achieve an adequate standard of living for its members; a society based on the pursuit of happiness, not the pursuit of profit.

consumer activism Direct participation in the protection of consumer rights by a continual scrutiny of market behaviour by corporations and agencies in both the private and public sectors. The term is associated with the name of Ralph Nader (b. 1934) in the USA whose book *Unsafe at any Speed* (1965) led to US car-safety legislation.

consumer protection Laws and measures to provide safeguards for consumers when buying goods and services in respect of quality, safety, suitability, specification, presentation, advertising, conservation of resources, and value to ensure fair and responsible trading.

consumers Organisms that feed on organic matter or other organisms and include herbivores, carnivores, omnivores, parasites, and scavengers.

contaminated sites Sites that present a concentration of chemical substances that are likely to pose an immediate or long-term hazard to human health or the ENVIRONMENT. A site may be unsafe or unfit for occupation or habitation by people or animals, or degraded in its capacity to support plant life. Historically, contaminated or degraded land has been recognized in public health and related legislation, such as sites used for refuse disposal which should not be built on for many years. In more recent times, the problem has taken on more widespread proportions due to the greater diversity of toxic substances that may be deposited and due to well-publicized incidents such as the LOVE CANAL crisis in the USA, and other incidents of schools built on lead-contaminated sites or

contamination

houses constructed on radioactive soil. Further, potential purchasers of land do not wish to acquire unwittingly an environmental problem that may require subsequent remediation. In the US, the COMPREHENSIVE ENVIRONMENTAL RESPONSE, COMPENSATION, AND LIABILITY ACT 1980 (CERCLA) (known also as Superfund) was introduced to meet the costs of the clean-up of abandoned hazardous waste sites, and to eliminate the most serious threats to public health and the environment. Special taxes were introduced to finance this programme. (See ENVIRONMENTAL AUDIT.)

contamination A condition or state of land, water, or air which represents a danger to health because of the presence of pathogenic bacteria or toxic materials or because it otherwise detracts from human wellbeing.

continental drift The very slow movement of the continents over their underlying plates, these plates being large sections of the lithosphere, the rigid outermost shell of the Earth that is about 100 km thick.

continental shelf A comparatively shallow zone that borders the continents and separates them from the deeper oceanic regions. Shelves exhibit a wide range of variation, particularly in size, depth, width, topography, and gradient.

continental slope The steeply sloping part of the sea floor between the CONTINENTAL SHELF and the bottom of the ocean.

contingent liability A liability that may occur but not with certainty; something possibly involving obligations. For example, a guarantee given for another company's borrowings to be invoked in the event of a default by that company; or a risk that current legal proceedings may result in a judgement against the company. The remediation costs that may arise under the US Superfund provisions for the clean-up of contaminated sites is a liability that affects every industry, from chemical and manufacturing companies to banks and insurers. Under Superfund, a company may be held responsible for the entire cost of cleaning up a site even if it contributed only a small fraction of the waste so that a liability is difficult to identify and even more difficult to quantify. (See CONTAMINATED SITES; SECURITIES AND EXCHANGE COMMISSION.)

contingent valuation A method of inferring the value of costs or benefits in the absence of a market. Individuals are asked to estimate what they would be willing to pay or willing to accept if a market existed for the good. (See WILLINGNESS TO PAY; EXTERNALITY.)

contour farming Ploughing, seeding, cultivating and harvesting across the slope rather than along it. Ploughing is horizontal, following the curvature of the hills with every furrow acting as a reservoir to retain water. Siltation, a common cause of water pollution, is also reduced.

contour-strip cropping The planting of crops on contour strips as an effective erosion deterrent; crops are alternated along the contours. Strip cropping is combined frequently with crop rotation.

controlled tipping See LANDFILL.

conurbation The English term for MEGALOPOLIS, a designation for a number of coalescing metropolitan areas to form a huge massing of people and economic activities. The term 'conurbation' was coined by Sir Patrick Geddes (1854–1932), the Scottish sociologist, in his book *Cities in Evolution* (1915). He described the waves of population inflow to large cities, followed by overcrowding and the creation of slums, and then the wave of backflow, the whole process resulting in amorphous sprawl, waste, and unnecessary obsolescence. As a planner, Geddes laid out some 50 cities in India and Palestine.

convection currents The movements that occur when liquids or gases are heated or cooled, due to changes in density. (See ANABATIC WIND; GREENHOUSE EFFECT; KATABATIC WIND; TEMPERATURE INVERSION; THERMAL POLLUTION.)

Convention for cooperation in the protection and development of the marine and coastal environment of the west and central African region Signed in Abidjan in 1981, an agreement by the countries of west and central Africa to adopt a comprehensive approach to pollution control and resource management. A related protocol deals with cooperation in pollution emergencies.

Convention for cooperation in the protection of the marine environment from pollution, Kuwait region An agreement by the countries of the Persian Gulf signed in Kuwait in 1978 providing for a comprehensive approach to pollution control and resource management. It is supplemented by a protocol concerning cooperation in combating pollution by oil and other harmful substances in cases of emergency.

Convention for the conservation of the Red Sea and Gulf of Aden environment An agreement, drawn up by the countries of the region and as part of the UNEP regional seas programme to monitor, prevent and control pollution from land-based sources in Saudi Arabia, Yemen, Jordan, Sudan, and Somalia entering the Red Sea, Gulf of Aqaba, Gulf of Suez, the Suez Canal, and the Gulf of Aden. It was signed in Jiddah in 1982.

Convention for the prevention of marine pollution by dumping from ships and aircraft (the Oslo convention) A convention signed in Oslo, Norway in 1972 by 12 European nations pledging to take all possible steps to prevent the pollution of the sea by dumping in the northeast Atlantic. The signatory nations were Belgium, Britain, Denmark, Finland, France, Germany, Iceland, the Netherlands, Norway, Portugal, Spain, and Sweden. A commission was created to administer the convention.

Convention for the prevention of pollution from ships (the Marpol convention) A convention concluded in London, England in 1973 at the end of the conference convened by the International Maritime Organization. The convention effectively superseded the previous convention for the prevention of the pollution of the sea by oil, but it does not deal with dumping from ships and aircraft covered by the Oslo convention. The Marpol convention aims at eliminating pollution of the sea by both oil and noxious substances. The convention applies to any ship of any type.

Convention for the protection and development of the marine environment of the Wider Caribbean Region (the Cartagena convention) An international agreement signed in 1983 by the representatives of 27 countries at Cartagenas de Indias, Columbia, to protect the marine environment of the Caribbean. From this emerged a Caribbean Action Plan. In the same year, a protocol to the convention was adopted to combat oil spills in the region and in 1990 at Kingston, a protocol was adopted concerning specially protected areas and wildlife.

Convention for the protection of the Mediterranean Sea against pollution (the Barcelona convention) A convention adopted by all Mediterranean countries in 1976 for the purposes of pollution control and resource management. Since the signing by the 15 Mediterranean countries, UNEP has been able to develop a Mediterranean Action Plan. The signatory countries undertook to develop programmes, conduct monitoring, prepare inventories and pursue relevant research. The convention has been reinforced by protocols relating to dumping by ships and aircraft, emergency situations, and the control of land-based sources. (See ATHENS TREATY ON LAND-BASED SOURCES OF POLLUTION.)

Convention for the protection of the natural resources and environment of the South Pacific Region (the Noumia convention) Adopted in 1986, an agreement to take appropriate measures in preventing, reducing, and controlling pollution in the South Pacific region (including the Australian east coast). In the same year two protocols were adopted: a protocol concerning cooperation in combating pollution emergencies in the South Pacific Region and a protocol for the prevention of pollution of the South Pacific Region by dumping. The former includes requirements for the exchange of information, reporting of pollution incidents and mutual assistance. The latter lists substances, the dumping of which is prohibited and substances for which special and general permits are needed. It also sets out implementing, enforcing, reporting, and emergency procedures.

Convention for the protection of the ozone layer (the Vienna convention) A convention adopted by 21 countries together with the whole of the European Union in 1985 as a first step in protecting human health and the environment from the adverse effects of activities likely to modify the ozone layer. The convention was reinforced by the Montreal protocol of 1988 aimed at protecting the OZONE LAYER by controlling the emission of CHLOROFLUOROCARBONS and HALONS, with a specific programme.

Convention for the protection of the world cultural and natural heritage (the Paris convention) A convention initiated by UNESCO and adopted in 1972. The convention established a World Heritage Committee which met for the first time in 1977. The convention also created a World Heritage Fund to allow financing of conservation projects. A World Heritage List was also created which includes the Grand Canyon in the US, the Galapagos Islands, the Pyramid Fields of Egypt, Chartres Cathedral, the Convent of Santa Maria delle Grazie, the Taj Mahal, and a whole range of national parks and places of natural beauty, including the Great Barrier Reef.

Convention for the regulation of Antarctic mineral resource activity (CRAMRA) A convention concluded initially in 1988 by the parties to the ANTARCTIC TREATY. The convention was to create a Minerals Resource Commission to collect data on the environmental effects of oil and mineral exploration and development proposals on the Antarctic continent. On the basis of this data, the commission would decide whether mining, drilling, or development might be undertaken and where. Such activity would require a licence with attached conditions. Australia opposed the convention, objecting to all mineral exploitation in Antarctica. In 1991, a protocol (the Madrid protocol) on environmental protection of Antarctica was adopted. The objective of the protocol was to provide 'comprehensive protection of the Antarctic environment and dependent and associated ecosystems'. Section 7 of the protocol prohibits 'any activity relating to mineral resources, other than scientific research'. Mining was banned by the protocol for a period of 50 years.

Convention for the regulation of whaling A convention, signed in 1946, creating the International Whaling Commission with responsibility for the conservation, development and optimal utilization of whale resources. The management of whale stocks remained very unsatisfactory over the years, with some species being hunted to virtual extinction. In 1982, the commission voted to ban commercial whaling worldwide beginning in 1986.

Convention of the World Meteorological Organization A convention that commits the many participating nations to international cooperation in monitoring, research, and data exchange in respect of the atmosphere, oceans, and inland waters. A UN agency, the WORLD METEOROLOGICAL ORGANIZATION (WMO) was created in 1947, evolving from its predecessor the International Meteorological Organization. WMO has participated in the UN Global Atmospheric Research Program, World Weather Watch, the World Climate Program, and the Global Environment Monitoring System.

Convention on biological diversity See CONVENTION ON PROTECTING SPECIES AND HABITATS.

Convention on Climate Change A framework convention endorsed by the UN CONFERENCE ON ENVIRONMENT AND DEVELOPMENT in 1992, to protect the world's climate system most notably against the effects of greenhouse gases and their warming influence. Developed nations are required to reduce their emissions of carbon dioxide and other GREENHOUSE GASES to 1990 levels. Several UN agencies are involved in the implementation of this convention. (See BERLIN CLIMATE CONFERENCE.)

Convention on environmental impact assessment in a transboundary context (the Espoo convention) A convention adopted by the ECONOMIC COMMISSION FOR EUROPE (ECE) at Espoo, Finland, in February 1991. The convention was the first multilateral treaty to specify the procedural rights and duties of countries with regard to the transboundary effects of their proposed activities. The convention stipulates the obligations of the parties to carry out EIA, as early as possible in respect of activities likely to cause significant adverse transboundary effects. Under the convention, countries have a general obligation to notify and consult each other about major projects under consideration, ranging from nuclear power stations to deforestation projects. Public participation is to be encouraged.

Convention on fishing and the conservation of living resources in the Baltic Sea and the Belts (the Gdansk convention) Adopted in 1973, an agreement between the riparian countries of the Baltic, also embracing the Danish belts and the Sound, for the purposes of the rational management of the marine resources of area.

Convention on international trade in endangered species of wild fauna and flora (CITES) (the Washington convention) A convention adopted in 1973, providing for the regulation of trade in whole plants and animals, dead or alive, and their readily recognizable parts and derivatives. Excessive international trade has contributed directly to the decline of many wild populations of rare and endangered species. The convention states that a protected species may not be the subject of international trade without a permit granted under the CITES procedures.

Convention on long-range transboundary air pollution A convention initiated by the UN ECONOMIC COMMISSION FOR EUROPE (UN ECE) for the purpose of curbing the incidence of ACID RAIN arising from the long-range transportation of pollutants, most notably SULPHUR DIOXIDE. The convention was signed in 1979 and came into effect in 1983. Basically, it was a statement of desirable objectives without numerical limits. In 1983, the Nordic countries proposed a 30% reduction in sulphur dioxide emissions in all member countries, to be achieved by 1993. The move was welcomed by most member countries. Regulations began to be introduced requiring power plants to be equipped with high-performance gas scrubbers, unless being retired by 1993. A protocol to the 1979 Convention was adopted in 1985 (the HELSINKI PROTOCOL) to require the reduction of sulphur emissions or their transboundary fluxes by at least 30%. In 1988 another protocol to the 1979 Convention was adopted (the SOFIA PROTOCOL), concerning the control of emissions of nitrogen or their transboundary fluxes. The objective here is to stabilize the emission of nitrogen oxides at their 1987 level by the end of 1994.

Convention on protecting species and habitats (Convention on biological diversity) A convention endorsed by the UN CONFERENCE ON ENVIRONMENT AND DEVELOPMENT in 1992, to conserve world biological diversity, that is, the variability among living organisms. Biological diversity is valuable for ecological, genetic, social, economic, scientific, educational, cultural, recreational, and aesthetic reasons. Initially treated with reserve by the US, it was subsequently endorsed in 1993, the US establishing a National Biodiversity Center.

Convention on the conservation of Antarctic marine living resources (the Canberra convention) Adopted in 1980, an agreement regulating the harvesting of krill, squid, and birds of the Southern Ocean, an area much larger than that embraced by the ANTARCTIC TREATY. It requires that harvesting decisions should take into account the possible effects on dependent species. The parties to the convention are Argentina, Australia, Belgium, Brazil, Britain, Chile, France, Germany, India, Japan, New Zealand, Norway, Poland, Russia, South Africa, and the USA.

Convention on the conservation of European wildlife and natural habitats (the Berne convention) Adopted in 1979 and coming into force in 1982, an agreement among members of the European Union requiring members to protect indigenous endangered species.

Convention on the conservation of migratory species of wild animals (the Bonn convention) Adopted in Bonn, Germany in 1979, a convention which obliges the parties to protect endangered migratory species and to endeavour to conclude agreements for the conservation of migratory species whose status is 'unfavourable'. The purpose is to protect wild animals, a significant proportion of whose numbers cyclically and predictably cross one or more national jurisdictional boundaries.

Convention on the conservation of nature in the South Pacific (the Apia convention) Adopted in 1976, an agreement aiming to establish a broad framework for nature conservation in the South Pacific region.

Convention on the control of transboundary movement of hazardous wastes and their disposal (the Basel convention) An international convention finalized in 1992. Its aim is to encourage countries to minimize the generation of hazardous wastes and the transboundary movement of such wastes. The convention was negotiated following concern about 'dumping' by industrialized countries of hazardous wastes on to African and other developing countries. The convention embodies the principle of prior informed consent, whereby the export of hazardous wastes is only allowed following the prior written approval of the importing country. However, in 1994, at a second meeting of the parties to the convention it was decided to prohibit immediately the export of hazardous wastes for final disposal from OECD to non-OECD countries and to phase out by 31 December 1997 all exports of hazardous waste destined for recycling or recovery operations from OECD to non-OECD countries. These decisions were strongly supported by the G77, the original group of non-OECD countries who were represented at the first Basel convention meeting in 1992.

Convention on the law of the sea (the Montego Bay convention) A UN convention which was adopted in 1982 and came into force in November 1994. It was the outcome of international discussions beginning in the early 1970s aimed at establishing a new legal regime for the oceans and their vast resources. The discussions embraced the concept of a 200-mile economic zone and have been concerned with the management and conservation of minerals and food supplies.

A subsequent protocol (the MADRID PROTOCOL) on environment protection, qualifying the ANTARCTIC TREATY, designates Antarctica as 'a natural reserve devoted to peace and science'.

Convention on the means of prohibiting and preventing the illicit import, export, and transfer of ownership of cultural property An agreement initiated by UNESCO and adopted in 1970. It provides a number of steps to protect moveable cultural property and to arrange for the return of illegally exported material. Each signatory should: maintain a national inventory of protected national property; set rules for professional conduct; and supervise and protect archaeological sites.

Convention on the prevention of marine pollution by the dumping of wastes and other matter (London dumping convention) Adopted in 1972, an agreement among 25 nations concerned with the effects of disposing of wastes at sea.

Convention on the prevention of marine pollution from land-based sources (the Paris convention) Adopted in 1974, by Britain, Denmark, France, Germany, Iceland, Luxembourg, the Netherlands, Norway, Spain, and Sweden to augment earlier agreements controlling the dumping of wastes at sea. The convention is implemented by a Commission, which meets annually.

Convention on the prohibition of fishing with long drift nets in the South Pacific (the Wellington convention) Adopted in 1989, a convention emerging from the Tarawa Declaration of the 20th South Pacific Forum, held earlier in the same year, aimed at banning the practice of ecologically damaging drift-net fishing in the South Pacific. The declaration was the result of an Australian proposal.

Convention on the protection of the environment between Denmark, Finland, Norway, and Sweden (the Nordic convention) Adopted in 1974 on the initiative of the Nordic Council following the UN CONFERENCE ON THE HUMAN ENVIRONMENT in 1972. The convention is an important element in Nordic cooperation in the field of environment protection ensuring that activities in one jurisdiction are not detrimental to other members and providing for compensation in the event of environmental damage.

Convention on the protection of the marine environment of the Baltic Sea (the Helsinki convention) Adopted in April 1992, an agreement by Denmark, Estonia, Finland, Germany, Latvia, Lithuania, Poland, Russia, and Sweden on the protection of the marine environment of the Baltic Sea with an action plan to reduce pollution in areas around the Sea.

Convention on wetlands of international importance, especially as waterfowl habitat (the Ramsar convention) Adopted in 1971 and coming into force in 1975, an agreement by the parties to take action to create reserves and otherwise protect wetlands that are internationally important for reasons including their habitat value for rare or migratory birds. Japan, for example, has concluded agreements for migratory bird protection with the US, Russia, and Australia to protect migratory birds facing extinction. Australia has designated three areas in the Northern Territory, including KAKADU NATIONAL PARK, as wetlands of international significance. A protocol adopted in 1982 in Paris strengthened the convention.

Convention to combat desertification   A UN treaty signed in Paris, October 1994, and ratified later by individual countries aimed at tackling the spread of deserts globally. The convention has as a central plank the voluntary implementation of a National Action Program to deal with the decay of farmland into desert. About one-fifth of the world's farmlands suffer from the effects of desertification. (See DESERTIFICATION CONTROL; UN CONFERENCE ON DESERTIFICATION.)

Conventions for the conservation of Antarctic living resources   Conventions agreed by the parties to the ANTARCTIC TREATY 1959. These have included: (1) Agreed measures for the conservation of Antarctic fauna and flora 1970; (2) Convention for the protection of Antarctic seals 1972; and (3) Convention on the conservation of Antarctic marine living resources 1980.

Conventions for the protection of the northeast Atlantic against pollution   Agreements adopted each dealing with separate aspects of pollution control in the northeast Atlantic region, including the North Sea. These agreements include: (1) Agreement for cooperation in dealing with pollution of the North Sea by oil (the Bonn convention) 1969; convention for the prevention of marine pollution by dumping from ships and aircraft (the Oslo convention) 1972; and the convention on the prevention of marine pollution from land-based sources (the Paris convention) 1974.

Conventions for the protection of the Rhine against pollution   (the Bonn conventions) Two conventions signed in 1976 by the five riparian countries and the European Union agreeing to eliminate pollution of surface waters by dangerous substances, identified as organohalogens, organophosphorus, carcinogenic compounds, mercury, cadmium, petroleum hydrocarbons, and chlorides. The parties also agreed to restrict a whole range of other undersirable substances. National programmes were to be established to deal with all these pollutants, which had caused the Rhine to be rated between 'massively' and 'critically' polluted.

cooling pond   (1) A large tank of water in which irradiated fuel elements from a nuclear reactor are stored to allow the short-lived fission products to decay. (2) An artificial lake used for the natural cooling of condenser-cooling water serving a conventional power station.

cooling tower   A device for cooling water by evaporation in the ambient air. A tower requires a flow of air and this may be induced by natural or mechanical means. The use of large cooling towers at power stations to dissipate recovered waste eliminates the problem of THERMAL POLLUTION in masses of water. However, the make-up water for a cooling-tower system is taken from a river or lake, thus eliminating its subsequent use downriver. After being used several times for cooling purposes, it is finally dissipated to the general atmosphere as steam.

Corbett National Park   In 1935, a national park established in Uttar Pradesh, India, near the foothills of the Himalayas. It is a region of hills and open plains. It is particularly rich in fauna, including the elephant, tiger, leopard, hyena, jackal, sloth, bear, wild pig, deer, and crocodile. The flora includes sal, silk-cotton, and kusum trees.

coriolis force   The apparent force caused by the Earth's rotation which serves to deflect a moving body on the surface of the Earth clockwise in the northern hemisphere and anticlockwise in the southern hemisphere.

Cornucopians   Those who believe that economic growth, technological advances, innovations and discoveries, will result in a world enjoying SUSTAINABLE DEVELOPMENT, minimal pollution, and an improving QUALITY OF LIFE for all upon the Earth, both now and in the future. Cornucopians reject the concept of limits to growth or the need for the rich to reduce consumption to help the poor, save in the short-term, when accelerated growth is needed.

cosmic rays   A complex, very penetrating system of radiation incident upon the Earth from outer space.

cost   (1) The expenditure incurred in obtaining the services of a FACTOR OF PRODUCTION. (2) The opportunities foregone in making a purchasing decision. (See OPPORTUNITY COST.)

cost-benefit analysis   (CBA)   A procedure for comparing alternative possible courses of action by reference to the net social benefits that they are likely to produce. The term 'net social benefits' refers to the difference between social benefits and social costs, the one being subtracted from the other. The results may be positive or negative. As far as possible, the costs and benefits are measured in money terms: where costs and benefits cannot be readily assigned dollar figures (the 'intangibles') they are separately identified and described for assessment in a wider context by the decision-maker. Cost-benefit was a technique developed in the USA in response to a legal requirement imposed in 1936 on federal water-resource projects that projects be undertaken only 'if the benefits to whomsoever they may accrue are in excess of the estimated costs' (US Flood Control Act 1936). Hence the concept of a benefit-cost ratio emerged. The principles of cost-benefit analysis (known as benefit-cost analysis in the USA) have since been applied or 'extended' to a wide range of projects, and to the design of public policies in various areas, for example, electric power generation, irrigation, airports, road projects, rail services, shipping, urban development and new towns, health services, education, and welfare and to social issues such as equity and social justice, income distribution, and employment opportunities. It has also been extended to a whole range of environmental issues ranging from pollution control, global strategies, parks and open spaces, environmental planning, and within the framework of EIA and SUSTAINABLE DEVELOPMENT. The project or activity that offers the greatest 'net social benefits', or the best benefit-cost ratio, may not be the one selected when larger considerations, of a political, social, funding, or international nature are taken into account. (See ECONOMIC SYSTEM, FUNCTIONS OF; ECONOMICS.)

cost-effectiveness analysis (CEA) An approach adopted when the benefits from a project or activity cannot be readily measured in monetary terms, or in some respects at all, yet the project or activity has political or corporate approval. In this situation the application of cost-benefit analysis (CBA) may be impracticable, yet a cost-effectiveness analysis (CEA) remains highly relevant. With the objectives given, such an approach examines the costs of alternative ways of meeting those objectives with a view to achieving the maximum value for the dollar invested. CEA cannot be used to compare projects with different objectives. CEA is often used in the difficult area of social infrastructure.

costs of pollution The social costs and penalties imposed on a community or population that are not recorded in the accounts of polluters. Various monetary estimates of the losses have been made, all of them of a high order of magnitude. When measures to control pollution are adopted at cost to the polluters, the costs are then said to be 'internalized'.

costs of production Expenditure incurred by way of payments for interest on loans, mortgages, rents, dividends, salaries and wages, buildings, plant, inventories, development, and marketing of a good or service. These are essentially accountancy or recorded costs internal to the enterprise. There may be other direct costs borne by the community by way of expenditure on roads and other infrastructure. Also, if the enterprise is environmentally deficient, and additional burden, unrecorded, will be borne by the community.

Council of Europe A body created in 1949 to secure a greater measure of unity between the European countries. It has a Parliamentary Assembly, a Committee of Foreign Ministers, and a European Commission investigating human rights. It has 21 member nations. It has been the task of the Assembly to propose actions to bring the European nations closer together, while the role of the Committee is to translate the Assembly's recommendations into action. Proposals have related to economic, legal, social, public health, educational, scientific, and environmental matters. The result has been over 70 conventions concluded among the members. In 1970, the Council adopted a 'Declaration on the Management of the Natural Environment of Europe', stressing the interdependence of continental land-use systems. In 1980, the Council initiated the European Campaign for Urban Renaissance. Regular ministerial conferences on the environment are held.

Council on Environmental Quality US (CEQ, US) A council created by the NATIONAL ENVIRONMENTAL POLICY ACT 1969, the members being appointed by the US President. The responsibilities of the CEQ include formulating policy recommendations on environmental matters, managing EIS and EIA procedures at the federal level, and publishing an annual report on the state of the US environment. The Council has groups studying various aspects of the environment. In 1977, the President authorized the CEQ to issue regulations which would ensure the effectiveness of EISs while reducing unnecessary paperwork, and to prescribe procedures to help resolve inter-agency conflicts regarding projects being reviewed under the Act. The regulations became effective in 1979.

coupe An area of forest felled in a single operation.

CRAMRA CONVENTION FOR THE REGULATION OF ANTARCTIC MINERAL RESOURCE ACTIVITY.

crop rotation An agricultural technique in which, season after season, each field is sown with crop plants in a regular rotation, each crop being repeated at intervals of several years. Crop rotation minimizes the risks of depleting the soil of particular nutrients. In rotation systems, a grain crop is often grown the first year, followed by a leafy-vegetable crop in the second year, and a pasture crop in the third. The last usually contains legumes; such plants can restore nitrogen to the soil. Notwithstanding, high yields tend to depend upon the continued addition of chemical fertilizers to the soil.

cropping   A form of agriculture that involves the growing of crops rather than the raising of livestock.

crown-of-thorns starfish (*Acanthaster planci*)   A multi-armed starfish, reaching a size of 60 cm in diameter, that has caused extensive damage to coral reefs in tropical waters in the Indian and Pacific oceans. It has caused damage to the GREAT BARRIER REEF.

crude birth rate   The annual number of live births per 1000 persons in a population.

crude death rate   The annual number of deaths per 1000 persons in a population.

cryosphere   The Arctic and Antarctic ice caps, the sea ice in the polar regions, and the many mountain glaciers. The cryosphere forms the Earth's most important freshwater reservoir, of which the inland ice of Antarctica accounts for about 80%. It also has much influence on the weather patterns of the world.

cryptic damage   Damage and impairment without visible signs, for example, reduced growth in plants.

Cryptosporidium   A genus in the Protozoa that is an enteric pathogen. It may contaminate water from pasture runoff and human waste, and cause diarrhoeal illness. It is extremely resistant to disinfection.

cullet   Recycled crushed glass used in glass-making to speed up the melting of silica sand.

culm and gob banks   Low-grade fuel and waste of no commercial value discharged from coal-processing plants onto land where it has formerly accumulated into hills or banks of culm (anthracite) and gob (bituminous coal). Such banks used to disfigure the landscape and were easily ignited.

cultch   Materials (such as oyster shells and pebbles) dumped into the sea to furnish places of attachment for the larval stage of the oyster (or any other shellfish).

cultivars   Varieties of plants or other organisms that have been developed during cultivation. They are distinct subspecies that do not occur naturally in the wild.

cultural environment   A concept which includes: rock-art sites, ceremonial grounds, sacred sites, camp sites, and shell mounds; historical and archaeological sites such as old missions and cemeteries, historic buildings and structures, historic towns and precincts; current evidence of distinctive characteristics of communities and their ambiance; and evidence of the degree of development of the arts in a society, viewed in a domestic and world setting. The cultural environment is one aspect to be considered in EIA.

cumulative effects   Progressive environmental degradation over time arising from a range of activities throughout an area or region, each activity considered in isolation being possibly not a significant contributor. Such effects may arise from a growing volume of cars and trucks, multiple sources of power generation or incineration, the increasing application of chemicals to land or substantial urban development. There are many ecological, social and economic factors to be taken into account in measuring cumulative effects including synergistic and threshold effects, time lags, effects on social infrastructure, and any effect on prices and quantities of goods and services. The solution to unacceptable cumulative effects is often better regional planning and control.

cyanides   The salts of hydrocyanic acid which are virulent poisons. The objections to cyanides in effluents include: (1) the toxic effects which they have on fish and other life in rivers, and on bacteria in sewage works; (2) their ability to absorb oxygen; and (3) the formation of extremely poisonous hydrogen cyanide gas which can endanger the lives of persons working in sewers or other confined areas. In some areas, significant contributions are made to the toxicity of rivers by cyanides derived from metal finishing processes such as electroplating, cleaning, and case-hardening, or from the scrubbing of coke-oven or blast-furnace gas. Cynanides may be removed from an effluent by oxidation with CHLORINE to produce cyanates which are relatively harmless, or by the precipitation of the cyanides after treatment with lime and ferrous sulphate.

cyanobacteria   Bacteria (formerly classified as BLUE-GREEN ALGAE), which occur in all natural waters, though of no concern in small numbers. Some species can cause skin symptoms, asthma, gastroenteritis, neuromuscular disorders, and liver damage. Blue-green algae are known as cyanophytes.

cyanophytes   See BLUE-GREEN ALGAE.

cyclic hydrocarbons   See HYDROCARBONS.

cyclodiene insecticides   A subgroup of the CHLORINATED HYDROCARBONS, including CHLORDANE, heptachlor, aldrin, dieldrin, and endrin.

cyclones, dust removal   Devices for removing particulate matter from the waste gases of industrial processes. A simple cyclone comprises a cylindrical upper section and a conical bottom section. The dust-laden gases enter the cylindrical section tangentially. Centrifugal action throws the grit and dust particles to the outer walls. These particles fall by gravity to the dust outlet at the bottom. The relatively clean gases leave through a centrally situated tube within the upper section. Cyclones may be used singly or in groups to achieve higher collecting efficiencies. Cyclones are inadequate to meet the demanding standards for emission

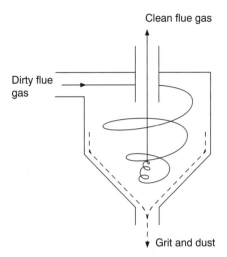

CYCLONES, DUST REMOVAL: Figure 5. Principle of a simple cyclone.

control on large installations such as power stations, where electrostatic precipitators or bag filters will be required. (See Figure 5.)

cytology The study of cells as the fundamental units of living things. It is a separate branch of BIOLOGY.

cytoplasm The contents of a cell, comprising about 70% water and 15% proteins, with the rest mostly composed of sugars, fats and salts. All cells are similar in composition, form, and function. However, the evolution of specialized cell features is responsible for the great diversity of life forms.

cytotype A member of a population composed of individuals with essentially similar KARYOTYPES or characteristics (in respect of sizes, shapes, and numbers), in contrast with those of individuals in other populations.

D

daily calorie supply per capita A statistic appearing annually in the *World Bank Atlas*. The calorie content of the food produced and imported for domestic consumption is divided by a country's population and expressed as a daily amount. Consumption varies from less than 2300 calories to more than 3300 calories per capita per day. The information comes primarily from the UN FOOD AND AGRICULTURE ORGANIZATION, supplemented by data from WORLD BANK sources.

Daisetsuzan National Park Established in 1934 in Hokkaido, Japan, a national park embracing a volcanic range with many peaks, cliffs, gorges, lakes, and waterfalls, with vast alpine meadows and virgin coniferous forests. Fauna include the Asiatic black bear, northern pika, chipmunk, and Japanese macaque. The park is the only habitat in Japan for the pika.

damper A manual or automatic device to regulate the rate of flow of combustion gases through the flues of a fuel-burning appliance, thus regulating the rate of combustion and heat output. Types include barometric, butterfly, guillotine, and sliding.

damping (1) In ACOUSTICS, the dissipation of energy with time or distance; the absorption of noise. The noise from machines may be reduced by the use of damping techniques, panels, and enclosures, sometimes by 10 dB(A) or more. Damping is extensively employed in the design of cars. The use of damped bits in road drills may give a reduction of 7 dB(A) in noise. (2) In ECONOMICS, a reducing effect. Thus, when the demand for labour is low, some people withdraw from the labour market, damping the rise in unemployment figures.

dams Structure constructed to restrict the flow of tidal or river water for the purposes of hydroelectricity or irrigation, or as a means of containing industrial wastes. Dams often attract opposition on visual grounds, on the grounds of interference with the local ecology, on the grounds that the formation of a new lake will involve loss of fertile land or the displacement of people and homes, perhaps in large numbers. (See ASWAN HIGH DAM; GORDON-BELOW-FRANKLIN DAM; INTERNATIONAL COMMISSION ON LARGE DAMS; THREE-GORGES WATER CONSERVATION AND HYDROELECTRIC PROJECT; WORLD BANK.)

dangerous goods Goods often classified as such by transport authorities; for example, explosives, gases, flammable liquids and solids, poisonous substances, toxic, infectious, and radioactive substances, corrosive and oxidizing agents. Vehicles carrying such loads are normally required to be clearly identified, indicating for rescue and fire crews the precise nature of the load.

Darling Harbour Located in Sydney, Australia, a major rehabilitation project completed substantially by 1990, transforming a derelict industrial district.

Darling, Sir Frank Fraser One of the world's pioneer ecologists. His work *Wilderness and Plenty*, being a series of Reith Lectures for the British Broadcasting Corporation, was published in 1970. It had a great effect on environmental thinking in Europe and the USA.

Dartmoor National Park Established in England in 1951, a high plateau with granite hills and tors (rocky pinnacles or peaks), bogs and moorland. Typical fauna of the park includes deer, badger, mink, vole, and squirrel. It has many relics of prehistoric Bronze Age civilization.

Dasmann, Raymond F. Author of *Environmental Conservation* (1984), a general textbook on the conservation of natural resources, stressing their interrelationships and uses.

DCF DISCOUNTED CASH FLOW.

DDT (dichlorodiphenyltrichloroethane) A chlorinated hydrocarbon insecticide remarkable for its high toxicity to insects at low rates of application and for the persistent effectiveness of its residual deposits. Developed in the 1940s, the initial results were spectacular: in some countries where DDT was used to combat the mosquito, malaria was reduced from being an important source of human illness and mortality to one of manageable proportions. Similarly, agricultural pests were reduced and crop yields soared in many areas. Eventually, it was recognized that DDT was having unexpected and severe consequences for the ENVIRONMENT. By the 1970s, the use of DDT anywhere for any purpose was open to debate. The use of DDT is now virtually banned worldwide. Its very strength of toxicity and persistence proved its greatest fault, particularly as it became accumulated through the food chain, for example, through plankton, fish, and fish-eating birds. In the US, there was a steady decline of predators such as the bald eagle, osprey, and peregrine falcon, whose prey accumulated DDT. Birds began to lay eggs with much thicker or thinner eggshells than usual. DDT began to appear in meat for human consumption. Also, in the early 1960s immune strains of mosquitoes and other disease-carriers emerged. (See BIOACCUMULATION; BIOMAGNIFICATION; CHLORINATED HYDROCARBONS.)

Dead Sea A salt lake between Israel and Jordan, the lowest body of water on Earth; it is situated in a desert. The waters are extremely saline; bathers float on it easily. The salinity excludes any animal or plant life other than bacteria. Fish carried in from the Jordan River die instantly.

Death Valley National Monument Established in 1933 in California–Nevada, USA, a large low-lying desert surrounded by mountains, the highest being Telescope Peak. There are several hundred kinds of desert plants. Fauna include desert bighorn sheep, deer, cougar, fox, coyote, and badger.

debt-for-nature swaps The promotion of nature conservation projects out of the vast indebtedness incurred by many Third-World countries during the 1980s. International debt became a global problem during that period, as many lenders and investors found that they would never be repaid. It then became possible to buy fairly worthless debt from banks and institutions at prices much below face value. Lenders were glad to receive something, rather than lose all. Conservation organizations found that it was possible to acquire some of this debt cheaply, with subscribed money. These bodies then negotiated a settlement of the debt with the debtor country, again well below face value, on the understanding that the funds received to settle the debt would be invested in the country concerned on conservation projects. The indebted country thus shed debt at a substantial discount, while the proceeds did not leave the country and boosted the local economy. Depending on the costs of the transaction, conservation organizations have been able to achieve significant increases in the resources available for conservation in the debtor countries. Increasingly, swaps involve SUSTAINABLE DEVELOPMENT projects. Creditor banks gain by converting their non-performing loans, though at substantial discounts. Debtor countries benefit from the reduction of their foreign currency debt, and add to their expenditure invested at home. The conservation investor receives a premium on its investment, through favourable rates of exchange. Debt-for-nature swaps have been endorsed by the WORLD BANK, the GLOBAL ENVIRONMENTAL FACILITY, and the G-7 countries.

decentralization A conscious policy of relocating some parts of an organization in outlying regions away from metropolitan areas, or guiding new organizations into those areas. The term may also be applied to organizations which delegate some decision-making from head office to district or regional offices. As an aspect of government policy, it often involves providing incentives to firms to locate in centres which would not otherwise have been chosen. General incentives have often been found to be ineffective, and the tendency in some countries is to offer incentives in relation to selected centres only. This is known as selective decentralization. Incentives are often of a temporary character.

decibel (dB) The unit of measure of sound-pressure level. The dB scale is logarithmic and hence can cover a wide range of sound. A sound of 0 dB is just audible to a person with good hearing and 120 dB causes pain. (See DECIBEL (A-SCALE).)

decibel (A-scale) An internationally weighted scale of sound levels which attenuates the upper- and lower-frequency content and accentuates the middle frequency, thus providing a good correlation for the human ear. Nearly all audible sound lies between 0 dB(A) and about 140 dB(A). An increase of 10 dB(A) means that the noise perceived by a listener has about doubled in loudness. The B, C, and D scales are used for more specialized noise measurements, for instance, the D scale is used for the measurement of jet engine noise at airports. (See DECIBEL (dB); Figure 6.)

deciduous trees and shrubs Trees and shrubs that shed their leaves before the onset of winter or a dry season. Examples of deciduous trees are oak and beech.

Declaration on the Human Environment A declaration of common outlook and principles adopted and issued at the conclusion of the UN CONFERENCE ON THE HUMAN ENVIRONMENT held in Stockholm, Sweden, from 5 to 16 June 1972. (See UN ENVIRONMENT PROGRAM; WORLD ENVIRONMENT DAY.)

decomposers Organisms, usually bacteria and fungi, that break down the organic compounds of dead organic material into simple compounds that primary producers (most being green plants) can utilize.

decomposition The separation of organic material into simpler compounds.

deep ecology A perspective that recognizes an inherent right to existence among all living organisms and which sees humans as not more or less important than any other species. (See ANTHROPOCENTRIC.)

deep freshwater marshes Marshes which generally remain inundated throughout the year, water depths being usually one or two metres. (See WETLANDS.)

deep repositories A solution offered for high-level RADIOACTIVE WASTES, the idea being to put nuclear waste far out of reach, and hopefully out of mind. GREENPEACE and others prefer to keep waste readily accessible, while guarding it carefully. Their argument is that the science of deep repositories is not fully understood and would make the waste hard

deflagration

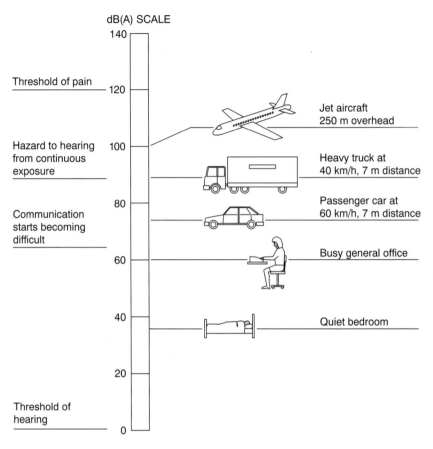

DECIBEL (A-SCALE): Figure 6. Level of common sounds on the dB(A) scale.

to get at should anything go wrong. Britain's only current plan for high-level waste is to leave it for 50 years to cool down, before making further decisions. In November 1994, it was revealed by Russia that waste had been injected directly into the ground at three sites, without packaging. The waste has spread much more widely and quickly than anyone expected, borne by water seeping through the ground towards the Ob and the Volga. The USA has its own plan for a deep repository for high-level waste, at Yucca mountain in Nevada. Disposal problems add to the economic case against NUCLEAR POWER PROGRAMMES, while renewable resources become cheaper. (See RENEWABLE RESOURCE.)

deflagration The chemical reaction of a substance in which the reaction front advances into the unreacted substance at less than sonic velocity. Where a blast wave is produced with a potential to cause damage, the term 'explosive deflagration' is usually used.

defoliant Any substance or mixture of substances causing the leaves or foliage to drop from a plant, with or without causing abscission. (See AGENT ORANGE.)

deforestation The removal or CLEAR-FELLING of forest and undergrowth to increase the amount of arable land or to use the timber for sale, for construction or industrial purposes. Forest and its undergrowth possess a very high water-retaining capacity, inhibiting runoff of rainwater. Removal of vegetation gives rise to erosion which can take on several forms: sheet erosion, rill erosion, gully or ravine erosion, sliding or slip erosion.

degradation A process by which a chemical is converted to another, usually less complex compound. This process can be the result of the activities of microorganisms, water, air, sunlight, or other agents.

demand In economics, the amount of a commodity or service that will be bought at any given price per unit of time. In society, the two main components of demand are consumption and investment. The market system only responds, however, to effective demand. It does not respond to need not accompanied by purchasing power, or to a community demand or need for the protection of environmental assets. It is the role of government to respond to needs and demands not matched by the market system and PRICE MECHANISM.

demand management The ways in which the character of effective demand can be influenced by pricing and other policies. For example, peak-load and off-peak pricing arrangements in the supply of electricity, by curbing the peaks and boosting the use of idle plant, can reduce the amount of generating plant required to serve a market, thus reducing the need for new plant and its demands on the ENVIRONMENT.

demersal fish Fish that live on, or adjacent to, the sea bottom, not to be confused with BENTHIC SPECIES.

demographic transition A fundamental change in the characteristic trends of a population, for example, a transition from high birth rates and high death rates, to low birth rates and low death rates. Much of Europe and North America has passed through this demographic transition, while Japan and other Asian countries are experiencing a similar transition. (See WORLD POPULATION.)

demography The study of the structure of human populations, for example, the size of a population and its distribution by age, sex, occupation, and location, and trends in these characteristics. (See CENSUS OF POPULATION.)

Denali National Park Established in 1917 in Alaska USA, an area of mountain peaks on the northern flank of the Alaska Range. The flora is comprised mostly of alpine tundra with mosses, lichens, and sedges; the fauna includes barren-ground caribou, grissly bear, wolf, and red fox. A special feature is Mount McKinley, the highest peak in North America.

denitrification The process by which specific bacteria break down nitrates and release nitrogen.

dense gas cloud A gas cloud whose density exceeds that of the air at the same temperature, as distinct from a buoyant gas cloud, which is lighter than the surrounding air, and the neutral density gas cloud, which has a density equal to that of the surrounding air.

Department for policy coordination and sustainable development, UN A department created in 1992 in response to the UN CONFERENCE ON ENVIRONMENT AND DEVELOPMENT. The Department is one of the prime moving forces behind the COMMISSION FOR SUSTAINABLE DEVELOPMENT.

depleted uranium Uranium in which the content of the fissile isotope uranium-235 becomes less than the 0.71% normally found in natural uranium. (See RADIOACTIVITY.)

deposit gauge An instrument placed out-of-doors to capture and measure the fall-out of dust and grit in the atmosphere, probably falling from a nearby industrial source. Some deposit gauges are directional, giving some indication of the likely source of the nuisance.

depurate Become free from contamination.

desalination The removal of dissolved salts from sea water and in some instances from the brackish waters of inland water bodies, or highly mineralized ground water. The process produces water fit for human consumption, irrigation, or industrial applications. Existing desalination technology consumes a great deal of energy and so the process is expensive.

desert Any large extremely dry area of land with fairly sparse vegetation. One of the Earth's major types of ecosystem, supporting distinctive plants and animals specially adapted to the harsh ENVIRONMENT. (See CONVENTION TO COMBAT DESERTIFICATION; UN CONFERENCE ON DESERTIFICATION.)

design checklist In hazard analysis, a method used for the auditing of new plant design from a safety aspect, at the detailed design stage.

designated development Proposed developments and activities appearing in national legislation in many countries in respect of which an ENVIRONMENTAL IMPACT STATEMENT must be prepared.

desorption The process of removing a sorbed substance; the reverse of ADSORPTION and ABSORPTION.

destructive distillation See PYROLYSIS.

desulphurization of flue gases The removal of sulphur compounds from stack or flue gases. Methods offering removal efficiencies in excess of 90% include: limestone scrubbing with additives; limestone scrubbing; dual alkali scrubbing; and regenerative processes. Desulphurization may be assisted also by: physical coal cleaning; chemical coal cleaning; low-sulphur fuel substitution; and limestone injection with a multistage burner. Thus, sulphur oxide emissions may be reduced directly by scrubbing or indirectly through the reduction of sulphur in the fuel.

detention period The average amount of time that each unit volume of liquid or gas is retained in a tank or chamber in a flow process. A minimum detention period may be essential to complete a given stage of a process; for example, the destruction of certain classes of bacteria, the settlement of certain size fractions of solid particles, or the digestion of organic material.

detergent A surface-active agent used for removing dirt and grease from a variety of surfaces and materials. Early detergents contained ALKYL SULPHONATES, which proved

detrimental level

resistant to bacterial decomposition, causing foaming in rivers and difficulties in sewage treatment plants. These hard detergents were replaced generally during the 1960s with soft BIODEGRADABLE detergents. Apprehension continues to be expressed about the use of phosphates in detergents, helping to promote the process of EUTROPHICATION. No satisfactory substitute has yet emerged.

detrimental level That concentration of any pollutant or indicator at or above which a substantial proportion of the exposed population may be adversely affected, or where significant changes are likely to be caused to some segments of the ENVIRONMENT.

detritivore A consumer organism that directly consumes dead organisms and the cast-off parts and organic wastes of other organisms. (See DECOMPOSERS.)

detritus Unconsolidated sediments formed at the bottom of a body of water composed of both inorganic and dead and semidecayed organic materials.

developing countries Generally, the poorer nations of the world often referred to as the Third World. Taken as a group, the developing nations (over 130 in number), account for roughly three-quarters of the world's population. But they are not homogeneous for the economies of the 'little dragons' (i.e. the outstanding economic performers among the smaller economies of Asia, such as South Korea, Taiwan, Singapore and Hong Kong) now have more in common with the First-World industrial economies and the Second-World eastern European countries than with the poorest economies in South Asia and black Africa. In 1995, South Korea, once one of the poorest nations on earth, stands in sharp contrast to India. It may be appropriate to introduce a concept of a Fourth World to accommodate Sub-Saharan Africa, Tanzania, Ethiopia, Peru, and the Philippines.

development The application of human, physical, natural, and financial resources to meet effective or prospective market demands and other human needs. The breadth of the concept is not always appreciated as it applies not only to the growth of industry, commerce, trade and financial institutions, but to the provision of INFRASTRUCTURE, sanitation, educational facilities, hospitals and public health services, housing, national parks, and tourist and recreational facilities. In all countries, development is a function of both the private and public sectors of the economy and voluntary activity. Activity of this nature may be profit-motivated, a matter of public policy, or unpaid. The general effect may be an increase in PER CAPITA INCOME with perhaps benefit or detriment to the QUALITY OF LIFE locally, regionally, nationally, or globally. Development may be sustainable if the resources used are renewable, or non-sustainable if the resource base is exhaustible. (See SUSTAINABLE DEVELOPMENT.)

development analysis An assessment of whether or not a proposed development or activity is likely to be beneficial in terms of meeting fundamental human needs. This is a two-stage analysis. The first stage reviews choices at a conceptual level, while the second analyses the resources to be consumed in the process.

development application A formal application to a planning body, authority, or department seeking to obtain development consent or planning permission for a proposed development or activity, with or without conditions, which may be of a residential, commercial, or industrial character. Such an application may need to be accompanied by an ENVIRONMENTAL IMPACT STATEMENT and undergo a full ENVIRONMENTAL IMPACT ASSESSMENT. Conditions imposed on development consents may range over many of the matters referred to in Box 7.

development control Considered broadly, all those statutory and regulatory elements which provide a framework for DEVELOPMENT, including all ENVIRONMENTAL PLANNING instruments, including ENVIRONMENTAL IMPACT STATEMENT and ENVIRONMENTAL IMPACT ASSESSMENT procedures, and pollution-control approvals and licences. However, the term may extend to much more detailed requirements stipulating, say, restrictions on the heights of buildings, facilities for car parking, landscaping and streetscape requirements, density and plot ratios, heritage conservation measures, facilities for disabled persons, design and location of signs, arrangements for refuse removal and recycling, pedestrian shelter and access, protection of trees, prohibited features, provision of public spaces, pedestrian arcades, aesthetics, shadow and wind-tunnel effects, and procedures for future changes and alterations.

development rights Traditional legal rights associated with the ownership of property, including (but not limited to) the rights to exclude trespassers and poachers, cut timber, mine or quarry, grow crops or keep animals, or to build structures for private use or for profit. These traditional rights have been progressively modified by planning legislation including planning schemes, zoning, DEVELOPMENT APPLICATION, and DEVELOPMENT CONTROL requirements, the outcome of public inquiries and appeals to the courts. No one can any longer do what they like with their own.

development standards (building standards) A more detailed extension of DEVELOPMENT CONTROL governing the quality of: construction; building location and frontage; side

development standards

> **Box 7 Typical matters dealt with in proposed conditions of development consent**
>
> 1. Approvals, licences, and permits to be obtained from statutory authorities, boards, departments, and local councils.
> 2. Conformity with certain specifications contained in the environmental impact statement (EIS).
> 3. Compatibility with all applicable planning instruments.
> 4. Control of air, water, and noise pollution including discharges to catchments, protection of aquifers, control of leachates, blasting controls, incineration, waste disposal, oil contamination, sewage treatment, drainage, storm water and runoff management, and dust suppression.
> 5. Life of project.
> 6. Location of buildings and individual items of equipment.
> 7. Sequence of mining, quarrying, and extractive operations.
> 8. Working hours.
> 9. Buffer zones.
> 10. Access roads, junctions, traffic, rail, pipeline, and transmission routes.
> 11. Risks and hazards.
> 12. Emergency procedures: fire-fighting, evacuation.
> 13. Water supplies and storage.
> 14. Heritage items.
> 15. Visual amenity, trees, vegetation, screening.
> 16. Rehabilitation.
> 17. Social and economic effects of proposal.
> 18. Housing of workforce.
> 19. Monitoring and recording of results.
> 20. Environmentally hazardous chemicals.
> 21. Heights of buildings and stacks.
> 22. Acquisition of properties.
> 23. Protection of wetlands, parks and reserves, oyster leases, mangroves, rainforest, and other natural resources.
> 24. Subsidence.
> 25. Closure of existing plant and replacement of less efficient plant.
> 26. Effects on residents, schools and hospitals, industries.
> 27. Lodging of guarantee funds in respect of performance, payment of levies towards future management of the site, contributions towards infrastructure costs and road improvements.
> 28. Appointment of environmental management officers by proponent.
> 29. Independent auditing of risks and hazards.
> 30. Annual reports to the Department of Planning.
> 31. Arrangements for continuous liaison with the public, local councils, conservation bodies, and resident action groups.
> 32. Matters relating to sustainable development, sustainable yield, or cross-frontier or global environmental protection.
> 33. Compliance with international conventions.

accesses and distances from other buildings; the proportion of the site that may be occupied; the bulk, scale, size, and height of structures; the external appearance of a building; the provisions for vehicles entering and leaving, standing, and parking; drainage and earthworks; sunlight, daylight, and shadows; services, facilities, and amenities; control of noise, air, and water pollution; access for public services; and landscaping and aesthetics.

dichlorodiphenyltrichloroethane

Development or building standards are a tailoring of development control standards to a particular project.

dichlorodiphenyltrichloroethane DDT.

dieback A general term for a decline in the vigour and health of trees, culminating in death. Dieback may be caused by a parasitic infection such as a fungus, or by insect attack, or air or water pollution, or other changes in the environment.

dieldrin One of the CHLORINATED HYDROCARBONS, used widely during the 1960s. Highly persistent and fat soluble, it contributed to BIOMAGNIFICATION or bioaccumulation along the food chain. Its use is now prohibited or severely restricted in most countries.

diffuse sources Sources of pollutants, contaminants or wastes which cannot be traced to a single source such as a sewer outlet or chimney stack. A diffuse source could be a number of properties contributing fertilizer runoff to a river system, or urban stormwater runoff to a wetland. The diversity of sources make regulation and control difficult. The solution often lies in ENVIRONMENTAL PLANNING. (See POINT SOURCES.)

diminishing returns A generalization that while an increase in some inputs relative to other fixed inputs may raise total output, there will come a point when the extra output is likely to diminish. In the extreme, the process of additions becomes counterproductive. An example might be the increasing application of fertilizer to a fixed area of land or an increase of staff in a business without a subsequent growth in sales. The principle applies also to everyday life when good things are consumed to excess.

dioxins A family of chlorinated organic compounds that are undersirable by-products in the manufacture of certain classes of herbicides, disinfectants, bleaches, and other agents. Dioxins are among the most toxic substances, the most notorious being one specific dioxin, 2,3,7,8-tetrachlorodibenzo-p-dioxin (TCDD) This and other dioxins are also formed when chlorine compounds such as PVCs are burnt at low temperatures in badly designed or operated domestic refuse and industrial waste incinerators. Another dioxin formed in these conditions is polychlorinated-dibenzo-p-dioxin (PCDD). The toxic properties of dioxins are enhanced by the capacity that they can pass into the body through all major routes of entry, including the skin, the lungs and the mouth. Dioxins can produce chronic skin diseases, muscular dysfunctions, bodily inflammations, impotency, birth defects, genetic mutations, nervous disorders, and cancers. In the early 1970s, the use of dioxin-contaminated oil to suppress dust on hundreds of roads in Missouri led to the permanent evacuation of whole communities. Dioxin was also present in AGENT ORANGE and prominent in the SEVESO INCIDENT, ITALY.

disability The absence, loss, or reduction of functional ability. It embraces various incapacities in respect of walking and movement generally, manual skills and intellectual impairment. The UN dedicated 1981 as the International Year for Disabled Persons, urging everyone to break down the barriers and help disabled people to enjoy a better life. A world previously designed for able-bodied people has been progressively modified to facilitate movement by the disabled.

disasters See ABERFAN; AGENT ORANGE; AMAZON RAINFOREST; AMOCO CADIZ DISASTER; ARAL SEA; BHOPAL DISASTER; 'BRAER' INCIDENT; CHERNOBYL NUCLEAR CATASTROPHE; EXXON VALDEZ DISASTER; FLIXBOROUGH DISASTER; HARRISBURG NUCLEAR PLANT INCIDENT; HENAN DAMS CATASTROPHE; IXTOC I INCIDENT; KOMI PIPELINE OIL DISASTER; LPG ROAD TANKER INCIDENT, SPAIN; MEUSE VALLEY INCIDENT; MEXICO CITY INDUSTRIAL DISASTER; PINATUBO, MOUNT; POZA RICA INCIDENT; SAINT HELENS MOUNT; SANTA BARBARA BLOW-OUT; SEVESO INCIDENT; STAVA TAILINGS DAMS DISASTERS; THREE-MILE ISLAND NUCLEAR INCIDENT; WINDSCALE NUCLEAR REACTOR INCIDENT.

discounted cash flow (DCF) A method of comparing the profitability of alternative projects. The 'rating' of a prospective investment is determined by the present cash value of sums to be received over future years. The present value or worth of a sum to be received at some future date is such an amount as will, with compound interest at a prescribed rate, equal the sum to be received in the future. Some environmentalists object to this approach on the grounds that the future is discounted compared with the present, to the disadvantage of rising and future generations. (See INTERGENERATIONAL EQUITY.)

diseases and afflictions See ANTHRAX; ARSENIC POISONING; ASBESTOSIS; ASTHMA; BACTERIA; BLACK DEATH; BOTULISM; BRONCHITIS; CADMIUM; CARCINOGENIC COMPOUNDS; CHEMICAL POLLUTANTS; CYANIDES; FOOT-AND-MOUTH DISEASE; HEAVY METALS; HEPATITIS; LEGIONNAIRES' DISEASE; MALARIA; MERCURY; MICHIGAN EPISODE, US; ONCHOCERCIASIS; SCHISTOSOMIASIS; SMALLPOX; TYPHOID; TYPHUS; VIRUSES; YELLOW FEVER.

disinfection The destruction of microorganisms by the use of such substances as creosote, alcohol, iodine, or chlorine; or by the application of heat through boiling or autoclaving. (See CHLORINATION.)

dismal swamp A marshy region on the coastal plain of southeastern Virginia, USA. The swamp is densely forested with cypress, black gum, juniper, and water ash. It is noted for fishing and hunting; deer, bear, raccoon, and opossum are plentiful. The swamp is the habitat of many rare birds, including the ivory-billed woodpecker. The swamp is nearly 70 km in length.

dispersal In BIOLOGY, the dissemination or scattering of organisms over periods within a given area. The two disciplines involved in the study of dispersal are systematics and evolution. Systematics is concerned with the relationships between organisms including the classification of life into ordered groups, while evolution is concerned with the processes of natural selection.

dispersion (1) The break-up of light into components, usually according to wavelength; any phenomenon associated with the propagation of individual waves. (2) The dilution of pollutants in either air or water. The dispersal of air pollutants is a function of meteorological influences, while the dispersal of water pollutants is subject to hydrological influences. In both cases there may be concurrent chemical and biological interactions.

disposable income That portion of an individual's income over which the recipient has complete discretion; it consists of gross income minus direct taxes and other compulsory deductions, plus social-security payments and pensions.

dispute resolution See ALTERNATIVE DISPUTE RESOLUTION; MEDIATION.

dissolved oxygen (DO) The concentration of oxygen dissolved in water or effluent in milligrams per litre (mg/l). The colder the water, the greater the amount of oxygen that can be dissolved in it. Fish and other aquatic organisms require generally more than 4.0 mg/l of DO to survive. The discharge of an organic waste to a water body imposes an 'oxygen demand', the oxidizing activities of bacteria withdrawing oxygen from the water. The extent to which the DO will fall depends upon the rate of oxidation and the rate of re-aeration. The oxygen balance of a river determines its degree of health and its ability to support a balanced aquatic ECOSYSTEM. When the flow of a river is at its lowest (the 'dry weather flow') the DO is also likely to be lowered as a result of sluggish water movement, while the concentration of pollutants may be high through diminished dilution. Low concentrations allow growth of nuisance microorganisms causing taste and odour problems, staining, and corrosion.

district heating A scheme in which both heat and hot water are provided from a central boiler plant to an entire housing estate or group of buildings. The consumer enjoys house heating at a comfortable level and a constant supply of hot water. The combustion of fuel takes place at one point in the most efficient manner, with the best facilities for dealing with the products of combustion. Many point sources of pollution are thus displaced. (See COGENERATION.)

diurnal Daily, or occurring every day; the diurnal cycle of AIR POLLUTION according to meteorological influences.

diversification An increase in the variations exhibited by a population of organisms or community over time.

diversity See BIODIVERSITY.

diversity factor The probability of a number of pieces of equipment being used simultaneously. For example, if 100 electrical appliances of 10 kW each rarely produce a maximum demand in excess of 200 kW, the diversity factor is said to be one in five, or 20%.

DO DISSOLVED OXYGEN.

dolphins Aquatic creatures noted for grace, intelligence, playfulness, and friendliness to humans. The most widely recognized species are the common and bottle-nosed dolphins (*Delphinus delphis* and *Tursiops truncatus*). Both are of the family Delphinidae being widely distributed in warm and temperate seas. The family Delphinidae comprises 14 genera and about 32 species of dolphins distributed throughout the oceans. Other types and variations of dolphin have been identified. The term 'dolphin friendly' is often attached to fish products caught in a manner non-injurious to dolphins.

domestication The process by which wild animals are modified and adapted to human conditions, for example, crops domesticated from wild plants, dogs domesticated from wolves, and the domestic cat.

domestic sewage Wastes consisting of human bodily discharges carried by flushing water: sullage water from kitchens, bathrooms, and laundries; and other water-borne material discarded as a result of regular household and human sanitary activities. In developed countries such wastes are often discharged into drains and sewers, reaching sewage treatment works before discharge to the ENVIRONMENT. In less developed societies, domestic waste may pass directly into the nearest stream. Intermediate arrangements involve the use of septic tanks and night-soil collection.

dominance hierarchy A social structure in which a linear or nearly linear ranking exists, with each animal dominant over those below it and submissive to those above it in the hierarchy. Dominance hierarchies are best known in social mammals, such as wolves and baboons, and in birds which have a 'pecking order'. Pecking orders may also be recognized in human societies: business structures, bureaucracies, political parties, class and caste systems, academic institutions, and residential locations. It is said that the individual's prospects in life are often determined by the care that was taken in selecting one's parents' post code.

Donora smog incident An air pollution episode occurring in Donora, Pennsylvania, USA in October 1948, involving morbidity and mortality due to excessive concentrations of SULPHUR DIOXIDE and particulate matter in the atmosphere. About 42% of a total population of 14 000 suffered some illness, while 18 deaths resulted.

Doppler effect The apparent difference between the frequency at which sound or light waves leave a source and that at which they reach an observer, being caused by the relative motion of the wave source and the observer. It was first described by the Austrian physicist Christian Doppler (1803-53). This phenomenon is used in astronomical measurements, radar, and modern navigation. As an example, as one approaches a blowing horn, the perceived pitch is higher until the horn is reached, and then becomes lower as the horn is passed. A shorter wavelength means a higher frequency.

dose The quantity of a substance or an amount of energy received either in a single exposure or over a prescribed period of time which results in an adverse effect; for example, the quantity of radiation delivered to a given area or volume, or to the whole body. It is usually measured in rems. A lethal dose (LD_{50}) is that quantity of material administered orally or by skin absorption or injection, which results in the death of 50% of the test group within a 14-day observation period.

dose-response approach See EXTERNALITY.

double-glazed windows A technique for heat and sound insulation, if properly designed and installed. An acoustic double-glazed window, for example, must have an airspace of 100 mm or more, have a resilient mounting for at least one pane, and a sound-absorbent lining to the reveals.

downdraught A region of severe turbulence formed on the leeward side when wind flows around and over a building. Chimney emissions discharged into a downdraught zone will be brought rapidly to the ground.

downstream processing The products of subsequent stages of processing, starting with the initial process. For example, the raw material bauxite is processed into alumina, which is then processed into aluminium, followed by the manufacture of aluminium products. From bauxite, it is all downstream.

downwash The drawing down of chimney gases into a system of vortices or eddies which form in the lee of a chimney when a wind is blowing. In extreme circumstances, it may assist in bringing flue gases prematurely to ground level. (See DOWNDRAUGHT.)

drainage basin See CATCHMENT.

drift net fishing The use of fishing nets of great length and depth, aptly described as 'walls of death' because of the huge numbers of marine mammals, birds, and turtles that became ensnared in them. The Tarawa Declaration of 1989 formulated at the 20th South Pacific Forum, aimed at banning drift netting in the South Pacific. In June 1992 the UN banned drift netting in all the world's oceans.

drinking-water quality guidelines Guidelines issued by the WORLD HEALTH ORGANIZATION. Essentially, drinking water should be both clean and safe, bright, colourless, adequately aerated, without discernible taste or odour, suspended matter, or turbidity. It should be pleasant to drink, free from harmful organisms and should not contain any excessive amounts of chemical substances.

drinking-water supply and sanitation decade, UN The decade 1980-90 devoted by the UN to bringing clean water and adequate sanitation to everyone by 1990. This involved bringing a water supply to 1800 million people, and providing sanitation for 2400 million people. The WORLD HEALTH ORGANIZATION estimated at that time that poor sanitation was responsible for some 25 million deaths a year.

drought Insufficiency of rain for an extended period with associated water shortages, crop impairment or failure, stream-flow reduction, and depletion of GROUND WATER and soil moisture. It occurs when TRANSPIRATION exceeds PRECIPITATION for a considerable period. In the driest of areas, agriculture may be impossible without continuous irrigation. With seasonal droughts, crop planting and harvesting must coincide with the rainy season. Droughts of an unpredictable kind, of indefinite duration, may be countered to some extent by reserve water storage and cattle feed, with adjustment arrangements with unaffected areas. Borderline water deficiency may simply diminish crop yields.

dry farming Or dryland farming, the cultivation of crops without irrigation in regions of limited moisture. Such farming depends upon the efficient storage of the limited moisture in the soil, and the selection of crops and growing methods that make the best use of that moisture.

dryland salinity The accumulation of salt in soil and water in non-irrigated areas, caused by clearing trees and vegetation. The uptake of water by vegetation reduces the water table with the concentration of soluble salts rising, killing what vegetation remains and creating bare erodible areas.

drylands project A project of the UN ENVIRONMENT PROGRAM, resulting in the 1987 report

Drylands Dilemma, an important document on the degradation of the world's drylands and the economic, social, and environmental consequences.

dry weather flow (DWF)   The rate of flow of sewage, together with infiltration if any, in a sewer in dry weather; or the rate of flow of water in a river channel, in dry weather.

dumping   See CONVENTION ON THE CONTROL OF TRANSBOUNDARY MOVEMENT OF HAZARDOUS WASTES AND THEIR DISPOSAL; CONVENTION ON THE PREVENTION OF MARINE POLLUTION BY THE DUMPING OF WASTES AND OTHER MATTER.

dune lakes   Clear freshwater lakes that may be found in elevated sand dunes, particularly where there is an impervious layer underlying the sand.

dust arrestor   Any device for removing the grit and dust from the gas stream of an industrial plant. It may work by sedimentation, inertial separation, precipitation, or filtration. (See BAG FILTERS; CYCLONES, DUST REMOVAL; ELECTROSTATIC PRECIPITATOR.)

dust bowl   Any area so dry and parched that surface soil may be raised and carried away by the wind. A large agricultural region of central USA experienced dust bowl conditions in the 1930s, some areas losing 60 to 90 cm of topsoil.

dust burden (dust loading)   The concentration of solid particles in combustion or other waste gases entering or leaving a gas-cleaning plant, often measured on a weight/volume basis.

dust loading   See DUST BURDEN.

Dutch Elm disease   A fungus disease of elm trees, spread by the elm bark beetle. The disease is widespread, Britain having a major outbreak in 1927. The disease was first identified in the Netherlands in 1918.

DWF   DRY WEATHER FLOW.

duty of care   A legal concept embodied or implied in much environmental legislation that those occupying positions of responsibility and trust have a duty to discharge those responsibilities with the utmost care.

dynamic equilibrium   A situation in which a system may be stable, that is in equilibrium, even though substances are constantly moving in and out, there being a balance in these movements.

dynamometer   A machine for measuring the brake horse-power of a prime mover and also used in connection with test cycles for assessing emission rates from the exhausts of motor vehicles.

E

EAD EQUIVALENT AERODYNAMIC DIAMETER.

Earth Day Initiated by the US Government in 1970, an opportunity for students, conservationists, and members of the public to demonstrate concern over the degradation of the ENVIRONMENT. Earth Day is celebrated in April, every year. (See WORLD ENVIRONMENT DAY.)

earthquake Any abrupt disturbance within the Earth, or the shaking of the Earth's surface. The scientific study of earthquakes is called seismology. Earthquakes result from the build-up of stresses that are tectonic or volcanic; they are measured on the RICHTER SCALE. Earthquakes have taken up to one million lives at a time, although historically the eruption of Vesuvius in AD 79, which eliminated the towns of Pompeii and Herculaneum in Italy, has attracted the most attention.

earth satellite See REMOTE SENSING.

earth sciences The fields of study concerned with the solid planet earth, its waters, and the atmosphere that envelops it. The earth sciences include the geological, hydrological, and atmospheric sciences and their various disciplines and subdisciplines.

Earth Summit See UN CONFERENCE ON ENVIRONMENT AND DEVELOPMENT.

earth tide The deformation of the planet Earth as it rotates within the gravitational fields of the Sun and the Moon; the Earth deforms because it has a certain elasticity, although the actual movement of the surface may be no more than 30 cm. Earth tides are similar to ocean tides.

earthwatch A plan for global environmental assessment arising out of the UN CONFERENCE ON THE HUMAN ENVIRONMENT in 1972. Earthwatch has four closely linked components: evaluation and review, research, monitoring, and exchange of information. The plan, administered by the UN ENVIRONMENT PROGRAM, contains important operational elements such as the GLOBAL ENVIRONMENTAL MONITORING SYSTEM (GEMS); the INTERNATIONAL REFERRAL SYSTEM (INFOTERRA); and the INTERNATIONAL REGISTER OF POTENTIALLY TOXIC CHEMICALS (IRPTC).

easements Contractual requirements between government planning agencies and landowners to protect the public's interest in a parcel of land. One such requirement permits public access to, or through, private land; another may maintain the scenic open quality of a piece of land; while a third category may ensure a path for the laying of public services such as water, gas, and electricity.

EC_{50} The effluent concentration (EC) or toxicant producing an effect (not mortality) in 50% of exposed organisms in a specified time.

ECE ECONOMIC COMMISSION FOR EUROPE.

ECN ENVIRONMENTAL CHANGE NETWORK.

ecobalance A condition of equilibrium between ecosystems, ensuring their coexistence. By removing or introducing flora and fauna, by destroying habitats, by polluting the ENVIRONMENT, and through population growth, human beings cause major changes, some of which may be irreversible. (See BALANCE OF NATURE; DYNAMIC EQUILIBRIUM.)

ecoclimate The climate within a plant community; the microclimate.

ecocline A gradient of ecosystems associated with an environmental gradient; a directional variation in the characteristics of a population or community along a gradient.

ecodevelopment A term to describe ecologically sound DEVELOPMENT; positive management of the ENVIRONMENT for human benefit and the natural world.

ecofeminism A view that androcentrism, with its associated domination of women, ethnic minorities, intolerance, and war, stands also at the root of the domination and exploitation of nature. The modification of androcentrism through ATTITUDINAL CHANGES would result in a more caring society, liberating much of society from bondage, with the evolution of more caring feministic influences in ecological management. The goals of production would be subverted to those of the reproduction of life, reversing present priorities. Ecofeminism draws on the accumulated wisdom of womankind to heal our relationship with the Earth. (See DEEP ECOLOGY.)

ecofundamentalism A view that environmental protection and ECOLOGICALLY SUSTAINABLE DEVELOPMENT require fundamental changes to the structure of society such as: the restructuring of the tax system to ensure much reduced disparities in the net incomes of individuals and to ensure greater INTRAGENERATIONAL EQUITY; having greater regard for the effects of all developments and activities for future generations ensuring greater INTERGENERATIONAL EQUITY; having a society governed by the principles of ECOFEMINISM; a progressive reduction, not increase, of international trade contrary to the GENERAL AGREEMENT ON TARIFFS AND TRADE (GATT) and other free-trade agreements; the progressive reduction of international tourism; a massive redistribution of wealth and income from the richer to the poorer nations; and a return to simpler lifestyles with the promotion of local, district, regional, and national self-suffiency. (See DEEP ECOLOGY; PRINCIPLES OF SUSTAINABILITY; WORLD POPULATION.)

eco-labelling The certification of a product so that it meets a published specification relating to environmental performance. For example, the British Eco-Labelling Board has introduced this kind of certification for washing machines and dishwashers. Well-established schemes operate in Germany and the Nordic countries.

ecological capacity See CARRYING CAPACITY.

ecological efficiencies The ratios between the quantities of energy flow at different points along a FOOD CHAIN.

ecological environmental impact assessment (EEIA) An assessment of the potential ecological impacts of a proposed DEVELOPMENT. It comprises the following minimum steps: (1) The identification of the ways in which a proposed development will directly or indirectly affect the ecology of the proposed site and adjoining area. (2) The quantification, wherever possible, of the value of these disturbed ecosystem elements in a local, regional, and national context, both in the short- and long-term. (3) The modifications flowing from specified mitigation measures. (4) A proposed monitoring programme should development proceed. Many of the above stages of an ecological assessment rely on accurate data gathering and the correct identification of species in the field. (See ENVIRONMENTAL IMPACT ASSESSMENT; ENVIRONMENTAL IMPACT STATEMENT; FAUNA IMPACT STATEMENT.)

ecological factor Any environmental factor, ABIOTIC or BIOTIC, that influences living organisms.

ecologically sustainable development (ESD) A term sometimes used to emphasize the natural environment aspects of SUSTAINABLE DEVELOPMENT.

ecological indicators Organisms whose presence in a particular locality or circumstance indicates particular soil, water, and microclimate conditions.

ecological niche See NICHE.

ecological pyramid A pyramid with the producer (green plant) level at the base with successive trophic or feeding levels (herbivores, carnivores, detritivores) above.

ecologism ECOLOGY viewed as a distinctive doctrine, system or theory; it implies a placing of ecological considerations at the centre of things.

ecology The study of the relationships between organisms and their ENVIRONMENT or the study of ecosystems. Ecology is concerned frequently with general principles that apply to both fauna and flora. Autecology is concerned with single organisms or species and synecology with communities of species, although these artificial partitions are no longer accepted by all ecologists. Human ecology is the study of the structure and development of human communities and societies in terms of the processes by which human populations adapt to their environments. As a subject it applies the perspectives of the biological sciences to the investigation of topics included in the social sciences. (See ECOSYSTEM.)

econometrics The use of mathematics and statistics in ECONOMICS to define economic issues, broadly defined, and seek solutions to those problems. The development of mathematical models is central to the discipline, major advances having been made in model formulation, data gathering, estimation, hypothesis testing and computing. The application of econometric tools to economic and some environmental problems is now a sophisticated business.

Economic and Social Commission for Asia and the Pacific (ESCAP) A regional economic commission created by the Economic and Social Council of the United Nations in 1978, ESCAP investigated early on several environmental problems such as industrial pollution in the Asian area, determining various means of tackling these problems and estimating the costs of pollution control. Since then, the activities of ESCAP in environmental matters have broadened considerably. In 1990, ESCAP produced a report on the state of the environment in Asia and the Pacific, resulting in a ministerial declaration on environmentally sound and SUSTAINABLE DEVELOPMENT in Asia and the Pacific. Within this strategy, the alleviation of poverty was clearly a most urgent task with emphasis on basic necessities such as food, shelter, safe drinking water, fuel, sanitation, primary and preventive health care, and education. ESCAP has strongly supported the principles of EIA coupled with public participation, urging the integration of environmental considerations into national economic policies.

Economic Commission for Europe (ECE) A body set up in 1947 by the Economic and Social Council of the United Nations initially to raise the level of post-Second World War European economic activity, and for maintaining and strengthening the economic relations of Europeans, not only among themselves but with other countries. The membership comprises European and North American countries. The environmental activities of the UN ECE have broadened considerably over the years and member countries now submit biennial monographs on their environmental policies and strategies. The UN ECE pioneered the CONVENTION ON LONG-RANGE TRANSBOUNDARY AIR POLLUTION in 1979, becoming operational

economic costs

in 1983. In 1991, a CONVENTION ON ENVIRONMENTAL IMPACT ASSESSMENT IN A TRANSBOUNDARY CONTEXT was adopted at Espoo, Finland.

economic costs Costs falling into three categories: (1) explicit costs in respect of which actual financial liabilities are incurred, also known as accounting costs; (2) implicit costs, being the cost of using resources already at the disposal of an organization and which could be used in other ways; and (3) costs, social and environmental, to society. (See EXTERNALITY; OPPORTUNITY COST.)

economic development The process whereby relatively poor countries are transformed into much richer industrial economies, the changes being both qualitative and quantitative. Not all the results have been beneficial, for example, when labour-intensive industry is replaced by capital-intensive industry with restricted alternative opportunities, or when progress involves the intensive use of non-renewable resources or the destruction of forests. Other structural imbalances may persist. Alternative patterns of development are now being sought to speed progress but with much less disruptive social effects. (See GROSS DOMESTIC PRODUCT.)

economic efficiency The efficiency with which scarce resources are organized and used to achieve nominated economic, social and environmental objectives within a framework of SUSTAINABLE DEVELOPMENT. The aim is to achieve best value for the resources committed. The concept of efficiency may ignore, however, considerations of INTRAGENERATIONAL EQUITY and INTERGENERATIONAL EQUITY.)

economic geology A scientific discipline concerned with the distribution of mineral deposits, an assessment of the reserves available, and the economic considerations involved in their recovery. Economic viability tends to vary over time depending on market prices, domestic and international, and the techniques used to recover minerals. A new technique may make low-grade ores and reserves, and even waste dumps from previous eras, economically viable.

economic growth An increase in GROSS DOMESTIC PRODUCT (GDP) of the community over time. An increase may correspond simply with population growth or it may represent an increase in per capita GDP, representing an improvement in material welfare. An improvement in material welfare may, however, be offset by a loss in QUALITY OF LIFE.

economic impact assessment Consideration of the economic implications alone of a proposed policy, plan, programme project, or activity. It is essentially narrower than a more comprehensive ENVIRONMENTAL IMPACT ASSESSMENT, but may be justified if the social, health, and environmental consequences are likely to be small. The main purpose of an economic impact assessment is to measure everything in financial terms from the point of view of the initiator, whether government or private sector.

economic indicators Statistics which are sensitive to changes in the state of the economy generally, notably industry, trade, and commerce. These statistics relate to such matters as: GROSS DOMESTIC PRODUCT (GDP), employment and unemployment, the participation rate, the budget deficit, the balance of payments, overseas debt, investment by the private and public sectors, housing construction, production, exports and imports, capital inflow and outflow, inventories, savings, inflation, exchange rates, and consumer and business confidence. Indicators may be classified into three types: 'leaders', offering advance pointers; 'coincidents', moving in tandem with business conditions; and 'laggers' showing delayed results.

economic instruments Economic or fiscal measures to influence environmental behaviour. (See CARBON TAX; CERTIFICATES OF FINANCIAL RESPONSIBILITY; DEBT-FOR-NATURE SWAPS; ECOTAXES; EFFLUENT CHARGE; EMISSION TRADING PROGRAM; ENERGY TAXES; ENVIRONMENTAL ECONOMICS; FISCAL POLICY; PERFORMANCE BOND; POLLUTER PAYS PRINCIPLE; POLLUTION CHARGES; POLLUTION-CONTROL CHARGES; POLLUTION CREDITS; PRODUCT CHARGES; TAX INCENTIVES; USER PAYS.)

economic rationalism A widely held and fashionable belief in the early 1990s that the private sector with competitive markets can always do everything better than governments. Hence, the aim of government should be to reduce the size of the public sector through the privatization of almost everything, move taxation away from business to the community at large, deregulate business and finance, stimulate personal effort and provide greater financial incentives, and reduce welfare spending while tolerating a higher natural level of unemployment.

economics A social science concerned with how people, either individually, in groups, or collectively, attempt to accommodate scarce resources to their demands for goods and services through the processes of exploration, investigation, research, investment, DEVELOPMENT, production, distribution, conservation, substitution, and exchange, within a framework of SUSTAINABLE DEVELOPMENT. More specifically, the science has been concerned with: the allocation of the resources of society among alternative uses and the means of arranging that allocation; the distribution of income and reward to the FACTORS OF PRODUCTION; the ways in which production and distribution change over time; the efficiencies and inefficiencies of economic systems; and the alternative routes by which a society might resolve its economic, social and environmental problems. (See COST-BENEFIT ANALYSIS; ECONOMIC SYSTEM, FUNCTIONS OF.)

economic system, functions of Functions may be identified as follows: (1) to match supply to effective demand for goods and services in an efficient manner and to an acceptable quality; (2) to allocate scarce resources among industries and activities producing and distributing goods and services; (3) to organize wholesaling, retailing, and trading facilities for all kinds of goods and services; (4) to provide channels for fruitful capital investment; (5) to fully utilize the human resources of society, ensuring adequate employment opportunities; (5) to utilize the natural resources of society within a framework of SUSTAINABLE DEVELOPMENT; (6) to convey to consumers information on which goods and services are available for best value, within a socially responsible framework; (7) to promote research and development aiming at improved, safer environmentally friendly products and services better designed to meet society's objectives; (8) to review, rationalize, and upgrade continuously the structure of the economic system; (9) to increase productivity in all sectors; (10) to conserve raw materials, achieve cleaner production, reduce waste, encourage reuse and recycling, and to minimize waste-disposal problems; (11) to control all emissions from activities to a socially acceptable level; (12) to adjust the behaviour and performance of the economic and financial system to meet the larger objectives and expectations of society; (13) to promote satisfactory industrial relations; (14) to promote fair trading in all markets; and (15) to conform with environmental standards and expectations.

economic welfare Defined by the Cambridge economist A.C. Pigou (1877-1959) as 'that part of social welfare that can be brought directly or indirectly into relation with the measuring rod of money'; in other words, those aspects of social welfare that are concerned with material as distinct from bodily, moral, or spiritual wellbeing, although obviously these are interrelated in many ways. Pigou stressed that there is no precise line between economic and non-economic satisfaction. He also warned that economic welfare does not serve as a barometer or index of total welfare because an economic benefit may affect non-economic welfare in ways that cancel its beneficial effect. Non-economic welfare, for example, is liable to be modified by the manner in which income is earned and also by the manner in which it is spent. A society may boost real income in the short term by cutting down all its trees and selling the timber, without any attempt at reafforestation. Thus, in the absence of a substitute product that society destroys in the end both its economic and non-economic welfare.

economies of scale The gains by way of reduced average and marginal costs of production per unit of output often arising from increasing the scale of a plant, business or industry. In appropriate circumstances, large-scale production leads to significant economies in the use of the FACTORS OF PRODUCTION, namely land and natural resources, capital, and labour; and in marketing, finance, and research. However, above a certain point diseconomies may appear as management becomes more top-heavy and inflexible, with a loss of coordination and a growth in inertia. Further, the potential markets for the increased scale of production may become saturated. A social consequence is that the capital-intensive nature of the expansion may lead to a smaller total workforce than before, without retrenched workers having viable alternative opportunities.

ecopolitics The influence of ecological considerations in local and national policies, expressed in policies, and programmes by incumbent governments of all shades, and in the presence of 'green' parties and candidates in many parliaments and councils. Governments may hold power, or lose power, on the movement of the 'green' vote in parliament, council, or community. 'Greens' generally embrace the PRINCIPLES OF SUSTAINABILITY and the concept of SUSTAINABLE DEVELOPMENT. (See NON-GOVERNMENT ORGANIZATIONS.)

ecosphere The sum total of the Earth's ecological systems, similar to BIOSPHERE though implying a concern with improved environmental management of the Earth's resources. (See ECOSYSTEM.)

ecosystem The complexity of living organisms, their physical environment, and their entire interrelationships within a defined unit of space, through which matter and energy flow. There are innumerable ecosystems: marine, freshwater, and terrestrial. Reference may be made to forest ecosystems, grassland ecosystems, and wetland ecosystems. The word ecosystem appears to have been coined by A.G. Tansley in 1935 in an article in the journal *Ecology* entitled 'The use and abuse of vegetational concepts and terms'. (See ECOSPHERE.)

ecotax Any tax or charge, usually on goods and services, intended to influence users and consumers so that a desirable shift in use and consumption occur. Taxes on cigarettes, alcohol, and gambling (sin taxes) were earlier attempts at this. Ecotaxes may include emission and effluent charges, user charges, and levies for the treatment of waste, and special taxes such as a CARBON TAX.

ecotone A transition area of vegetation between two different plant communities, such as between rainforest and woodland or grassland. Some animals have their greatest abundance at the ecotone or edge, while some are restricted to it. The influence of the bordering communities on each other is known as the edge effect.

ecotourism The planning, development, and management of tourism in natural environments while maintaining environmental quality. The concept implies enhancing visitor appreciation of natural and cultural values, while contributing to the long-term conservation and management of environmental resources.

ecotoxicology The field of science dealing with the adverse effects of chemicals, physical agents, and natural products on populations and communities of plants, animals, and human beings.

ecumenopolis The universal city; the logical consequence of unlimited population growth. A term attributable to C.A. Doxiadis, the Greek architect and city planner, who traced city development through the metropolis, conurbation, megalopolis, urban region, urban continent, and ecumenopolis, the all-encompassing urban city.

edaphic Of or pertaining to SOIL, especially with regard to its influence on plants and animals.

eddy A current of air or water moving contrary to the main current or stream, especially one having a rotary or whirlpool motion.

eddy diffusion The process by which gases diffuse in the ATMOSPHERE; molecular diffusion is extremely slow by comparison and is usually ignored at this scale.

EEA EUROPEAN ENVIRONMENTAL AGENCY.

EES ENVIRONMENTAL EFFECTS STATEMENT.

effective height of emission The height above ground level at which a plume of waste gas is estimated to become approximately horizontal. It is usually higher, perhaps considerably more so, than the height of the chimney due to the initial upward momentum and the buoyancy of the hot gases.

effluent Strictly, anything flowing forth to air or water; however, usually, any liquid which is being discharged as sewage, trade or industrial waste, untreated or treated. Sewage effluent is the liquid finally discharged from a sewage treatment plant.

effluent charge (See ECOTAX; CARBON TAX; POLLUTION CHARGES.)

effluent standard The maximum amount of specified pollutants allowed in a waste stream as permitted by regulatory bodies seeking to achieve acceptable standards in receiving waters.

efflux velocity The speed, measured in, say, metres per second, at which flue gases are discharged from a stack or chimney.

egocentric traits Traits which favour the survival of the individual organism.

EHIA ENVIRONMENTAL HEALTH IMPACT ASSESSMENT.

Ehrlich, P. R. Professor of biology at Stanford University, published in 1970, with A. H. Ehrlich, *Population, Resources, Environment: Issues in Human Ecology*; and in 1971 *The Population Bomb*, a popular book that sparked off the population debates of the early 1970s.

EIA ENVIRONMENTAL IMPACT ASSESSMENT.

EIS ENVIRONMENTAL IMPACT STATEMENT.

ekistics The science of human settlements. The word was coined by the Greek architect and city planner, C. A. Doxiadis and first used in his lectures of 1942 at Athens Technical University.

elasticity (1) The ability of a material to yield to stress and to recover, without permanent deformation. (2) In ECONOMICS, the responses of prices to changes in demand in various circumstances, and the rates at which substitution occurs in responses to changes in prices.

electric vehicles Battery-powered or solar-powered cars and vehicles often regarded as the ultimate replacement for the internal combustion engine on the grounds of both pollution abatement from the vehicle itself and the conservation of petroleum. Battery-driven vehicles have been in use for milk rounds and the like for many years; their relative silence is appreciated by residents in the early hours. They have low running costs and require little maintenance. In the role of passenger vehicles a number of difficulties have to be overcome. There is slower acceleration and a limited top speed and range, while batteries occupy much space and need regular recharging. Solar-driven vehicles carry little weight. However, research continues into all aspects of the problem with prospective wider employment of such vehicles.

electrolysis The chemical decomposition of a substance when electricity is passed through it in solution or in the molten state. The electrolysis of water produces hydrogen and oxygen as separate components. The electrolysis of brine is the main process for the manufacture of CHLORINE. Electrolysis is one of the processes by which ores and minerals are refined to yield pure metals; it is most often used in the refining of aluminium.

electromagnetic radiation Energy that is propagated through space or matter in the form of electromagnetic waves such as radio waves, visible light, and gamma rays. Electromagnetic radiation spans an enormous range of frequencies or wavelengths and there are many sources of such radiation, both natural and human-made. In recent years, electromagnetic radiation from power lines has been suspected of causing leukaemias.

electrostatic precipitator A device for the arrestation and removal of dust from a gas stream. It utilizes the general principle that if a gas is passed between two electrodes, one of which is supplied with a very high negative voltage and the other earthed, the gas and any particles of dust in suspension become electrically charged. The dust is attracted to the earthed electrodes, where it collects. A mechanical rapping device dislodges the particles and they fall into a collecting hopper beneath the precipitator. Units installed at large coal-fired power stations are designed to operate at collecting efficiencies of over 99%. (See BAG FILTER.)

element (1) A substance that cannot be decomposed (split up) by chemical changes into simpler substances. A compound consists of two or more elements combined together chemically in definite proportions. An element is defined by its atomic number. (2) In relation to the ENVIRONMENT, any of the principal constituent parts of the environment including water, the atmosphere, soil, vegetation, climate, sound, odour, aesthetics, fish, and wildlife.

El Niño southern oscillation An abnormal appearance every few years of unusually warm ocean conditions along the tropical west coast of South America. The warm ocean conditions cause large-scale anomalies in the ATMOSPHERE, rainfall increasing in Ecquador and Peru with adverse effects for agriculture and fishing. Conversely, strong El Niño events are associated with droughts in Indonesia, Australia and northeastern South America. Ocassionally, the effects of an El Niño event are felt worldwide. The change in the tropical atmosphere is known as the southern oscillation, involving dramatic changes in wind direction.

Elton, C. S. An English biologist, born 1900, credited with framing the basic principles of modern animal ECOLOGY. Elton's first book *Animal Ecology* (1927) became a landmark for the basic principles established, notably the concepts of food chains and the nutrition cycle, the characteristics of foods, ecological niches, and the 'pyramid of numbers'. In 1930, his provocative book *Animal Ecology and Evolution* appeared. He challenged the concept of the BALANCE OF NATURE and stressed the role of the migration of animals in which the ENVIRONMENT is chosen by the animal as opposed to the natural selection of the animal by the environment. In 1932, Elton established his Bureau of Animal Population at Oxford University, which became a world centre for data and research.

elutriation The classification or grading of particles affected by movement relative to a rising fluid. A known weight of dust is placed in a receptacle at the base of the apparatus and is then subjected to an upward current of air or water. By varying the velocity of the upward stream, it is possible to obtain a series of fractions expressed in terms of the falling velocity of the dust. (See AIR POLLUTION.)

EMAP, US ENVIRONMENTAL MONITORING AND ASSESSMENT PROGRAMME, US.

EMAR EUROPEAN ECO-MANAGEMENT AND AUDIT REGULATION.

eminent domain The superiority of the sovereign power over all the property of the nation, whereby it is entitled to appropriate any part required for the public advantage, compensation being made to the owner. Thus, all levels of government often have authority to acquire land and property for public purposes through compulsory-purchase procedures. The power of compulsory purchase is embodied in much US and British planning, transport, and housing legislation. Land may be acquired for public housing, a needed highway, an urban renewal scheme, or a redevelopment programme.

emission A liquid or gaseous EFFLUENT, or noise, emitted to the general ATMOSPHERE. An immission is effluent or noise received. (See IMMISSION.)

emission charge See EFFLUENT CHARGE.

emission inventory See POLLUTANT RELEASE INVENTORY, CANADA; TOXIC RELEASE INVENTORY, US.

emission standard The amount of pollutant permitted to be discharged from a pollution source. Emission standards are commonly described in one or more of the following ways: weight of pollutants per unit volume of discharged gas, for example, micrograms per cubic metre ($\mu g/m^3$); volume of pollutants per unit volume of discharged gas, for example, parts per million (ppm); weight of pollutants per weight of materials processed, for example, 10 kilograms per tonne (10 kg/t); weight of pollutants over a certain time, for example, kilograms per hour (kg/h), tonnes per day (t/day). Limits for noise are set in the DECIBEL (A-SCALE), or dB(A).

emission trading program An attempt to introduce some free-market principles into the use of environmental resources introduced into the USA in the early 1980s. The amount of pollution allowable from individual firms in a region is fixed by US EPA permits to operate. Once limitations are set, emissions trading programs allow firms to trade among themselves. The limitation on the total quantity of specific pollutants in the atmosphere of a region remains, and may even be tightened progressively. The region as a whole must continue to meet ambient air quality goals. Under a trading program, there is an incentive to achieve reductions in emissions below the legal requirements enabling a firm to expand or to sell the resultant credits to other firms needing them. Sulphur credits are traded on the Chicago Board of Trade. (See BUBBLE CONCEPT.)

endangered species Fauna and flora likely to become extinct due to direct exploitation by humans, intrusion into highly specialized habitats, threats from other species, interruption of the food chain, pollution, or a combination of such factors. Although the process of extinction has been a natural process, the process has been accelerated by human influences. Such losses are irretrievable. An endangered species may become rare. However, all rare species, are not necessarily endangered. Most countries have taken legal and physical measures to protect endangered species from extinction.

endemic Descriptive of a species that is restricted naturally to a particular region or locality; for example, kangaroos and eucalypts are endemic to Australia.

endosulphan See CHLORINATED HYDROCARBONS.

endrin See CHLORINATED HYDROCARBONS.

energetics The abstract study of the ENERGY relationships in physical, chemical, and biological changes.

energy The capacity of matter or radiation to do work. It may not always be possible to effect the transition from energy to work: not all potential energy is available energy. The unit of energy is the same as that of work, the Joule. There are several forms or descriptions of energy. Kinetic energy is the ability of matter to do work by virtue of its motion, for example, the energy of flowing water may be used to generate electricity. Potential energy indicates the ability to do work by virtue of the position or configuration of matter, for example, a bent spring possesses potential energy for it is capable of work in returning to its natural form. Other types of energy are: chemical energy, electrical energy, nuclear energy, radiant energy, and heat energy.

Energy and Environment, UNEP Collaborating Centre on Based in Denmark at the Risø National Laboratory, a research centre funded by UNEP, the Danish International Development Agency (DANIDA) and the Risø National Laboratory to promote the incorporation of environmental considerations into energy policy and planning, especially in developing countries. Established in 1990, the centre collaborates with institutions and experts worldwide, providing direct support to UNEP headquarters in the furtherance of SUSTAINABLE DEVELOPMENT.

energy audit A systematic review of the use of energy within an organization, with a view to the avoidance of unnecessary waste and any needless contribution to the GREENHOUSE EFFECT. Energy audits have tended to receive much less attention than the ENVIRONMENTAL AUDIT, as energy costs are often not major items in the accountancy costs of a company. However, with greenhouse gases becoming a major target, nationally and internationally, the emphasis on energy audits may well substantially increase.

energy cascading The utilization of ENERGY, otherwise wasted, in the performance of work matched to the characteristics of the energy, for example, the use of waste heat from an industrial process for a variety of purposes in sequence, performing work at a lower temperature at each stage. The sequence might include air heating, steam raising, and water heating.

energy conservation The better and more efficient use of energy with proper regard to the related costs and benefits, whether these be economic, social, or environmental. Measures may be aimed at achieving higher thermal efficiencies, higher load factors, better heat insulation, and more effective use of wastes through ENERGY CASCADING, all within a pricing structure designed to modify peak demands and curb the needs for additional INFRASTRUCTURE. Energy conservation also has regard to the longer view in respect of SUSTAINABLE DEVELOPMENT, the GREENHOUSE EFFECT, and the conservation of resources of the nonrenewable kind.

energy, environmental effects of See Box 8.

energy farming The growing and harvesting of BIOMASS, such as trees and field crops, for energy production, especially for liquid fuels needed in transport.

energy flow In ECOLOGY, the total assimilation at a particular trophic level which equals the production of BIOMASS plus respiration.

energy-intensive industry Industry in which the consumption and costs of energy are significant as in electricity generation and aluminium production.

energy pyramid In ECOLOGY, a diagram showing the rate of energy flow or production at successive trophic levels.

energy stocks Stocks and shares in companies and corporations operating in the energy sector of the national economy, notably in oil, shale oil, natural gas, uranium, coal mining, and power. Companies may be engaged also in exploration, drilling, mining, treatment, processing, transmission, generation, storage, transport, and distribution.

energy taxes Taxes on fossil fuels such as petrol and oil. (See CARBON TAX.)

Box 8 Principal environmental effects of energy sources

1. Fossil Fuels

	Impact on lithosphere	Impact on hydrosphere	Impact on atmosphere	Impact on biosphere
1. Coal extraction	Land subsidence in underground mining, land disturbance in strip mining	Acid mine drainage effects on underground and surface waters	Dust in mining areas	Health hazards in underground mining
Processing	Storage areas	Waste water of coal processing	Coal-dust	Acid mine drainage effects on fish and wildlife
Use (in power plant)		Thermal discharges	Emissions from power plants (particulate SO_x, NO_x, CO, HC) Possible local meteorological effects (pollutants, thermal discharges)	Health effects of pollutants Impact of thermal discharges on aquatic life Effect on plants
2. Oil Extraction	Oil pollution Disposal of brines	Oil pollution in offshore activities		Effects of oil spills on marine life
Transportation	Pollution from pipelines Disturbance of land (permafrost, etc.)	Marine oil pollution		Effect on marine life Effect on wildlife (pipelines)
Processing	Oil pollution	Refinery wastes and their effects on groundwater and/or coastal waters	Emissions of SO_x, HC, NO_x, CO	Health effects of pollutants Effects of industrial wastes on aquatic life

continued overleaf

Box 8 (continued)

	Impact on lithosphere	Impact on hydrosphere	Impact on atmosphere	Impact on biosphere
Use (in power plants)		Thermal discharges	Emissions of SO_x, HC, NO_x, CO Possible local meteorological effects (pollutants, thermal discharges)	Health effects of pollutants Effects of thermal discharges on aquatic life Effect on plants
Use (in mobile sources)	Gasoline and oil leakages Lub-oils	Oil leakage from ships, boats, etc.	Pollutants from automobiles Pollutants from aircraft and their effects on ozone layer	Health effects of pollutants Effects on aquatic life Effects of pollutants on plants in forest areas (highways)
3. Natural Gas				
Extraction	Same as oil	Same as oil		
Transportation	Accidents (fire, etc.) in liquefaction plants	Marine spills	Air pollution in case of accidents	Health effects in liquefaction and regasification plants Effects on marine life in case of spills
Use (in power plants)		Thermal discharges	Emission of particulate SO_x NO_x	Health effects of pollutants Impact of thermal discharges on aquatic life

II. Renewable Sources of Energy

	Impact on lithosphere	Impact on hydrosphere	Impact on atmosphere	Impact on biosphere
1. Geothermal Energy				
Extraction	Land subsidence, local disturbance of land (drilling, pipelines)		Noxious gases	Effects on plants
Use		Disposal of brines Thermal discharges	Noxious gases (e.g. hydrogen sulphide), thermal discharges	Health effects of pollutants Effects on plants Effects on aquatic life

2. Hydroelectric			
Man-made lakes	Disturbance of land, possible earth-crust disturbances. Terrestrial effects around lakes (soils, etc.)	Changes in hydrological cycle, water quality, nutrients, etc.	Possible local meteorological effects
			Microclimatic changes due to evaporation (humidity)
			Aquatic life (fish, plants, etc.) Health effects
Downstream effects	Erosion of river banks Delta-shoreline erosion	Changes in hydrological cycle, water quality	Aquatic life (fish, plants, etc.) Health effects
3. Solar Energy	Land areas for solar farms	Thermal pollution from water-cooled turbines	Possible local meteorological changes
4. Tidal Energy	Land drainage (shore-lines), sediment budget Geomorphological changes		Possible effects on aquatic life
5. Wind Energy	Areas for windmills		Possible local meteorological changes
6. Sea-thermal Power		Possible changes in sea temperature	Possible effects on marine life
7. Renewable Fuel Sources	Land areas for photosynthetic material Disposal of wastes and storage	Disposal of wastes from digesters	Health effects of wastes (in handling) Effects on agriculture, etc.

Source: OECD (1992) Paris.

enriched uranium Processes that raise the content of the fissile isotope, uranium-235, from the 0.71% normally found in nature to a higher level. Low-enriched uranium, containing 2 to 4% of uranium-235, is used in many types of nuclear reactor. Highly enriched uranium, containing perhaps more than 90% of uranium-235, is used as fuel in some reactors and to make nuclear weapons.

enteritis Inflammation of the intestines caused by irritants, poisons, disease organisms, or other factors.

enthalpy The total amount of heat that water, steam, air, or other gas contains, measured from 0°C.

entomology The scientific study of insects; the largest single class of animals comprising about 700 000 known species and probably about as many yet to be classified. Entomology has numerous subdisciplines such as taxonomy, ecology, behaviour, and physiology. Pest insects tend to receive the most attention and the bulk of research funds, with a view to effective control measures.

entrainment The collecting and transporting of a substance by the flow of another fluid moving at a high velocity, For example, boiler water may become entrained in the steam leaving the boiler under certain conditions, or dust and flyash particles may become entrained in the flue gases and carried towards the stack.

entropy A measure of the amount of disorder in a system. As entropy increases, the amount of available energy in the form that can be used usually decreases. The processes of life reduce the entropy, that is, increase the order within living things, at the cost of increased entropy outside. The concept of entropy is sometimes presented as the second law of thermodynamics. According to this law, entropy increases during an irreversible process such as the spontaneous mixing of hot and cold gases, the combustion of a fuel, and the expansion of a gas into a vacuum.

environment A concept which includes all aspects of the surroundings of humanity, affecting individuals and social groupings. The European Union has defined the environment as 'the combination of elements whose complex interrelationships make up the settings, the surroundings and the conditions of life of the individual and of society, as they are or as they are felt'. The environment thus includes the built environment, the natural environment, and all natural resources, including air, land, and water. It also includes the surroundings of the workplace. The idea of a beneficial environment which renders more services than disservices relates also to that other concept, QUALITY OF LIFE. Often the word is used in other more narrower contexts, such as a business environment, a musical environment, a cultural environment, or a home environment. (See ABIOTIC; BIOTIC; ENVIRONMENTAL IMPACT ASSESSMENT; ENVIRONMENTAL PLANNING; ENVIRONMENTAL SCIENCE; SUSTAINABLE DEVELOPMENT.)

Environment Agency, Japan An agency created by the Japanese Government in 1971. It has jurisdiction over basic environmental planning; promotes the coordination of government pollution control policies, including pollution control expenditures; supports research and investigation; enforces national environmental quality standards; and is responsible for national parks and nature conservation. Other ministries and agencies also have environmental responsibilities. The agency prepared the Japanese Government submissions to the UN CONFERENCE ON THE HUMAN ENVIRONMENT 1972 and the UN CONFERENCE ON ENVIRONMENT AND DEVELOPMENT 1992.

environmental agencies Government bodies set up to administer the environmental planning, pollution control, and nature conservation activities, which are responsibilities of that government. Following the UN CONFERENCE ON THE HUMAN ENVIRONMENT 1972, perhaps over 100 governments created environment ministries, departments, and authorities. Since then practically all governments have environmental responsibilities embodied in single agencies, whereas previously, environmental functions were fragmented and embodied in several different ministries, such as departments of health or public works, or water and sewerage boards. The integration of environmental function has been the first step in the integration of environmental considerations into government decision-making and SUSTAINABLE DEVELOPMENT. The record of environmental departments and agencies has varied according to the priorities of government, the introduction and enforcement of appropriate legislation, the political will of individual ministers, and the ranking of environment and conservation ministers within the pecking order of cabinets. Further, these considerations reflect to a marked degree the creation and activities of voluntary conservation groups and the participation of individuals in the community. Environmental outcomes often reflect community inputs, and from time to time governments retain or lose office on environmental issues. (See ENVIRONMENT AGENCY, JAPAN; ENVIRONMENT, DEPARTMENT OF THE, UK; ENVIRONMENTAL PROTECTION AGENCY, USA.)

environmental audit An assessment of the nature and extent of any harm or detriment, or any possible harm or detriment, that may be inflicted on any aspect of the environment by any activity process, development programme, or any product, chemical, or waste substance. Audits may be designed to: verify or otherwise comply with environmental requirements; evaluate the effectiveness of existing environmental management systems; protect the organization against external criticism; assess risks generally; or assist in planning for future improvements in environment protection and pollution control. An audit may be conducted in-house, though an independent audit may sometimes be required by the authorities. Occasionally, an audit or post-project analysis may be conducted by an enforcement agency, or it may be undertaken when a transfer of ownership or control of land is taking place and it is necessary to declare the degree of contamination of the land.

environmental change network (ECN) A network for monitoring the changes which may be taking place in British ecosystems, distinguishing natural variations from human-made changes and giving early warning of undesirable effects. ECN was launched in 1992. The operation of the network depends on the voluntary collaboration of the sponsoring agencies in providing sites and staff. Twenty-four sites in the system represent a broad range of climate, soil, habitat, and land management in Britain The project is managed by the NATURAL ENVIRONMENTAL RESEARCH COUNCIL.

environmental economics A recognized field of specialization in the discipline of ECONOMICS, with a number of associated academic journals. It embraces the issues of pollution control and environment protection, in which costs and benefits are difficult or impossible to estimate, much of the subject matter falling outside the competitive market system. Yet, it is an area in which immense common property resources need to be allocated sensibly to the overall public good. The subject is also very much concerned with ways and means to achieve this sensible allocation such as emission and effluent charges, user charges for the treatment or disposal of waste, environmental taxes, product charges, deposit refunds, tradeable pollution rights, performance bonds, natural resource accounting, and the economic implications of SUSTAINABLE DEVELOPMENT.

environmental education A concept ranging from media coverage of environmental issues to formal environmental education at university level in such subjects as environmental engineering, resource management, and environmental law. The subject received recognition from the Belgrade Charter which emerged out of the Belgrade International Workshop on Environmental Education convened in 1975 as part of a UNESCO-UNEP programme; the charter formulates guidelines which can be applied internationally. The subject of environmental science has found its place in most schools and colleges in many countries. Many universities now have centres focusing on environmental research topics. (See EARTH DAY; WORLD ENVIRONMENT DAY.)

environmental effects statement (EES) A term preferred by some environmental agencies to the more usual term ENVIRONMENTAL IMPACT STATEMENT. The argument is that most of the adverse environmental consequences of activity are not hammer blows or 'impacts' but essentially slow and progressive 'effects' over time. Therefore, 'environmental effects' is a more accurate description. Victoria, Australia, has an Environmental Effects Act.

environmental engineering The application of the principles and findings of engineering and ENVIRONMENTAL SCIENCE to the protection of the ENVIRONMENT and the improvement of the QUALITY OF LIFE. Environmental engineering is concerned with water-pollution control, air-pollution control, noise control, solid- and liquid-waste processing and disposal, the safe disposal of hazardous chemicals and intractable wastes, soil erosion and rehabilitation, the mitigation of visual pollution, monitoring, ENVIRONMENTAL PLANNING, the preparation of environmental impact statements and assessments, and the achievement of SUSTAINABLE DEVELOPMENT. It is particularly relevant to pollution-control engineering in the chemical and metallurgical industries; power generation; transport systems; mining and quarrying; oil and gas exploration and production; and nuclear developments. Environmental engineering is a subject of great complexity which owes its origin and development to the growth in public concern over environmental issues with ever-increasing statutory and licensing requirements.

environmental forecasting A forecasting programme capable of timely and effective warning of technologically induced perturbations of any health-welfare parameter of the population. The need for environmental forecasting has grown with the pace of technological and community change. In making detailed forecasts concerning community hazards, models need to take account of: emission levels; the transport, storage, and reaction of pollutants in the ENVIRONMENT; and the resulting exposure of the community and its response. An example of the need for forecasting arises when substantial urban developments are planned for areas with limited assimilative capacity for pollutants.

environmental geology The application of geological data and principles to the solution of problems likely to be created by human occupancy and use of the physical ENVIRONMENT; geology oriented towards the planned utilization of resources and the safeguarding of the environment. The areas of interest include: metallic and non-metallic minerals; mineral

environmental health impact assessment

fuels; ground water; soils and soil conditions unsuitable for septic systems or sanitary landfill use; land use; and beach preservation.

environmental health impact assessment (EHIA) An assessment of the effects on the ENVIRONMENT and people of aspects of a project recognized as having potentially adverse health effects; the health component is often insufficiently addressed in environmental impact assessments (EIAs). In 1982, the WORLD HEALTH ORGANIZATION recommended that EHIA studies should be conducted for all major development projects. Health assessment embraces the following: risks and hazards (direct and indirect), involving explosion, fire, shock, heat, blast, vibration, and destruction of property beyond the boundaries of the plant; biological factors such as parasites, helminths, protozoa, bacteria, mycobacteria, rickettsia, and viruses; toxic, carcinogenic, or mutagenic chemicals and heavy metals; ionizing and non-ionizing radiation; noise; dust and other irritants; and excessive temperature or humidity. The possible implications for human health may be measured in terms of mortality and morbidity. (See ASWAN HIGH DAM.)

environmental heritage Those buildings, works, relics, or places of historic, scientific, cultural, social, archaeological, natural, or aesthetic significance for the city, region, province, state, or nation.

environmental impact assessment (EIA) The critical appraisal of the likely effects of a policy, programme, project, or activity on the environment. To assist the decision-making authority, assessments are carried out independently of the proponent, who may have prepared an environmental impact statement (EIS). The decision-making authority may be a level of government (local, state, or federal) or a government agency (at local, state, or federal level). Assessments take account of: any adverse environmental effects on the community; any environmental impact on the ecosystems of the locality; any diminution of the aesthetic, recreational, aesthetic, scientific, or other environmental values of a locality; the endangering of any species of fauna or flora; any adverse effects on any place or building having aesthetic, anthropological, archaeological, cultural, historical, scientific, or social significance; any long-term or cumulative effects on the environment; any curtailing of the range of beneficial uses; any environmental problems associated with the disposal of wastes; any implications for natural resources; and the implications for the concept of SUSTAINABLE DEVELOPMENT. EIA extends to the entire process from the inception of a proposal to environmental auditing and post-project analysis. (See Box 9.)

environmental impact statement (EIS) A document, prepared by a PROPONENT, describing: a proposed activity or development and

---

Box 9 National principles for EIA in Australia

PRINCIPLES FOR ASSESSING AUTHORITIES

(1) Provide clear guidance on types of proposals likely to attract environmental impact assessment and on levels of assessment.

(2) Provide proposal-specific guidelines (or procedure for their generation) focussed on key issues and incorporating public concerns; and a clear outline of the EIA process. Amendments to guidelines should only be based on significant issues that arise after guidelines have been adopted.

(3) Provide guidance to all participants in the EIA process on criteria for environmental acceptability of potential impacts including such things as the principles of ecologically sustainable development, maintenance of environmental health, relevant local and national standards and guidelines, codes of practice and regulations.

(4) Negotiate with key participants to set an assessment timetable on a proposal-specific basis and commit to using best endeavours to meet it.

(5) Seek and promote public participation throughout the process, with techniques and mechanisms tailored appropriately to specific proposals and specific publics.

(6) Ensure that the total and cumulative effects of using or altering community environmental assets (for example air, water, amenity) receive explict consideration.

(7) Report publicly on the assessment of proposals.

continued overleaf

environmental impact assessment

Box 9 *(continued)*

(8) Ensure predicted environmental impacts are monitored, the results assessed by a nominated responsible authority and feedback provided to improve continuing environmental management of proposals.
(9) Monitor properly the efficiency and effectiveness of the environmental impact assessment process to learn from the past, streamline requirements and help maintain consistency.
(10) Review, adapt and implement techniques and mechanisms which can improve the process and minimise uncertainty and delays.
(11) Ensure that educational opportunities inherent in the EIA process are actively pursued.

PRINCIPLES FOR PROPONENTS

(1) Take responsibility for preparing the case required for assessment of a proposal.
(2) Consult the assessing authority and the community as early as possible.
(3) Incorporate environmental factors fully into proposal planning, including a proper examination of reasonable alternatives.
(4) Agree on a proposal-specific evaluation timetable and commit to using best endeavours to meet it.
(5) Take the opportunity offered by the EIA process to improve the proposal environmentally.
(6) Make commitments to avoid where possible and otherwise minimise, ameliorate, monitor and manage environmental impacts; and implement these commitments.
(7) Amend environmental management practices responsibly, following provision dissemination of environmental monitoring results.
(8) Identify and implement responsible corporate environmental policies, strategies, and management practices, with periodic review.

PRINCIPLES FOR THE PUBLIC

(1) Participate in the evaluation of proposals through offering advice, expressing opinions, providing local knowledge, proposing alternatives and commenting on how a proposal might be changed to better protect the environment.
(2) Become involved in the early stage of the process as that is the most effective and efficient time to raise concerns. Participate in associated and earlier policy, planning and programme activities as appropriate, since these influence the development and evaluation of proposals.
(3) Become informed and involved in the administration and outcomes of the environmental impact assessment process, including:
 - assessment reports of the assessing authority
 - policies determined, approvals given and conditions set
 - monitoring and compliance audit activities
 - environmental advice and reasons for acceptance or rejection by decision-makers.
(4) Take a responsible approach to opportunities for public participation in the EIA process, including the seeking out of objective information about issues of concern.

PRINCIPLES FOR GOVERNMENT

(1) Provide policy and planning frameworks which set contexts for the environmental assessment of proposals.
(2) Base decisions on proposals having potentially significant environmental impact on advice resulting from the EIA process and include provisions for effective protection and management of the environment.
(3) Apply the EIA process equally to proposals from both the public and private sectors.

continued overleaf

environmental impact statement

Box 9 *(continued)*

(4) Within each jurisdiction (Commonwealth, State/Territory) provide for a coordinated government decision-making process to which the outcomes of EIA can be directed; and develop mechanisms to synchronise processes for decision-making such that, where possible, the opportunity exists for decisions to be made in parallel rather than sequentially for proposals requiring multiple approvals.
(5) Ensure assessment reports are available to the public before or at the time of decision-making.
(6) Establish one national agreement to ensure a single orderly process is in place where the EIA responsibilities of several governments are involved.
(7) Provide support, if and when appropriate, to participants in the process to enable better and informed involvement.
(8) Provide opportunities for reasonable public and proponent objections, on decisions made other than at Ministerial level, regarding the requirement for and level of assessment, adherence to due process, and environmental advice given to decision-makers.
(9) Implement this national approach including, where appropriate, progressive amendment of statutory provisions, to increase consistency in the process.
(10) Maintain the integrity of the EIA process.
Source: Australian and New Zealand Environment and Conservation Council (1991), *A National Approach to Environmental Impact Assessment in Australia*, ANZECC, Canberra.

---

identifying the possible, probable, or certain effects of the proposal on the ENVIRONMENT; examining the possible alternatives; setting out the mitigation measures to be adopted; proposing a programme of environmental management; provisions for monitoring, post-project analysis, or auditing; and plans for decommissioning and rehabilitation. An EIS should be prepared following SCOPING exercises to identify the key issues. It should be objective, thorough, and comprehensive, but without superfluous material. EISs are usually prepared by consultants working for the proponent presenting what has been described as an ethnical dilemma. However, the ultimate test is not pleasing the proponent in the short term, but achieving development consent after rigorous examination by a government agency and the public. This has ensured an increasing degree of integrity in the preparation of EISs. An EIS is often a key document in the EIA process.

Box 10 outlines the structure of a good EIS, though such a structure needs to be adapted

---

Box 10 Structure of an environmental impact statement (EIS)

(1) A summary of the EIS, intelligible to non-specialists and the public, should precede the main text.
(2) Acronyms and initials should be defined; a glossary of technical terms may be relegated to an appendix.
(3) The list of contents should permit rapid location of the main issues.
(4) The authors of the EIS should be clearly identified.
(5) A brief outline of the history of the proposed development should be given, including details of early consultations.
(6) A full description of the proposed project or activity, its objectives and geographical boundaries; its inputs and outputs and the movement of these; also the inputs and outputs specifically during the construction phase. Diagrams, plans, and maps will be necessary to illustrate these features, with a clear presentation of the likely appearance of the finished project.

continued overleaf

environmental impact statement

Box 10 *(continued)*

(7) A full description of the existing environment likely to be affected by the proposal; the baseline conditions; deficiencies in information; data sources; the proximity of people, other enterprises, and characteristics of the area of ecological or cultural importance.

(8) The alternative locations considered, or alternative processes, resulting in the preferred choice of site; evidence of credible studies will be needed here.

(9) The justification of the proposal in terms of economic, social, and environmental considerations; the consequences of not carrying out the proposal for the proponent, the locality, the region, and the nation.

(10) The planning framework, relevant statutory planning instruments, zoning, planning and environmental objectives.

(11) The identification and analysis of the likely environmental interactions between the proposed activity and the environment.

(12) The measures to be taken in conjunction with the proposal for the protection of the environment and an assessment of the likely effectiveness of those measures, particularly in relation to pollution control, land management, erosion, aesthetics, rehabilitation, ecological protection measures, and decommissioning. Measures to achieve clean production and recycling; the management of residuals.

(13) The effect on the transport system of carrying people, goods, services, and raw materials, to and from the project.

(14) The duration of the construction phase, operational phase, and decommissioning phase; housing the workforce, both construction and permanent.

(15) The implications for public infrastructure such as housing, schools, hospitals, water supply, garbage removal, sewerage, electricity, roads, recreational facilities, fire, police, emergency services, parks, gardens and nature reserves; the implications for endangered species and threatened ecological features and ecosystems; the prospective financial contributions of the proponent.

(16) Any transfrontier or transborder implications of the proposal.

(17) Any cumulative effects from similar enterprises should be considered, being either short-term or long-term, permanent or temporary, direct or indirect.

(18) Proposals for annual reporting to the decision-making body on the implementation of the conditions of consent; post-project analysis and environmental auditing.

(19) Arrangements for consultation with the relevant government agencies, planners, the public and interested bodies during the concept, preliminary, screening, scoping phases, the preparation of the EIS, the EIA stage, the construction, operational, and decommissioning stages; the communication of results.

(20) Any unique features of the proposal of national or community importance such as technology, employment characteristics, training, contributions to exports or import replacement, defence, landscaping, recreational; facilities, foreign investment, or marked multiplier effects.

(21) The contribution towards the containment of global environmental problems such as the 'greenhouse effect' and the protection of the ozone layer.

(22) The implications for renewable and non-renewable resources.

(23) The implication for intragenerational and intergenerational equity.

(24) The relevance of the precautionary principle.

(25) The implications for the conservation of biological diversity.

(26) Any other implications for sustainable development or sustainable yield generally.

to the individual case. (See ENVIRONMENTAL IMPACT ASSESSMENT; PROGRAMME IMPACT STATEMENT.)

environmental law A separately identified branch of law, being primarily concerned with the development of legislation relating to pollution control (air and climate change, water, noise, land contamination, and environmentally hazardous chemicals); national parks, wildlife, fauna and flora, wilderness, and biodiversity; environmental and occupational health; environmental planning; EIA of policies, plans, programmes and projects; heritage conservation; ecologically SUSTAINABLE DEVELOPMENT; resource management generally; and a large number of international conventions relating to the environment. Environmental law has developed a considerable literature with regular conferences conducted by environmental law associations.

environmental management A concept of care applied to localities, regions, catchments, natural resources, areas of high conservation value, cleaner processing and recycling systems, waste handling and disposal, pollution control generally, landscaping and aesthetics, enhancement of amenities. In general, it means the efficient administration of environmental policies and standards. It involves the identification of objectives, the adoption of appropriate mitigation measures, the protection of ecosystems, the enhancement of the quality of life for those affected, and the minimization of environmental costs. The two most important environmental management standards in Britain are BS 5750 (quality management systems) and BS 7750 (environmental management systems). Certification against both standards involves assessment by professional independent assessors. (See EUROPEAN UNION.)

Environmental Monitoring and Assessment Program, US (EMAP, US) A national monitoring and research programme to determine the status of critical ecological resources, and identify trends in the condition of those resources. EMAP uses seven resource categories for its fundamental research and monitoring units: forests, arid ecosystems, agro-ecosystems, estuaries, the Great Lakes, inland surface waters, and wetlands. A demonstration project has been completed in estuaries in the mid-Atlantic region and on the Gulf Coast of the United States. In international applications, EMAP has been used to determine the spatial decline of Brazil's rainforests, and to characterize Australia's arid lands.

environmental planning The identification of desirable objectives for the physical environment, including social and economic objectives, and the creation of administrative procedures and programmes to meet those objectives. Matters embraced include: city and regional planning; land use; transportation; employment; health; growth centres and new towns; population and national settlement policies; locational problems; industrial and urban development; national, regional, and local environmental policies; planning permission and development consent procedures; zoning ordinances and development controls; subdivision regulations; building codes; housing standards; urban renewal; community development programmes; welfare policies; living-resource conservation; landscape conservation; heritage conservation; wilderness, national and marine parks; pollution-control strategies; environmental impact statements and assessments; public hearings and inquiries; appeal mechanisms and procedures; and the application of international conventions and agreements.

Environmental Protection Agency, US (EPA, US) A federal agency created by the US Government in 1970; its purpose is to protect the ENVIRONMENT to the fullest extent possible under the laws enacted by the US Congress. Its mandate was to mount an integrated, coordinated attack on environmental pollution and degradation in conjunction with state and local governments. The EPA was formed initially from the amalgamation of some 15 agencies or parts of agencies. It became at once responsible for the federal programmes for air and water pollution, solid-waste disposal, pesticide registration, toxic substances, the setting of radiation standards, noise control, and EIA procedures.

environmental protectional policies Policies developed by governments, agencies, associations, communities, groups, corporations, and companies, relating to the protection of the natural environment, the control of wastes, the improvement of the human-made environment, the protection of heritage values, the declaration of national parks and reserves, the protection of fauna and flora, the conservation of forests and landscapes, the protection of wilderness, the promotion of ENVIRONMENTAL PLANNING, and the implementation of international conventions and agreements.

environmental quality standards (1) Levels of exposure to pollutants which should not be exceeded, such as those suggested by the WORLD HEALTH ORGANIZATION; see also AMBIENT QUALITY STANDARDS. (2) Upper limits for exposure to noise at different times of the day. (3) Restrictions in respect of pesticides and detergents. (4) Specifications in relation to product lifetime performance. (5) Standards of performance, design, or appearance set within the framework of ENVIRONMENTAL PLANNING. (6) Standards set by permits and licences in respect of, say, the composition of effluents. (7) Restrictions on discharge of trade wastes to sewers.

environmental science A body of particular facts, together with organized methods of investigation, which aims at an improved understanding and interpretation of the various characteristics of the ENVIRONMENT. Environmental science draws on many established sciences and subjects including biology, geology, geomorphology, physics, chemistry, engineering, economics, and other social sciences. It is, however, unique in focusing its attention on the interaction between humans, both individually and collectively, and the ambient physical world, BIOTIC and ABIOTIC.

environmental sustainability See SUSTAINABLE DEVELOPMENT; SUSTAINABLE YIELD.

environmentally hazardous chemicals Chemicals and chemical wastes which pose a threat to the ENVIRONMENT, directly or indirectly, either in their immediate potential effects, or through long-term insidious adverse influences. These include: (1) toxic wastes which have mutagenic, teratogenic, or carcinogenic effects or other latent effects on health; (2) persistent substances disposed to land or the marine environment; (3) toxic or noxious materials applied in gardens, homes, and play areas; (4) agricultural pesticides, herbicides, fertilizers, and other products likely to be carried by runoff into the broader environment; (5) some products and by-products of the chemical and petrochemical industries. (See ARSENIC; ASBESTOS; BERYLLIUM; CADMIUM; CHLORINE; CUMULATIVE EFFECTS; HEAVY METALS; LEAD; MERCURY; PESTICIDES; RADIOACTIVE WASTES.)

environmentally sensitive areas In Britain, a scheme introduced by the Ministry of Agriculture in 1984 to protect 10 of the most beautiful areas in Britain from loss and damage from agricultural activity.

Environment, Department of the In Britain, a department formed in 1970 to bring together, for the first time, responsibilities for environmental planning and land use, pollution control, public building and construction, inner-urban and inner-city areas, new towns, regional affairs, and initially, transport. In 1976, the department was split, hiving off those matters concerned with transport. However, close collaboration between the Department of the Environment and the Department of Transport was established, with joint arrangements on urban transport and planning matters. Other ministries retain some environmental responsibilities, such as the Ministry for Agriculture remaining responsible for the control of pesticides used on land, the disposal of farm wastes, and the monitoring of contaminants in food. Other departments look after oil pollution at sea and noise around airports. In Scotland, Wales, and Northern Ireland environmental matters remain the responsibility of their respective departments. The Foreign Office remains responsible for international environmental conventions and policies.

enzyme A biological CATALYST that converts one chemical into another, without itself being destroyed or modified in the process. Enzymes are complex proteins which digest food, convert food energy into adenosine triphosphate (ATP) which drives many biological processes, assist in the manufacture of all the molecular components of the body, produce copies of deoxyribonucleic acid (DNA) when a cell divides, and control the movement of substances into and out of cells. Of many different kinds, enzymes promote the multitudinous reactions that take place in the living organism; for example, the hydrolysis of fats, sugars, and proteins, and their resynthesis, and the many forms of oxidation and reduction which provide energy for the cell. METABOLISM depends entirely upon enzymes, a very small amount producing a very great cumulative effect. Enzymes are susceptible, however, to a wide variety of substances which may act as poisons; they may be destroyed or inactivated, for example, by insecticides.

EPA, US ENVIRONMENTAL PROTECTION AGENCY, US.

ephemeral Descriptive of: (1) any organism or phenomenon that is essentially short-lived; (2) a plant that completes more than one life cycle, from seed to seed, within a year; (3) organisms adapted to making rapid use of favourable environmental conditions, such as plants in arid ecosystems adapted to surviving as seeds for a long time and yet, with adequate rain, germinating, growing, and producing new seeds within a very short period.

epidemiology A branch of medical science concerned with the study of the environmental, personal, and other factors that determine the incidence of disease; it concerns itself with groups of people rather than individual patients, using statistics. One of its chief functions is the identification of populations at high risk for a given disease, so that the cause may be identified and preventive measures adopted.

epilimnion The upper stratum of water in a lake that usually has the highest oxygen concentration; the warmer uppermost layer of water, lying above the THERMOCLINE.

epiphyte Any plant that uses another plant in some manner for physical support and not for nutrition, such as lichen on trees.

equilibrium (1) In physics, a condition in which the resultant, or vector sum, of all forces acting upon a point is zero. (2) In ECONOMICS, an economic system at rest and in which no one has an incentive to behave differently; it is not a state, however, that necessarily provides the community with the maximum possible economic, social, and environmental benefit;

(3) In ECOLOGY, stability. (See DYNAMIC EQUILIBRIUM.)

equivalent aerodynamic diameter (EAD) The most common and perhaps the most useful of the definitions of particle size; it is commonly defined as the diameter of a spherical particle of the density of water which has the same aerodynamic behaviour as the particle in question. The EAD is used extensively in AIR POLLUTION studies.

equivalent continuous sound (L_{eq}) A level of constant sound measured in dB(A) which would have the same sound energy over a given period as the measured fluctuating sound.

eradication The complete and final extinction of a species throughout its range; eradication programmes frequently fail in this absolute sense.

ergonomics The study of the relationship between people and the electronic equipment, machinery, tools, and furniture they use at work; the purpose is to improve work performance by removing or minimizing sources of stress and fatigue.

erosion A process by which rock particles and soil are detached from their original site, transported, and then deposited at some new locality. The main agents of erosion are water and wind. Geological or natural erosion operates at an extremely slow rate; it is with accelerated erosion arising from the activities of humans that the conservationist is primarily concerned. Soil erosion can occur when land is cleared and cultivated or when vegetation is destroyed by fire, overgrazing, or the development of conditions unsuitable for vegetation, such as high salinity.

ESCAP ECONOMIC AND SOCIAL COMMISSION FOR ASIA AND THE PACIFIC.

Escherichia coli (***E. coli***) A species of bacterium which lives in the alimentary canal of a number of animals, including humans. *E. coli* is normally harmless, occurring in large numbers in faeces and in water contaminated by faeces. Water is usually considered fit for domestic supply when there is no more than one cell per 100 ml. While *E. coli* may not in itself be dangerous, it is indicative that other bacteria of a harmful character may be present. (See COLIFORM BACTERIA.)

ESD ECOLOGICALLY SUSTAINABLE DEVELOPMENT.

Espoo convention See CONVENTION ON ENVIRONMENTAL IMPACT ASSESSMENT IN A TRANSBOUNDARY CONTEXT.

essential ecological processes Those processes that are necessary for food production, health, and other aspects of human survival and SUSTAINABLE DEVELOPMENT, such as the recycling of nutrients and the cleansing of waters.

estuaries Semi-enclosed coastal areas where salt and fresh water meet. Mangroves and salt-tolerant plants usually fringe the edges of estuaries, and seagrasses carpet the tidal mudflats of these areas. They tend to be extremely productive places and provide nursery grounds for many species of fish.

ethanol (ethyl alcohol) Derived from the fermentation of starch and sugar obtained from energy crops, including wheat, barley, rye, pearl millet, grain, sorghum, sugar cane, cassava, and sugar beet. Many cars may run on a blend of 80% petrol (gasoline) and 20% ethanol without significant engine modification. In industry, ethanol is mainly synthesized from ethylene, being used as a solvent or chemical raw material; it is the basis of alcoholic drinks.

ethnocentrism The evaluation of cultural traits against the background of a particular culture that is assumed to be superior.

ethology The study of animal behaviour; a combination of laboratory and field science. An ethologist may study one kind of behaviour in a number of different species.

Etosha National Park Established in 1907 in Namibia, southwest Africa, an area of broad semi-arid plains which merge into the Etosha Pan, which is sometimes largely submerged. The park is one of the world's largest game areas, with a considerable range of fauna and flora.

EU EUROPEAN UNION.

Eugenic The controversial study of human improvement by genetic means. Sir Francis Galton (1822-1911) in his work *Hereditary Genius* proposed that a system of marriages between men of distinction and women of wealth would eventually produce a gifted race. He coined the term eugenics in 1883. Others before and since have produced theories and policies for the elimination of the disadvantaged with a view to human improvement, the Nazis using eugenics to support the extermination of Jews, blacks, and homosexuals in the Second World War. US eugenists have argued for the sterilization of insane, retarded, and epileptic people. Since the 1950s, there have been many developments in the handling of genetic problems and disabilities which follow more humane and constructive paths.

euphotic zone The open water of the ocean, corresponding to the LIMNETIC ZONE of a lake. The zone has sufficient sunlight to support PHOTOSYNTHESIS and a considerable population of PHYTOPLANKTON. Usually, sunlight cannot penetrate deeper than 200 m in most marine habitats; this depth is frequently considered the lower border of the euphotic zone. Below the euphotic zone lies the BATHYAL ZONE and the ABYSSAL ZONE.

European Bank for Reconstruction and Development A bank created by the EUROPEAN UNION to assist the eastern European countries following the collapse of communism; it finances and promotes appropriate development. The bank has a responsibility through the whole range of its activities to promote environmentally sound and SUSTAINABLE DEVELOPMENT in central and eastern Europe, including Russia.

European Conference of Ministers of Transport An intergovernmental organization with 31 European member states, administratively attached to the OECD in Paris, working on international transport policy issues, with a strong focus on the protection of the ENVIRONMENT and the integration of the Eastern European countries into the European transport market. (See ORGANIZATION FOR ECONOMIC COOPERATION AND DEVELOPMENT.)

European Eco-Management and Audit Regulation (EMAR) Adopted by the EUROPEAN UNION in March 1993, a voluntary scheme encouraging companies to set up environmental targets, with external auditors verifying compliance. Companies fulfilling the requirements are entitled to a Statement of Participation. The scheme is without prejudice to pollution control and other related environmental legislation. A list of participating companies is published once a year in the official EU journal. A label can be used on company advertising, but not on products. Participating companies must introduce a scheme for ENVIRONMENTAL MANAGEMENT, setting pollution prevention targets, conducting self-assessments every one to three years with external validation. The scheme started in 1995.

European Environmental Agency (EEA) Established in 1993, an agency of the EUROPEAN UNION located in Copenhagen, Denmark. The EEA functions as a monitoring authority, implementing a public awareness and environmental education programme.

European Union (EU) An economic and political alliance formed initially by six European nations through the Treaty of Rome in 1957, by 1995, the EU (formerly the European Community) comprised Austria, Belgium, Britain, Denmark, Finland, France, Germany, Greece, Ireland, Italy, Luxembourg, the Netherlands, Portugal, Spain, and Sweden, with prospects of further expansion under the Single European Act 1987 and the Maastricht Treaty 1993. The EU is committed to further steps in economic and monetary integration. The combined population exceeds 370 million. The EU has come to adopt a broad approach to environmental policy formulation. The first environmental programme was adopted in 1973, shortly after the UN CONFERENCE ON THE HUMAN ENVIRONMENT, followed by directives, decisions, and regulations intended to be adopted by member countries and embodied in national legislation. Any member country failing to implement EU directives may be brought before the European Court of Justice. Directives have related to motor vehicle exhausts, intractable wastes, drinking-water standards, recreational-water standards, marine pollution, dangerous substances, packaging, air-quality standards, and environmental quality. Other significant directives have included the Birds Directive (79/409/EEC) and the Environmental Assessment Directive (85/337/EEC). The EA directive sought to a bring a measure of uniformity into the assessments, of projects, public and private, throughout the member countries. The directive lists those projects likely to have significant adverse effects on the environment and which should be subject to rigorous EIA procedures; it also lists other categories which should receive close attention. Another important directive is the Habitats Directive (officially the Directive on the Conservation of Natural Habitats and of Wild Fauna and Flora 92/43/EEC). The directive aims to establish a Europe-wide network of protected areas called Special Areas of Conservation (SACs) by the year 2000. The network will be known as Natura 2000. SACs will be both terrestrial and marine. They will require not only protection, but also positive management in the interests of habitats and species. In 1993, the EU introduced a EUROPEAN ECO-MANAGEMENT AND AUDIT REGULATION (EMAR); this involves the appointment of independent verifiers to check environmental statements and their compliance with EMAR. In October 1993, Copenhagen was chosen as the location of the new European Environment Agency. (See BRITISH STANDARD 7750; COUNCIL OF EUROPE; See also Box 11.)

eutrophic Describing an aquatic environment that is rich in dissolved nutrients and abundant in FLORA and FAUNA. (See OLIGOTROPHICS.)

eutrophication The process of nutrient enrichment of water that leads to enhanced organic growth causing, if carried too far, undesirable effects. In lakes, rivers, harbours, and estuaries the accumulation of nutrients is a natural process. These nutrients include carbon, sulphur, potassium, calcium, magnesium, nitrogen, and phosphorus. Thriving on these nutrients, algal blooms may be red, brown, or blue-green. In small concentrations these blooms contribute to the oxygen balance of the water body and also serve as food for fish. Industrial and domestic wastes often contribute greatly to the amount of nutrients entering water, phosphorus

eutrophication

Box 11   Environmental directives issued by the European Union and its predecessor, the European Community

1970   Directive against air pollution by exhaust fumes from motor vehicle engines; Directive on motor vehicle noise standards
1973   Directive on polychlorinated biphenyls (PCBs), polychlorinated terphenyls (PCTs) and vinyl chloride monomer; Directive on detergents; Directive on the testing of the biodegradability of anionic surfactants
1975   Directive on drinking-water standards; Directive on bathing-water quality standards; Directive on the sulphur content of gas oils; Directive relating to the preparation of inventories of harmful wastes and residues
1976   Directive on the discharge of dangerous substances
1977   Directive on biological standards for lead and on screening of the population for lead
1978   Directive on the discharge of titanium dioxide; Directive on measures to combat oil pollution; Directives controlling pollution from the wood and pulp mill and other industries; Directive setting the maximum permitted lead compound in petrol (0.4 g/l)
1979   Directive on the classification of packaging and labelling of dangerous substances (sixth modification to 1967 Directive)
1980   Draft Directive on environmental impact assessments; Directive on air quality limit values and guideline values for sulphur dioxide and suspended particulates; Directive on the protection of ground water against pollution; Directive on drinking-water standards
1981   Directive on an air quality standard for lead; Directive following the conclusion of the CONVENTION ON THE CONSERVATION OF ANTARCTIC MARINE LIVING RESOURCES
1982   Draft directive on the environmental effects of proposed development projects; Directive on the conclusion of the CONVENTION ON THE CONSERVATION OF EUROPEAN WILDLIFE AND NATURAL HABITATS; Directive on the conclusion of the CONVENTION ON THE CONSERVATION OF MIGRATORY SPECIES OF WILD ANIMALS. Directive on the implementation of the CONVENTION ON INTERNATIONAL TRADE IN ENDANGERED SPECIES OF WILD FAUNA AND FLORA (CITES)
1983   Directive on establishing a Community system for the conservation and management of fishery resources
1984   Directive in respect of emissions from industrial plant; Directive in respect of lead in petrol (gasoline)
1985   Directive on ENVIRONMENTAL IMPACT ASSESSMENT (EIA); Draft directive in respect of motor vehicle exhausts
1988   Directive on emissions from large combustion plants
1989   Directives on emissions from new municipal plants and the reduction of air pollution from existing plants
1990   Directive establishing the European Protection Agency; Directive on the freedom of access to information on the environment
1991   Directive on substances that deplete the OZONE LAYER; Directive on hazardous waste
1992   Directive on the conservation of natural habitats and of wild flora and fauna; Directive on air pollution by ground-level ozone; Directive on a Community system for fisheries; Directive on a Community eco-label award scheme
1993   Declaration on the protection of animals (Maastricht Treaty), Eco-management, and audit regulation

---

and nitrogen being the main contributions. Fertilizers, also containing phosphorus and nitrogen, may be carried by surface or ground water into lakes or streams. In an advanced stage, eutrophication may create problems of water colour, taste, and odour; the water may become less attractive for swimming, fishing, and boating; in the extreme, the more desirable types of fish may die. Possible remedies include: the removal of nutrients from waste water; removal of excessive weeds and debris; dredging of lake sediments; application of chemicals to destroy algal growths; and the by-passing of lakes, diverting waste waters to water bodies below. Catchment management has also much to contribute. (See ALGAE; BLUE-GREEN ALGAE; CATCHMENT.)

evaporation The conversion of a liquid into vapour, without necessarily reaching boiling point. The fastest-moving molecules escape from the surface of a liquid during evaporation, the average kinetic energy of the remaining molecules being reduced; in consequence evaporation causes cooling.

evapotranspiration The total amount of moisture that evaporates from any specific area of soil and vegetation in a particular ECOSYSTEM, being the difference between the total precipitation falling on the area and the amount which runs off as estimated by stream flow, assuming a constant soil-water content. It is the combined EVAPORATION of water from the soil surface and the transpiration or loss of water vapour from a plant, mainly through the stomata.

event-tree analysis A technique used in risk assessment to find the various possible outcomes arising from a given initiating event.

Everglades National Park Established in 1934 in Florida, USA, a large flat region, embracing most of Florida Bay, with many swamps, islands, and extensive freshwater and saltwater areas. The flora comprises subtropical wilderness, with open prairies, mangrove forests, palm trees, cypress, and sawgrass marshes. Fauna include manatees, crocodiles, alligators, turtles, panthers, raccoons, opossums, numerous snakes, and many kinds of bird. The Everglades is a refuge for many rare animals.

evo-economics A merging of ECONOMICS and Darwinian BIOLOGY, resulting in the discipline of evolutionary economics. The common ground between economics and evolution is a focus on the individual; in other words, an understanding of what happens to groups, populations, nations, or species depends upon an understanding of what motivates individual people and animals. In society, individuals tend to maximize their 'utility functions' or consumption; animals tend to maximize their 'fitness functions' to ensure the survival of their species. The process for humans or animals is much the same; individuals do things for the benefit of the larger group only if it is for the good of the individual. Animals rarely do selfless things for their species, while humans rarely do selfless things for the benefit of society. Yet, altruism abounds in human and animal life to somewhat modify this singular conclusion. Love of the ENVIRONMENT and active protection of it, for example, illustrates that humanity moves in a somewhat larger framework than theory might suggest. A satisfactory theory of human and animal behaviour has yet to emerge.

examination in public In Britain, a public inquiry directed by the Secretary of State for the Environment into a structure or regional plan, in its entirety or into selected issues. Such an inquiry is normally conducted by a panel of three. The inquiry takes the form of a discussion, with selected participants including the county planning authority responsible for the plan. (See PUBLIC LOCAL INQUIRY.)

exclusionary zoning The use of ZONING in a manner that discriminates against racial, social, and economic groups; for example, zoning that seriously impedes or absolutely prevents the construction of low-cost housing. In the US, exclusionary zoning has often been found illegal on the grounds that zoning ordinances must provide for a wide range of housing types suitable for all income groups.

existence value The benefits from knowing through reading, or watching films or viewing pictures that something of great geological, biological, or heritage value exists, even though the individual may never visit or see personally any of these assets. Thus, the existence of a wide range of fauna and flora throughout the world may be greatly appreciated and valued by many, while few will have direct experience of this BIODIVERSITY. While existence value is difficult to measure, it is argued that this value should be added to a nation's list of economic resources.

existing environment An essential component of the ENVIRONMENTAL IMPACT STATEMENT, a competent description of the ENVIRONMENT as it is and as it is likely to become, without the proposed project or activity. The concept includes: the history of the area; the land uses; the fauna and flora; the ecological features; the demography; existing industrial, commercial, and agricultural activities; the measured levels of air pollutants, water pollutants, and noise; the presence of contaminated land; unusual features topographically and biologically, including rare and endangered species; and the socio-economic features of the area. To gain an adequate and reliable impression of an existing environment may require data for some years past, against which, the possible effects of a new development may be considered.

existing use The use of premises immediately before the introduction of a new zoning or planning scheme; often these are allowed to continue under the new scheme, presenting conflicting and jarring activities, until the demise of the business.

exothermic Of a chemical reaction, one that takes place with the release of heat.

exotic A species not native to a region or locality; from another part of the world.

externality A benefit or cost falling on a third party who normally cannot pay or be compensated for it through the market mechanism. An external benefit is often termed a positive externality; an external cost, a negative one. An externality embraces any production or consumption process which 'spills over' such that other (third) parties are affected. An example of a positive externality is the beneficial effect of an increasing use of electricity for domestic heating in reducing the amount of pollution in the general atmosphere from the use of open fires using fossil fuels. An example of a negative externality is the adverse effect on a community of water pollution caused by an inadequate sewage treatment plant. Four types of techniques are widely used in measuring externalities: the related market (hedonic price) approach; the hypothetical market (contingent valuation) approach; the dose-response approach; and the minimum-maximum value approach.

Exxon Valdez disaster A major incident when the tanker *Exxon Valdez* ran aground in March 1989 in Prince William Sound, Alaska, on the Bligh Reef; 250 000 barrels of oil poured into the Sound. Some 2400 kilometres of beach were fouled by the spill. A jury decided that the owners, America's biggest oil company, and the captain, were reckless. Plaintiffs in the Federal Court case included more than 10 000 commercial fishermen, Alaskan Indians, and property owners, who claimed they had suffered economic harm after the spill.

F

fabric filter See BAG FILTER.

facadism The philosophy and practice of requiring developers to retain the frontages or facades of important heritage buildings, while allowing redevelopment behind the facades. The retention of facades is common practice throughout the world, most notably in Australia, Britain, Germany, Italy, the Netherlands, New Zealand, and the USA. Through this practice, a famous warehouse or bond store may become a shopping precinct or a block of home units, with the whole exterior preserved. While the primary purpose may be heritage preservation, substantial savings may also occur in avoiding the complete clearing of a site with entirely new construction.

factors of production The various agents which combine to produce goods and services. Land and natural resources, labour and capital are the traditional categories; organization or enterprise is sometimes added as a fourth category. Occasionally, all the factors are lumped together as capital: natural capital, renewable or non-renewable; human capital; and human-made capital. In this instance natural capital embraces all environmental assets including soil, minerals, atmosphere, forests, water, and wetlands; human capital embraces the workforce; and human-made capital consists of infrastructure, factories, ships, houses, and transportation. There are many types or categories of each factor of production and, at the margin, quite often one factor may be substituted for another.

facultative A condition where both AEROBIC and ANAEROBIC conditions occur; the surface of a pond may be aerobic and the bottom anaerobic. The term also refers to microorganisms which can survive and reproduce under both aerobic and anaerobic conditions.

faecal organisms A group of microorganisms normally found in the gut of warm-blooded animals, and whose presence in the ENVIRONMENT is used as an indicator that faecal pollution has taken place. (See COLIFORM BACTERIA; *Escherichia coli*.)

fail-safe fault A fault in an industrial plant, which does not result in the deterioration of safety standards.

fail-to-danger fault A fault which moves a plant towards a dangerous condition, or which limits the ability of a protective system to respond to a dangerous situation.

failure mode The manner in which a component in an industrial plant fails.

failure mode and effects analysis A method of hazard identification where all known failure modes of components or features of a plant or system are considered in turn, and undesirable outcomes eliminated.

fallowing The practice of leaving land in an uncultivated state for a period of time before sowing a crop or between subsequent crops, to allow land to recover naturally.

family In relation to taxonomy, a commonly used category of classification that incorporates a range of genera (below a family and above a species) possessing a number of characteristics in common. (See TAXONOMIC.)

family planning See UN CONFERENCE ON POPULATION AND DEVELOPMENT.

fanning Descriptive of the behaviour of a chimney plume when the air is very stable, the gases quickly reaching their equilibrium level and travelling horizontally with sideways meanderings but with very little dilution and mixing in the vertical direction. Fanning produces a thin but concentrated layer of pollution that may impinge on hillsides or tall buildings.

FAO See UN FOOD AND AGRICULTURE ORGANIZATION.

FAR FATAL ACCIDENT RATE.

fatal accident rate (FAR) The number of deaths (mortality) that have occurred, or are predicted to occur, in a defined group in a given ENVIRONMENT during a specified period of total exposure.

fault-tree analysis A method of examining complex engineering systems to determine the probable occurrence and consequences of an accident in any component of the total system. It involves tracing the sequence of events in all parts of the system that could result from any particular accident.

fauna Animal life; the animals of a particular region or period of time. Faunal regions, also called zoogeographic regions, are the six or seven areas of the world defined by animal geographers on the basis of their distinctive animal life. These regions differ only slightly from the floristic regions defined by botanists. Each region coincides more or less with a major continental land mass. (See FAUNA IMPACT STATEMENT.)

fauna impact statement (FIS) A specialized impact statement concerned with the possible effects and implications of a proposed policy, plan, programme, project or activity for fauna. An FIS might well be incorporated in a comprehensive ENVIRONMENTAL IMPACT STATEMENT.

faunal succession, law of Observations that assemblages of fossil animals and plants succeed or follow each other in time in a predictable manner. Faunal succession occurs

feedback

because evolution tends to progress from simple to complex in an orderly and non-repetitive manner. Faunal succession is a fundamental tool of stratigraphy and comprises the basis for the geological time scale.

feedback The return of a part of the output of a system or process to the input; the process whereby the output of a system affects the input. Positive feedback increases or reinforces something; negative feedback simply keeps a process within certain limits. Negative feedback is widely used in living organisms and in ecosystems to help maintain their existing conditions (a process called HOMEOSTASIS). Positive feedback can occur also in natural systems where it amplifies a very small effect, yet upsetting the previous equilibrium.

feral Descriptive of FAUNA and FLORA that are no longer domesticated or cultivated and have reverted to a wild state.

fertility rate The number of children born in the lifetime of each woman. In the developed world generally, the rate is less than two; while throughout almost the whole of Africa and the Middle East, the rate is five or more. Many countries now discourage large families and encourage wise family planning. China has a very restrictive policy on the size of families. (See DEMOGRAPHIC TRANSITION; WORLD POPULATION.)

fertilizer Natural or artificial materials containing chemical elements that improve the growth and productiveness of plants. Fertilizers enhance the natural fertility of the soil, compensating for the chemical elements taken from the soil by previous crops. Animal manures and vegetational wastes have been applied to land for thousands of years. Natural organic fertilizers applied today include manure, compost, and sawdust, while inorganic fertilizers include crushed limestone, gypsum, sulphur, and rock phospate. In addition, vast amounts of manufactured chemical compounds are applied containing nitrogen, potassium, phosphorus, and sulphur. Many of these manufactured fertilizers are quite soluble, and if not taken up by plants are easily leached away. Such runoff may have a detrimental effect on water quality. Thus, appreciable quantities of nitrates may be leached to water. Nitrate fertilizers may also pose a health hazard if excessive nitrates occur in food, particularly leafy vegetables, and in drinking water. (See EUTROPHICATION.)

filter bag See BAG FILTER.

filter ratio The number of cubic metres of gas that will pass through one square metre of filter surface per minute at a given filter resistance.

filter resistance The pressure drop across a filtering surface expressed in millimetres of mercury.

final settlement tank A tank through which the effluent from a percolating filter or aeration tank flows for the purpose of separating settleable solids; often called a humus tank.

Fiordland National Park Established in 1952 in New Zealand, a rugged area on the southwestern coast of the South Island, including the Cameron, Hunter, Murchison, Stuart, and Franklin mountains. The flora includes alpine meadow, with a snow covering on the higher slopes and subtropical rainforest on the lower slopes. The fauna includes kea (mountain parrots), kakapo (owl parrots), takahe (rare flightless rails), red deer, and opossums.

fire A process of combustion characterized by heat, smoke, or flame, or any combination of these.

fireball A fire burning sufficiently rapidly for the burning mass to rise into the air as a cloud or ball.

fire prevention Measures taken to prevent outbreaks of fire at a given location.

fire protection Design features, systems, or equipment that are intended to reduce the damage from a fire at a given location.

fire storm A fire covering an extremely large area, resulting in a tremendous inrush of air which may reach hurricane force.

FIS FAUNA IMPACT STATEMENT.

fiscal policy The policy adopted by government for raising revenue through various forms of taxation to meet government expenditure and to achieve economic management objectives. Revenue is commonly raised from personal taxes, company taxes, sales taxes, excise duties, customs duties, land taxes, revenues from public sector enterprises, payroll taxes, government charges such as stamp duties, and taxes on cigarettes, alcohol, and gambling. Carbon taxes have been urged as a means of discouraging to some extent the use of fuels with high carbon contents, thus attempting to curb the GREENHOUSE EFFECT. Taxes of various kinds are suggested from time to time to serve as economic instruments in ENVIRONMENTAL MANAGEMENT.

Fish and Wildlife Service, US (FWS, US) An agency to conserve, protect, and enhance fish and wildlife and their habitats. The FWS is responsible for the US National Wildlife System officially established in 1966, and is concerned with the development of this system beyond the year 2000. The overall aim is to provide, preserve, restore, and manage a national network of lands and waters sufficient in size, diversity, and location to meet society's needs for such areas.

fisheries Areas dedicated to the harvesting of fish, shellfish, and sea mammals as a commercial enterprise. Fisheries range from small family operations relying on traditional fishing methods to large enterprises using large fleets and sophisticated technology. Factory ships sail thousands of kilometres from home. Fish constitutes less than 1% of the world's diet while the various hazards of the industry such as weather, environmental pollution, the perishability of the harvest, and the high operating costs discourage much commercial growth. Of the world's total harvest, about a quarter is provided by the herring family, while the cod family accounts for about one-sixth. Since 1950, the world's fish catch has increased almost five-fold, fishermen turning to new seas and new species. Overfishing is a characteristic of many traditional fishing grounds and fish stocks. Changes in the size of fish populations can alter entire ecosystems. Many governments subsidize their fishing industries in various ways, encouraging an unsustainable level of activity. The world's marine fish catch peaked in 1989 and has since slowly declined: the expression 'Plenty more fish in the sea' is no longer true. (See AQUACULTURE; CONVENTION ON THE PROHIBITION OF FISHING WITH LONG DRIFT NETS IN THE SOUTH PACIFIC; DRIFT NET FISHING.)

fission   In physics, the process by which an atomic nucleus splits into two fragments plus several free nutrons, giving off large amounts of energy in the form of gamma radiation and heat. (See RADIOACTIVITY.)

fission products   The mix of nuclides resulting from FISSION. Fission products are often unstable and undergo radioactive decay, emitting radiation of a range of types and energies. In addition to uranium and plutonium, these may consist of more than 40 different radioactive elements; for example, arsenic, barium, cadmium, cerium, iodine, silver, and tin.

fixation   In SOIL, any process by which a nutrient essential for plant growth is changed to a soluble form that can be absorbed by plant roots from an unavailable and often insoluble form. Nature accomplishes nitrogen fixation, for example, by means of nitrogen-fixing bacteria.

fixed carbon   (1) Carbon which does not pass off in volatile matter (tarry matter and gases) when coal is heated, remaining in the coke or residue. (2) A measure of the primary productivity of an ECOSYSTEM based on the amount of carbon fixed by PHOTOSYNTHESIS per unit area. (See CARBON CYCLE.)

fixed costs   Production costs which tend to be unaffected by variations in the volume of output; also known as 'supplementary costs' or 'overhead costs'.

flammable   Capable of igniting easily and burning very rapidly; synonymous with 'inflammable'.

flammable limits   The concentration in air of a flammable material below or above which combustion will not propagate.

flash fire   The combustion of a flammable vapour and air mixture, in which flame passes through the mixture at less than sonic velocity; as a result negligible overpressure is generated.

flash point   The lowest temperature at which a liquid will form a flammable vapour/air mixture under standard conditions; it indicates the temperature at which a liquid becomes a fire hazard.

Flixborough disaster   A major explosion at the plant of Nypro Ltd, Flixborough, England in 1974, killing 28 of the plant personnel. The plant was situated on a bend of the River Humber in northeast England. A temporary dogleg pipe connection with stainless steel bellows at each end failed between two reactors containing mainly cyclohexane, at elevated temperature and pressure. The flooding liquid jets from the reactor openings formed a vapour cloud containing some 50 tonnes of flammable hydrocarbon. This finally ignited at a nearby furnace and exploded devastating the site. In response to public concern, the British Government created the Health and Safety Commission.

floating floor   A technique for obtaining high sound insulation for a floor. A basic concrete slab has a quilt of rock wool laid upon it and a screed reinforced with wire mesh cast on top of the quilt. It is also a highly effective way of insulating the people below from impact noises on the floor above.

floating plant   A free-floating or anchored plant adapted to grow with most of its vegetative tissue at or above the water surface.

floating value   A phenomenon whereby potential development value accrues to or 'floats' over a number of identical competing sites. Not all sites will be developed, but all the individual owners have an equal right to believe that theirs is the plot which will attract the development and the development value.

flocculation   The joining together of particles in a suspension; the coagulation of small particles into larger particles.

flood mitigation works   Levees, floodway schemes, drains, floodgates, river-bank stabilization, pumping facilities, flood-free mounds, diversions, dams, and dredging. Additional measures include zoning, land acquisition, land exchange, relocation, provision of flood-free land, notification of flood hazards, flood warnings, and floodproofing.

floodplain   A relatively smooth portion of a river valley, adjacent to the river channel, built of sediments carried by the river and covered with water when the river overflows its banks. Floodplains are described on the basis of how often they are likely to flood. Generally, that land which, on the average, is likely to flood once in every 100 years is known as the '100-year floodplain'. Floodplains are also

flora described in terms of the severity of a flood. Floodplains can be divided into two parts, the floodway and the flood fringe. (See FLOOD MITIGATION WORKS.)

flora Plant life; the plant life of a geological period or of a region. The term flora corresponds to FAUNA, used to denote animal life.

flora reserves Reserves which contain significant examples of native or exotic vegetation with considerable floral value. Reserves are set aside primarily to conserve species that may be rare or endangered, and other plants of particular conservation significance.

flotation A process in which water is aerated, usually after the addition of a frothing agent. Separation takes place by particles and solids floating on the froth.

fluidization A process in which a bed of solid particles has properties similar to those of a liquid, by virtue of air or fluid being forced through it from below and effectively holding the particles in suspension.

fluidized combustion A system of combustion in which the fuel particles are supported by a stream of air and are in rapid motion relative to one another. The bed has the appearance of a boiling liquid and shows many fluid-like properties.

fluorescein A liquid dye suitable, for example, for detecting condenser or boiler tube leaks, or tracing leaks from drains and sewers.

fluoridation The addition of FLUORIDE to public water supplies in appropriate cases as an additional precaution against dental decay. Water containing the optimal amount of fluoride appears to increase resistance to tooth decay, while water containing too much fluoride is capable of producing dental fluorosis, a mottling of tooth enamel associated with brittleness and general deterioration. In public water supplies, fluoride is often added up to 1 mg/l; anything above 1.5 mg/l may cause dental fluorosis.

fluorides Compounds containing fluoride, the lightest and most reactive of the halogens. Fluorides occur widely in rocks and soils, consequently, gaseous and particulate fluorides are emitted from many industrial operations using natural materials; for example, in electrolysis of cryolite during the manufacture of aluminium; when fluorspar is used as a flux in the steel industry; in sintering phosphate rock; in superphosphate manufacture; in thermal processing of electric furnace slag; in cement manufacture; in brick and pottery kilns; and in the combustion of coal. Very low concentrations of fluorides in the atmosphere can cause damage to sensitive crops and fluorosis in cattle.

flushing time In respect of bays and inlets, the time taken for the volume of water in one location to be replaced by fresher water from outside.

flux (1) The movement of substances in and out of different phases of a cycle; for example, water entering the atmosphere through EVAPOTRANSPIRATION and leaving the atmosphere as rain or precipitation. (2) The rate of flow of fluid or radiation across an area; for example, in nuclear physics, the number of particles crossing a unit area per second, the unit area being at right angles to the path of the particles.

flyash The non-combustible particles suspended in the flue gas produced from combustion. The combustion of pulverized coal in modern steam generators results in large quantities of fine ash particles being carried forward from the boiler, through the dust-arresting equipment such as the ELECTROSTATIC PRECIPITATOR or BAG FILTER. Formerly regarded as a waste product, flyash is often used in building materials.

fog Visible moisture in the ATMOSPHERE; by international agreement, fog is defined as visibility below 1000 m.

FOI FREEDOM OF INFORMATION.

Food and Agriculture Organization, UN See UN FOOD AND AGRICULTURE ORGANIZATION.

food chain A sequence of organisms including producers, herbivores and carnivores through which energy and materials move within an ECOSYSTEM. (See FOOD WEB.)

food poisoning Acute illness caused by harmful bacteria, natural poisons, or contaminated food. The most frequent cause of food poisoning is the Salmonella bacterium; this comes in many forms.

food technology The application of scientific and engineering knowledge to the commercial processing and manufacture of foodstuffs, especially in respect of preservation techniques such as refrigeration, deep freezing, dehydration, irradiation, canning, and various alternative forms of packaging.

food web The linking and interlinking of many food chains, as in a complex ecosystem with several TROPHIC levels. (See FOOD CHAIN.)

foot-and-mouth disease A contagious viral fever which causes deterioration of milk yield and abortions in cattle. It affects nearly all cloven-footed mammals, including cattle, sheep, goats, and swine; some wild animals are also susceptible. However, the horse is resistant to the infection. The virus can survive for relatively long periods of time in the air, in food and garbage, and even in hides, hair, and wool. Following the diagnosis of infected animals, the area must be quarantined and affected animals slaughtered, with their carcasses burned. The quarantined area must be left uninhabited for several months. The development of effective vaccines has helped control epidemics, but has not eliminated the disease.

footloose industry (mobile industry)   Industry which is not tied to any existing location because of its independence from specific transport links or markets.

foreground   The landscape that is perceived by an observer up to a distance of 400 to 600 m; within this range, the observer experiences maximum discernment of details such as shape, colour, and contrast.

foreign aid   Assistance given by wealthier countries to aid less fortunate countries by way of grants and loans. Such aid is often funnelled through the WORLD BANK and other multilateral development banks. The UN CONFERENCE ON ENVIRONMENT AND DEVELOPMENT 1992 endeavoured to raise the level of foreign aid to 1% of gross domestic product (GDP), a target long held up in the period following the Second World War as attainable and reasonable, though never achieved. (See AGENCY FOR INTERNATIONAL DEVELOPMENT, US; BRANDT COMMISSION.)

forest products   A wide range of commodities and services valuable to affluent industrial and poor rural communities alike: timber for housing, construction, and furniture; pulp for packaging and newspapers; timbers for mining and railways; fuelwood for villages; fodder, fruits, honey, and pharmaceuticals; meat, fibres, resins, gums, dyes, skins, waxes and oils; beauty, amenity, and recreational qualities; preservation of genetic diversity; and absorption of CARBON DIOXIDE as food. Such products and services help to sustain climatic change and contribute to a sustainable future.

Forest Service, US   An agency responsible for the US National Forest System. Recent statutes have directed the Forest Service to produce land and resource management plans for each of the national forest and national grassland management units, and to review and update these plans regularly. Such management plans must provide for the diversity of plant and animal communities. The Forest Service also conducts a research programme that includes studies of forest productivity, disease and pest control, wildlife management, and forest ecology.

forestry   The business of managing forests for commercial wood production and other FOREST PRODUCTS. Forests also have important roles as water catchments and as a means of preserving natural ecosystems and providing wildlife shelter; they serve as a genetic, scientific, educational, and landscape resource. Forest areas can be managed to provide a balanced supply of many services and benefits. However, the point of balance is often a matter of intense controversy. Concern is often expressed that forestry often means: deforestation accompanied by erosion; the permanent loss of OLD GROWTH FORESTS; the loss of HABITAT and diversity; the destruction of MANGROVES and RAINFOREST; and that reforestation often means replacing interesting and varied trees by uniform coniferous plantations. (See INTERNATIONAL TROPICAL TIMBER AGREEMENT.)

fossil fuel   A general term embracing coal, oil, and natural gas which are fuels derived from organic deposits laid down in past geological periods; such fuels comprise essentially carbon and hydrocarbons. Nuclear fuels and wood are not fossil fuels. All fossil fuels can be burned with air (or with oxygen) to provide heat. Since the late 18th century, fossil fuels have been consumed at an ever-increasing rate, supplying most of the energy needs of industrial countries. While new reserves are regularly discovered, they are, by their nature, finite. Predictions regarding amounts, reserves, technology, and market conditions are notoriously unreliable, however. Apart from new discoveries in a world little explored, technology also changes making known reserves more accessible, while price movements change the entire perspective. Nuclear power, now a major contributor in many countries, is regarded with due reserve as a consequence of hazardous incidents including the CHERNOBYL NUCLEAR CATASTROPHE. Opposition to hydroelectric schemes around the world has also increased. In the meanwhile, concern over climatic change contributed to by fossil fuels and the problems of ACID RAIN and GREENHOUSE GASES generally, adds to the central problem of energy sources for a sustainable future.

fractionation   A screening process where samples of waste water are chemically and physically separated to determine the cause of toxicity by chemical groups such as polar organics, metals, or chlorine, for example.

France, evolution of environmental policy   See Box 12.

franchise agreements   Agreements reached directly between developers and governments (often endorsed by an act of parliament). The effect is to settle the terms of the proposed development, often exempting the development, in whole or part, from meeting statutory pollution control and environmental requirements and procedures and making the basic development decision immune from review by the courts. The developer becomes exempt, not only from normal planning law, but also from the surveillance of the government 'watchdogs'. Any pollution-control requirements or environmental protection measures are confined to the clauses of the franchise agreement.

freedom of information (FOI)   The right of the individual, facilitated by legislation, to gain

free field

	Box 12 France: Evolution of environmental policy
1913	Historic Monuments Act
1921	Cultural Heritage Act
1930	Act for the protection of national monuments and sites of artistic, historical, scientific, traditional or picturesque interest
1942	Toxic and Hazardous Substances Act
1948–49	Noise and Air Pollution Control Acts
1957	Parks and Nature Reserves Act
1958	Fauna and Flora Act
1960	National Parks Act
1961	Air Pollution and Wastes Act
1962	Town and Country Planning Act; Urban Renewal Act
1963	First national parks established
1964	Water Supply and River Management Act
1971	Ministry for the Environment established
1972	Protection of mountain regions directive
1973	Amendments to Pollution Control and Water Acts
1974	Further amendments to Environment Protection Acts
1975	Wastes Act
1976	National Conservation Act; Marine Pollution Act; Environmental Impact Assessment Act
1977	Chemical Products Act
1980	Air Quality Agency set up
1983	Environment Protection Zone Act; Oil Pollution Act; Environmental Impact Assessment Act
1986	Coastal Protection Act
1991	Comprehensive review of pollution control and environment protection legislation
1992	*Akatsuki Maru* arrives to load reactor-grade plutonium for Japan
1993	Rigorous legislation on contaminated land

access to information held as records by government agencies, ministers, local government, and other public bodies. Any kind of personal or non-personal information may be requested. Generally, the right of access may be denied where there is a case for commercial security, national security, or another person's privacy may be invaded. Many democratically advanced countries have such legislation.

free field   A region in which no significant reflections of SOUND occur. An anechoic room (or dead room) is a room designed to simulate free-field conditions.

free goods   In ECONOMICS, those attributes of the natural world which are valued by society but are not included in individual ownership, and do not enter into the processes of market exchange and the price system. Notable among such resources are the general atmosphere, watercourses, ecological systems, and the visual properties of the landscape. In all economies, they are common property resources of great and increasing values, presenting society with important and difficult allocational problems which exchange in the market place cannot resolve.

freeway   A highway (or motorway) that has several lanes and no intersections or traffic lights providing for rapid, unhindered movement of large numbers of vehicles. Ramps provide access and exits.

frequency   (1) The number of occurrences per unit of time; the likelihood of an event occurring over a time period. (2) The number of times a vibrating system or particle completes a repetitive cycle of movement in a period of one second, expressed in hertz or 'cycles per second'. Certain wavelengths of ELECTROMAGNETIC RADIATION are harmful to all living things.

freshwater meadows   Wetlands including shallow (up to 0.3 m) and temporary (less than four months' duration) surface water, with soils generally waterlogged throughout the winter months.

functional pattern

Friends of the Earth International A voluntary conservation organization founded in 1971. With over 800 000 members, it is represented in 27 industrialized nations and 16 developing nations. Each national group is autonomous, with its own campaigns, structure, and funding base. Members are bound together by a common cause: the conservation, restoration, and rational use of the Earth's resources. A coordinating office is based in Amsterdam, The Netherlands.

front In METEOROLOGY, the interface between two air masses of different temperature or humidity. Fronts usually occur when warm air from one region of the Earth's surface meets cold air from another.

fuel A substance used to produce heat, light, or power, usually by COMBUSTION in air or oxygen. Most natural or primary fuels such as coal, wood, peat, oil, and natural gas are made up of compounds of carbon, hydrogen, and oxygen, in association often with mineral ash, moisture, sulphur, and nitrogen. Nuclear fuel stands in sharp contrast by virtue of its nature and mode of heat release. (See FOSSIL FUELS.)

fuel cell An electrochemical cell which operates by utilizing the energy of spontaneous chemical reaction, that is, the COMBUSTION of a carbonaceous, hydrogen or hydrocarbon fuel with oxygen from the air. The reactants are fed to the cell at rates proportional to the amount of electrical energy required. Fuel cells are silent and reliable.

fuel efficiency The proportion of the potential heat of a FUEL that is converted into a useful form of energy.

fuel-reduction burning The planned use of fire to reduce hazardous fuel quantities in forests, woods, and bushland; the aim is to reduce undergrowth and accumulated debris at ground level to help slow the spread of a fire.

fume Air-borne solid particles arising from the condensation of vapours or from chemical reactions; fume particles are generally less than 5 μ in size, respirable, and visible as a cloud. They may be emitted in the following processes: volatilization, sublimation, distillation, calcination, and chemical reactions of many sorts.

***Fumifugium, or the Smoke of London Dissipated* (1661)** A thesis presented to King Charles II by John Evelyn (1620-1706), English diarist and author. A work discussing the air pollution problem of London, and possible technical and town planning solutions.

fumigant Any volatile, poisonous substance used to kill insects, nematodes, and other animals or plants that damage stored foods or seeds, or infest dwellings, clothing, and stocks.

fumigation A rapid increase in AIR POLLUTION at ground level caused by turbulence of the ATMOSPHERE created by a rising morning sun following a nocturnal inversion in which pollutants have become concentrations at a high level. Very high ground-level concentrations may be experienced for an hour or more at a distance many kilometres from the source of pollution. (See Figure 7.)

functional pattern The overall physical organization of urban land uses in the town and city including: neighbourhoods; retail, commercial,

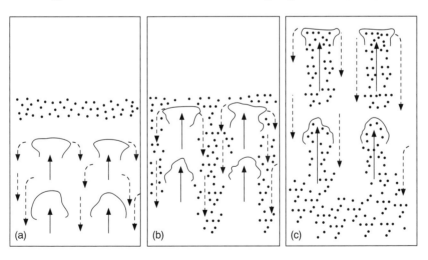

FUMIGATION: Figure 7. Fumigation
(a) Solar heating of ground initiates mixing by up-currents.
(b) Fumigation occurs as mixing involves layer of polluted air.
(c) Dilution, as clean air is introduced from above.

fungicide

or industrial concentrations; major institutions; and centres of government activity.

fungicide A chemical used to treat fungus diseases in plants and animals. Compounds containing sulphur are widely used.

fungus A group of organisms, separate from green plants, that lack leaves and roots. They contain no CHLOROPHYLL and reproduce by spores. Yeasts, moulds, and mushrooms are examples of fungi.

Fuji-Hakone-Izu National Park Established in Japan in 1936, a national park encompassing Mount Fuji, the Izu peninsula with its notable hot springs, seven active volcanic islands with crater lakes and lava cliffs, and several rivers. It incorporates special gardens, including one with over 2500 varieties of cactus.

fusion Short for thermonuclear fusion, a type of nuclear reaction in which two light nuclei such as deuterium fuse together to form one heavier nucleus, and in the process releases huge amounts of energy. Fusion reactions only take place at exceedingly high temperatures. These reactions are the source of the energy given off by the sun. (See RADIOACTIVITY.)

FWS, US FISH AND WILDLIFE SERVICE, US.

G

Galapagos National Park Created in 1968, a national park administered by Ecuador with the assistance of the Charles Darwin biological station based at Santa Cruz island. The park is renowned for its unusual animal life and takes its name from its giant land tortoises. The islands became internationally famous when visited by Charles Darwin, the English naturalist, in 1835; their unusual fauna contributed to the formation of his ideas on natural selection.

gamma radiation A form of electromagnetic radiation, similar to light or X-rays, distinguished by its high energy and penetrating power. Gamma radiation is emitted from many nuclei undergoing radioactive decay and in many other nuclear reactions. (See RADIOACTIVITY.)

Garamba National Park Created in 1938, a national park in northeastern Zaire, bordering on The Sudan; it is a continuation of the Sudanese savannah. Wild life includes the rare white rhinoceros, hippopotamuses, and giraffes.

garbage Solid, organic domestic waste, together with inorganic waste such as cans. Much household rubbish is now recycled, with householders separating paper, glass, cans, plastic bottles, and other items into separate containers for easy collection.

garden-city movement A movement in England arising from the writings of Sir Ebenezer Howard (1850-1928), which proposed preplanned new cities to be constructed on land held by the community and limited to a population of about 30 000, complete with employment opportunities, and surrounded by permanent green belts of rural land. Letchworth was started in the early 1900s and Welwyn Garden City in the 1920s. The movement was a great and continuing influence on efforts elsewhere to improve the urban environment.

gas cloud A mass of gas/air mixture. A buoyant gas cloud is lighter than the surrounding air, a dense gas cloud is heavier than the surrounding air, while a neutral-density gas cloud has a density equal to that of the surrounding air. (See BOILING LIQUID EXPANDING VAPOUR EXPLOSION; UNCONFINED VAPOUR CLOUD EXPLOSION.)

gasohol A blend of petroleum and organic alcohol in a ratio of 9:1, for use in cars. The carbohydrates in organic materials, such as grain, wood, whey, and sugar beet, can be fermented to produce alcohol for use as a fuel. Efforts have also been made to produce gasohol from sugar cane.

gasoline See PETROL.

GATT GENERAL AGREEMENT ON TARIFFS AND TRADE.

Gaussian distribution (normal distribution) A distribution that shows the maximum number of occurrences or events at or near the centre of mean point. A progressive decrease in the number of occurrences with increasing distance from the centre, with a symmetrical distribution of occurrences on both sides of the centre. A term used in statistics to describe the extent and frequency of deviations or errors. Gaussian distributions may be detected in the dispersal of air pollutants, water pollutants, thermal pollution, and noise.

GCP GOOD CONTROL PRACTICE.

Gdansk convention CONVENTION ON FISHING AND THE CONSERVATION OF LIVING RESOURCES IN THE BALTIC SEA AND THE BELTS.

GDP GROSS DOMESTIC PRODUCT.

GEF GLOBAL ENVIRONMENTAL FACILITY.

GEMS GLOBAL ENVIRONMENTAL MONITORING SYSTEM.

gene The unit of inheritance that is transmitted from parents to offspring and which controls the development in the offspring of a certain characteristic of the parents.

General Agreement on Tariffs and Trade (GATT) An international treaty which came into force in 1948 to promote world trade through a progressive reduction of trade barriers. By 1994, over 110 governments accounting for over 80% of world trade were parties to the agreement. Through rounds of talks over many years, tariff reductions have been achieved in the face of many impediments. The Uruguay Round of Talks for seven years between 1986 and 1993; the result was a commitment to a substantial reduction in tariff protection for industrial products, agricultural produce, and services. The agreement also created the World Trade Organization. Environmentalists have objected to the 1993 GATT agreement on the grounds that: trade liberalization encourages economic growth and so damages the environment; by limiting national sovereignty, countries cannot apply whatever environmental measures they choose; countries are not allowed to keep out a product because of the way it is produced or harvested; the use of trade measures to influence environmental policies abroad is disapproved of. The agreement also precludes export bans on, say, timber or wildlife and may undermine international environmental agreements. Countries may, of course incorporate environmental standards in their national legislation when ratifying and implementing GATT.

genetic Inherited; derived from parents by the transfer of heritable characteristics.

genetic diversity The range of organisms occurring in the world and the range of variation found within species or organisms. (See CONVENTION ON PROTECTING SPECIES AND HABITATS.)

Geneva convention See CONVENTION ON LONG-RANGE TRANSBOUNDARY AIR POLLUTION.

genotoxic A toxic effect on the chromosomes; for example, through mutation. (See PESTICIDE.)

gentrification The movement of people, often young professionals, into rundown neighbourhoods, who undertake the rehabilitation of what were once fine houses or apartments. Higher-standard homes are thus created in improving districts, often at a cost much lower than that of a home in the suburbs.

genus (plural, genera) A group of SPECIES that share similar characteristics; the taxonomic group above species in biological classification.

geodesic line The shortest distance between two points on the surface of a sphere; used to describe an imaginary line on the surface of the Earth, connecting two points on the surface by the shortest distance.

geographic information system (GIS) A computerized system for entering, storing, handling, analysing, and displaying data that can be drawn from different sources, both statistical and mapped. Remotely sensed data can be regularly fed into a GIS to keep it up to date. A GIS enables an environmental manager to: analyse spatial relationships, such as estimating the number of people living in a floodplain; identify regions that meet multiple criteria, such as identifying lands that would be suitable for specific agricultural crops; model the effects of policy options, such as the amount of erosion that may occur if an area is deforested; and monitor dynamic processes, such as crop yields or changes in a coastline.

geomorphology A scientific discipline concerned with the description, classification and analysis of the Earth's topographic features and the origin of landforms.

geophysics A major branch of the earth sciences that applies the principles and methods of physics to the study of the Earth. Geophysics analyses a wide array of geological phenomena, including the temperature distribution of the Earth's interior, the geomagnetic field, and large-scale features of the terrestrial crust.

geosphere The internal structure of the Earth, subdivided into crust, mantle, and core; the mineral non-living portion of the Earth. Other subsystems are the ATMOSPHERE, HYDROSPHERE, CRYOSPHERE and BIOSPHERE.

geosynclinal belts Large, elongated zones of the Earth's crust which have experienced a complex history of subsidence, sediment-infilling, volcanic activity, and deformation. With few exceptions the major mountain systems of the world have formed on such belts; for example, the Alps, Andes, Himalayas, and Rockies.

geothermal energy Power obtained by using heat from the Earth's interior, usually in regions of active volcanoes. Hot springs, geysers, pools of boiling mud, and fumaroles (vents of volcanic gases and heated ground water) are often sources of such energy. The Romans used hot springs to heat baths and homes. The greatest potential use for geothermal energy, however, lies in the generation of electricity. By the late 20th century, geothermal power plants were in operation in Iceland, Italy, Japan, Mexico, New Zealand, Russia, and the USA. Many more are now in operation. Water and steam hotter than 180°C are the most easily utilized for electricity generation, being used in most existing geothermal power stations. Geothermal power generation is non-polluting, competitive, and offers an alternative in certain locations to polluting fossil-fuel and nuclear power stations. Much depends on local geology and it is limited to a depth of about 5 km. (See Figure 8.)

germ theory The theory that certain diseases are caused by the invasion of the body by microorganisms which cannot be seen by the naked eye. Louis Pasteur (1822–95), Joseph Lister (1827–1912), and Robert Koch (1843–1910) are given much of the credit for the development of germ theory. In the 1880s, the organisms that cause tuberculosis and cholera were identified.

German UVP-Zentrum A centre for environmental impact assessment, based in Hamm, Germany. Privately funded, the centre decided in 1994 to strengthen its research into ENVIRONMENTAL HEALTH IMPACT ASSESSMENT, within the framework of ENVIRONMENTAL IMPACT ASSESSMENT.

Germany, evolution of environmental policy See Box 13.

Ghana, evolution of environmental policy in See Box 14.

Giardia A protozoan that may cause diarrhoeal illness. Drinking water may play a role in transmission if it is allowed to enter the distribution system.

Gir Lion National Park Established in 1965 in Gujarat, India, a somewhat hilly arid region, the last remaining natural habitat of the Asiatic lion, whose numbers in this sanctuary have increased to several hundreds. Other fauna include the leopard, wild pig, spotted deer, four-horned antelope, and chinkara. A large central waterhole contains a few crocodiles.

GIS GEOGRAPHIC INFORMATION SYSTEM.

glacial age See ICE AGE.

Global Environmental Monitoring System

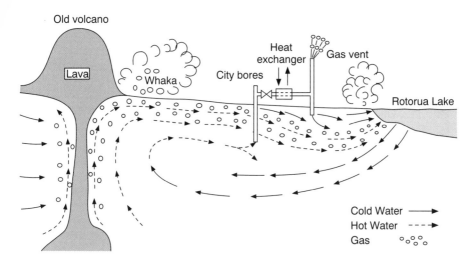

GEOTHERMAL ENERGY: Figure 8. City bores, Rotorua, New Zealand, providing heat and hot water to houses.

Glacier Bay National Park and Preserve Formerly Glacier Bay National Monument, established in 1925, enlarged and renamed in 1980, a national park in southeastern Alaska, USA. Among the notable features of the park are great tide-water glaciers, a dramatic range of flora, and an unusual variety of wildlife including brown and black bears and mountain goats.

Glacier National Park Established in 1886, a park in southeastern British Columbia, Canada, in the heart of the Selkirk Mountains. Snowcapped peaks such as Hermit, Cheops, Grizzly, Sifton, Grant, Avalanche, and Sir Donald, flanked by immense ice fields and glaciers present an impressive Alpine panorama with canyons, rivers, waterfalls, and flower-filled meadows.

glaciology The scientific discipline concerned with all aspects of ice on land masses, including the structure and properties of glacier ice, the dynamics of ice flow, and the interactions of ice accumulation with climate. (See GREENLAND ICE-CORE PROJECT.)

global energy and water cycle The HYDROLOGICAL CYCLE considered on a global scale. Water in all its forms is a vital component of the climate system and a contributor to the GREENHOUSE EFFECT. As cloud, it dominates the planetary ALBEDO and the radiative energy available at the surface of the Earth. The balance between precipitation and evaporation determines the availability of fresh water at ground level for living matter, and the entire body of salt water that covers about 72% of the Earth's surface. The global water system is a key component of the energetics of the Earth's system through atmospheric transport, latent and radiative heating, and through redistribution of water over the Earth's surface. Energy and water exchanges at the surface of the land are also important in the context of possible climatic change. European Conferences on the Global Energy and Water Cycle are held at intervals to review research in this major field.

Global Environmental Facility (GEF) A facility to provide additional grant and concessional funding for the achievement of agreed global environmental benefits. It finances activities in four areas: global warming, biological diversity, ozone depletion, and international waters. The first three areas are covered by international conventions. The GEF was established initially by the WORLD BANK in 1991; its establishment was subsequently endorsed by the Governing Councils of the UN DEVELOPMENT PROGRAM (UNDP) and the UN ENVIRONMENT PROGRAM (UNEP). Management procedures were agreed by the heads of those three bodies, known as the implementing agencies. Nations that have contributed financially to the GEF are known as 'participants'. At each of their regular meetings, the participants review and endorse the proposed work programme. The GEF participants are supported by a Scientific and Technical Advisory Panel (STAP) composed of 16 independent members. The GEF represents a highly innovative approach to environmental protection.

Global Environmental Monitoring System (GEMS) One of the components of EARTHWATCH, the UN Action Plan for the Human Environment. GEMS comprises five closely linked major programmes each containing various monitoring networks relating to pollution, climate, health, ecology, and oceans.

> **Box 13 Germany: evolution of environmental policy**
>
> 1869 Trade regulations restraining air pollution and noise
> 1935 Nature Conservation Act
> 1957 Federal Water Law
> 1969 Federal Government policy statement on the environment
> 1970 Environment Policy Act adopted by the German Democratic Republic
> 1971 Urban and rural renewal legislation introduced
> 1972 Disposal of Wastes Act; Traffic in DDT Act
> 1974 Federal Environmental Agency established
> 1975 Principles relating to environmental impact assessment adopted; Federal Forest Act; Federal Game Act; Plant Protection Act; Detergents Act
> 1976 Federal Nature Conservation Act
> 1977 Waste Water Act
> 1980 Amendment of the Federal Nature Conservation Act
> 1981 Federal Emissions Control Act, restricting air pollution, noise, and vibration
> 1983 Statutory restrictions on sulphur dioxide emissions
> 1986 Waste Avoidance and Waste Management Act; Gradual introduction of EIA throughout the Laender in implementation of the European Communities Directive
> 1988 Federal Government endorses the Vienna convention for the protection of the ozone layer; also the Montreal protocol on substances that deplete the ozone layer
> 1989 Amendment of Federal Nature Conservation Act; large combustion plants ordinance
> 1990 Environmental Impact Assessment Act; Federal Emissions Control Act; carbon dioxide reduction programme; Radiological Protection Act
> 1991 Environmental Impact Assessment Act applied to infrastructure projects; ordinances on the avoidance of packaging waste, on the prohibition of certain ozone-depleting substances, and the problem of discarded automobiles
> 1992 Environmental Impact Assessment Act applied to industrial projects; adoption of an environmental improvement plan for the 'new Laender' in eastern Germany
> 1993 National waste-recycling system introduced imposing separation duties on consumers; recyclables exceed recycling-plant capacity
> 1994 National plan to halve 1987 emission levels of methane, nitrous oxide, volatile hydrocarbons, and carbon monoxide by the year 2005

The WMO Background Air Pollution Monitoring Network (BARMON) and the WHO Program of Monitoring Air Quality in Urban and Industrial Areas have been developed as part of GEMS.

Global Ozone Observing System (GOOS) A system organized under auspices of the UN to monitor OZONE in the general atmosphere. The system combines ground-based stations and balloon-borne and satellite-borne ozone monitors. The most common instruments are the Dobson and Brewer spectrophotometers. A World Ozone Data Centre is operated in Downsview, Ontario, Canada on behalf of the WORLD METEOROLOGICAL ORGANIZATION (WMO). The measurements serve a UN programme called the World Plan of Action on the Ozone Layer.

Good control practice (GCP) The application of established pollution control methods, techniques, processes, or practices that a control agency considers capable of achieving an adequate degree of reduction and subsequent dispersion of wastes emitted from a particular source. GCP might be appropriate in a relatively isolated situation. (See BEST AVAILABLE CONTROL TECHNOLOGY; BEST AVAILABLE CONTROL TECHNOLOGY NOT ENTAILING EXCESSIVE COSTS; BEST PRACTICABLE MEANS; MAXIMUM ACHIEVABLE CONTROL TECHNOLOGY.)

GOOS GLOBAL OZONE OBSERVING SYSTEM.

Gordon-below-Franklin Dam A hydroelectric project proposed for southwest Tasmania, Australia. The successful campaign to stop the construction of the dam made conservation history. Prior to the Gordon-below-Franklin proposal, a major conservation battle had been waged and lost over the flooding of Lake Pedder. After two years of debate, the Tasmanian State Parliament decided to go ahead with

> **Box 14 Ghana: evolution of environmental policy**
>
> 1973 Environment Protection Council created by the Government of Ghana, bringing together for the first time all activities and efforts aimed at protecting and improving the quality of the environment; and to 'ensure the observance of proper standards in the planning and execution of all development projects'
> 1978 EIA procedures introduced for industries, but supporting legislation not passed
> 1985 Ghana Investment Code passed, under which the Ghana Investment Centre must have regard to any effect an enterprise is likely to have on the environment and the measures proposed for the prevention and control of any such harmful effects
> 1986 Environment Protection Council sets up an EIA committee to examine ways in which EIA can be put into operation as a management tool in Ghana; guidelines issues on EIA
> 1988 Government of Ghana gives environmental issues a priority on the development agenda
> 1989 Environmental Action Plan (EAP) introduced by government for the 10 years 1991–2000; attempts to make developments more environmentally sustainable; the EAP states that the aim of the plan is to ensure sound management of resources and the environment, avoiding exploitation of resources that will cause irreparable environmental damage; the EAP refers to the need for EIA
> 1990 Environmental concerns given prominence in the national budget for the first time; directive to all agencies that the Environment Protection Council was to be consulted formally on all development proposals, with an EIA to follow
> 1991 Environment Protection Council produces EIA guidelines; no development to proceed without an environmental impact certificate; EISs subject to review
> 1992 Ghana Environmental Resource Management Project (GERMP) developed with the participation of the World Bank, the Danish International Development Agency, and the British Overseas Development Administration; the aim is to have an effective national environmental management system for Ghana

the Gordon-below-Franklin scheme. However, the Australian federal government opposed the scheme, sought to include southwest Tasmania in the UN WORLD HERITAGE LIST, introduced appropriate legislation, and launched High Court proceedings for an injunction to stop construction of the new dam. The High Court endorsed the federal government's actions and construction work ceased. The campaign had been characterized by street marches and rallies all over Australia. What began as a state issue finished as a national issue.

gradient In ACOUSTICS, a variation of the local speed of sound with height above ground causing refraction (or bending) of sound; it may be caused by rising or falling temperatures with altitude or by differences in wind speed.

Gran Paradiso National Park Established in 1936, initially as a hunting zone, in northwestern Italy, a national park with a typically alpine terrain and numerous glaciers. Fauna include the ibex, ermine, weasel, hare, and golden eagle.

Grand Canyon National Park Created in 1919, a national park embracing the Grand Canyon in northwestern Arizona, USA, noted for its fantastic shapes and colouration. Awesome grandeur and beauty characterize the national park, which was greatly extended in 1975. The depth of the Grand Canyon is due to the powerful cutting action of the Colorado River, but its great width is due to rain, wind, temperature, and the chemical erosion of soft rock.

grassland An area dominated by grasses and grasslike plants. Grasslands, before extensive agriculture, were the largest single biome type in the world, perhaps half of the Earth's total surface. Today, the most extensive grassland areas of the temperate zone are the steppes of Eurasia, the prairies and plains of central and western North America, and the pampas areas of Argentina. Less extensive areas are the velds of South Africa, the mountain grasslands of South America, and those of Australia and New Zealand. Tropical and subtropical grasslands are located in central Africa and central South America.

gravimetric analysis A method of quantitative chemical analysis in which the constituent sought is converted into a substance that can be separated from the sample and weighed. The steps followed are: (1) preparation of a solution containing a known weight of the sample; (2) separation of the desired constituent; (3) weighing the isolated constituent; and (4) computation of the amount of the constituent in the original sample.

Great Barrier Reef Marine Park   A marine park established in 1975 by the Australian Government embracing substantial areas of the Great Barrier Reef which runs in an extended system along practically the whole east coast of Queensland. It is the largest assemblage of living corals and associated organisms in the world covering an area of 350 000 km^2. It has evolved over the last 8000 years. The park is managed by the Great Barrier Reef Marine Park Authority, which has prepared management plans for various sections of the reef. The reef was included in the WORLD HERITAGE LIST in 1991. The reef is now fully protected against inappropriate activity, including oil exploration. (See Figure 9.)

Great Lakes Water Quality Agreement, US-Canada   An agreement signed in 1972 (and amended in 1978 and 1987) by the USA and Canada to clean up pollution of the Great Lakes, particularly Lakes Erie and Ontario. The Great Lakes comprise 20% of the world's freshwater supply; 30 million people live within the drainage basins. Significant progress has been made since 1972 in restoring the ecological integrity of the Great Lakes, yet, significant problems still remain including the continued loss of habitat and the presence of persistent bioaccumulative toxic chemicals. Remedial Action Plans continue to be implemented.

Great Smoky Mountains National Park   Established in 1934, a national park in eastern Tennessee and western North Carolina, USA to preserve the last remaining sizeable area of southern primeval hardwood forest in the United States; it contains some of the highest peaks in the Appalachian Mountains. Important species of fauna in the park include black bears, white-tailed deer, foxes, bobcats, raccoons, grouse, and songbirds. The park was included in the WORLD HERITAGE LIST in 1983.

green algae   Member of the division Chlorophyta with about 6000 species; most green algae occur in fresh water, usually attached to submerged rocks or appearing as scum in stagnant water. There are also terrestrial and marine species.

Green Alliance   A London-based umbrella group for environmental bodies in some 19 countries; it is funded by the WORLD WIDE FUND FOR NATURE. In respect of environmental issues, the Green Alliance accuses the industrial nations of a persistent failure to act effectively on remedies which are well understood, widely communicated and for which there is strong public support.

Green bans movement   Originating in Sydney, Australia, in 1971 a movement based upon close cooperation between citizen groups and trade unions to prevent or change proposed developments considered to be socially and environmentally undesirable. Work bans and citizen blockades often resulted in police action, but in the end forced government reviews with improved outcomes. In Britain, a similar ban by the National Union of Seamen brought to an end the dumping of radioactive waste at sea.

green belt   A belt of land, usually several kilometres wide, around urban or metropolitan areas, preserved substantially as open space. Its purpose is to prevent further expansion and the coalescence of neighbouring cities and conurbations, and to bring fresh air and unspoiled countryside within reach of as many urban dwellers as possible. The concept has been employed in London, Paris, and Moscow, and other cities in Europe and North America.

green company   A company or corporation perceived by the public to be linked with environmental protection, for example, through investment in waste-recovery techniques, recycling processes, including paper recycling, or the manufacture of environmentally sound consumer products.

green consumerism   Purchasing which is influenced by environmental and conservation issues, directing buying preferences towards products which are less harmful to the environment; the buying of environment-friendly products.

green Great Wall   A desertification control measure introduced by the People's Republic of China in the late 1970s. The State Council decided to plant protective trees in the northern region and create a 'green Great Wall' for the benefit of present and future generations. The wall will provide much needed protection for the agricultural and pastoral districts menaced by sandstorms. (See Figure 10).

greenhouse effect   In the general ATMOSPHERE surrounding the Earth, the warming effect due to selective absorption by certain gases such as CARBON DIOXIDE, METHANE, NITROUS OXIDE, tropospheric OZONE, CHLOROFLUOROCARBONS (CFCs), and water vapour. These greenhouse gases prove transparent to incoming shortwave solar radiation but relatively opaque to long-wave radiation reflected back from the Earth, the result being a warming or greenhouse effect. Without the greenhouse effect, the Earth, certainly through the history of humanity, would have been very much

greenhouse effect

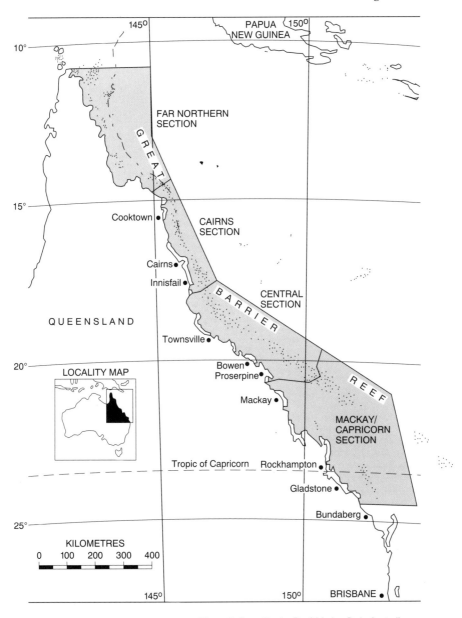

GREAT BARRIER REEF MARINE PARK: Figure 9. Great Barrier Reef Marine Park, Australia.

colder; life has depended, therefore, on the beneficial effects of the greenhouse effect. Concern has arisen in recent decades because of evidence that the levels of greenhouse gases in the atmosphere have been progressively increasing, notably the carbon dioxide level, due to the activities of humanity, both industrial and agricultural. Carbon dioxide enters the atmosphere largely as a waste product from the burning of fossil fuels, from naturally occurring fires, through the decomposition of plant and animal tissues and in exhalation from the lungs. It is removed by absorption through the leaves of plants and trees, while the oceans act as a sink. The concentrations of carbon dioxide in the atmosphere appear to have increased since 1890 from around 290 ppm to over 350 ppm. If this trend continues, climatic

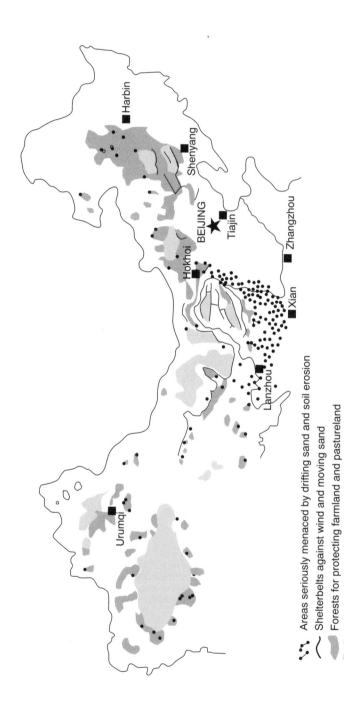

GREEN GREAT WALL: Figure 10. The green Great Wall of China.
Source: *Beijing Review* No. 36, 7 September 1979.

changes may occur which are of benefit to some regions but detrimental to others. The sea levels may rise due to the thermal expansion of the oceans. The Arctic and Antarctic polar ice caps might eventually melt. At several international conferences, measures to curb the increase in carbon dioxide have been urged through greater efficiency in the use of energy and in the conservation of forests. Measures against CFCs are already being taken to protect the OZONE LAYER. (See CARBON CYCLE; CONVENTION FOR THE PROTECTION OF THE OZONE LAYER; CONVENTION OF THE WORLD METEROLOGICAL ORGANIZATION; CONVENTION ON CLIMATE CHANGE; GREENHOUSE GASES; HAGUE CONFERENCE ON THE ENVIRONMENT; INTERGOVERNMENTAL PANEL ON CLIMATE CHANGE; LONDON CONFERENCE ON CLIMATIC CHANGE; MONTREAL PROTOCOL ON SUBSTANCES THAT DEPLETE THE OZONE LAYER; NITROGEN OXIDES; OZONE; TORONTO CONFERENCE ON THE CHANGING ATMOSPHERE; VILLACH CONFERENCE ON CLIMATIC CHANGE; WORLD COMMISSION ON ENVIRONMENT AND DEVELOPMENT; WORLD CONGRESSES ON CLIMATE AND DEVELOPMENT; WORLD METEROLOGICAL ORGANIZATION; Figures 11 and 12.)

greenhouse gases A collective expression for those components of the ATMOSPHERE that influence the GREENHOUSE EFFECT, namely CARBON DIOXIDE, METHANE, NITROUS OXIDES, OZONE, CHLOROFLUOROCARBONS (CFCs) and WATER VAPOUR. Some components are much more damaging than carbon dioxide, yet carbon dioxide is the most abundant, accounting for just over half of any global warming. The most potent greenhouse gases are the CFCs; however, they have such a low concentration in the atmosphere as to make a total lower contribution than other less effective, greenhouse gases. Water vapour, existing at a concentration of about 1% in the atmosphere, is the chief greenhouse gas, yet is rarely listed.

Greenland Ice-Core Project (GRIP) A programme of ice-core drilling in central Greenland to provide insight into the evolution of the Earth's climate and environment through the last glacial cycle. In 1994, there were eight European partners in GRIP, including the British Antarctic Survey. Ice cores drilled through the layered accumulations of snow and ice deposited in the Antarctic and Greenland ice sheets allow a very detailed reconstruction of past climate; the records in the ice extend back for up to several hundred thousand years. Evidence has been found that the whole of the

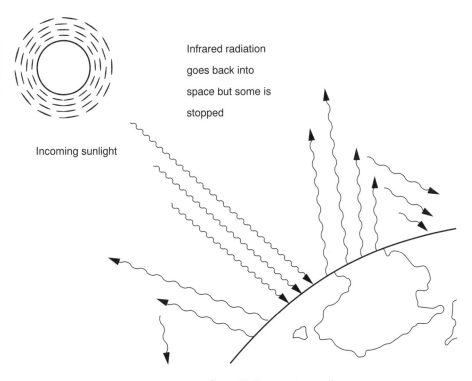

GREENHOUSE EFFECT: Figure 11. The greenhouse effect.

Green Lights Program, US

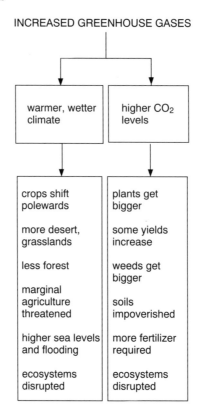

GREENHOUSE EFFECT: Figure 12. The social and environmental effects of increased greenhouse gases.

last glacial period was interrupted by a series of warm periods.

Green Lights Program, US An innovative programme, sponsored by the ENVIRONMENTAL PROTECTION AGENCY, US that encourages major US corporations to install energy-efficient lighting. Using these technologies, energy consumption can be substantially reduced while delivering the same or better-quality lighting.

Greenpeace Founded in 1971, the world's largest voluntary environmental organization with nearly five million supporters in 150 countries and observer status on 25 international bodies. Greenpeace seeks to protect BIODIVERSITY through the protection of rainforests and the protection of the marine environment from drift netting, overfishing, and commercial whaling. It also seeks: to promote the production of goods and services without involving toxic substances, such as CHLORINE, and to halt the international trade in toxic wastes; to protect the atmosphere from the burning of fossil fuels and threats to the OZONE layer; and to oppose uranium mining, reprocessing, and waste disposal.

In 1985, the Greenpeace flagship *Rainbow Warrior* was sunk in Auckland Harbour, New Zealand, by French terrorists acting for the French Government. Greenpeace was protesting against France's programme of nuclear testing at Mururoa Atoll in French Polynesia.

green revolution The achievement of higher crop yields through the adoption of new varieties of rice, wheat, and other grains associated with heavy inputs of fertilizers and improved irrigation. Since the 1960s yields in several countries have increased dramatically, the benefits being felt in India and Pakistan, the Middle and Far East, North Africa, and Latin America. The success of the green revolution in India, Pakistan, and elsewhere demonstrates that farmers are willing to learn new skills when they can gain an advantage for themselves in doing so.

Greens, The Environmentally or ecologically oriented political parties formed in European countries and elsewhere throughout the world during the 1980s. An umbrella organization known as the European Greens was formed in

Brussels, Belgium, in 1984 to coordinate the activities of the various European green parties, while Green representatives in the European Parliament have formed the Rainbow Group. The first green party was formed in Germany in 1979 arising out of the merger of about 250 ecological and environmental groups, becoming a national party in 1980. By the end of the 1980s, almost every European country had a party known as the Greens such as the Green List in Italy, the Green Alliance in Finland and Ireland, the Green Alternatives in Austria, the Green Ecology Party in Sweden, and the Ecological party in Belgium. Green parties have also emerged in countries such as Australia, New Zealand, Canada, Argentina, Chile, and the countries of eastern Europe. (See ENVIRONMENTAL AGENCIES; NON-GOVERNMENT ORGANIZATIONS.)

GRIP   GREENLAND ICE-CORE PROJECT.

gross domestic product (GDP)   The value of final goods and services produced within a nation during a specified period of time, usually one year. GDP may be measured at market prices (the expenditure method) or at factor cost (the income method). GDP at market prices is an estimate of the total market value of goods and services produced, excluding goods and services used up in the process of production, and before allowing for the depreciation of fixed capital. GDP at factor cost is an estimate of the total cost of producing those goods and services; it is the sum of the gross payments made to the FACTORS OF PRODUCTION. GDP per capita is calculated by dividing GDP by the population of the country, while GDP at constant prices may be measured over a series of years by discounting changes in the value of money. These modes of calculation have been criticized on the grounds that the value of domestic work, which is considerable, is excluded. Also, GDP does not reflect changes in quality, in the underlying resource base, or any adverse effects the getting of GDP may have on the environment or the prospects for SUSTAINABLE DEVELOPMENT. Still less does it measure at all accurately that more subtle concept, the QUALITY OF LIFE.

ground water   Subsurface water that fills voids in soils and permeable geological formations. Water-bearing formations are called aquifers. Ground water accounts for over 95% of the Earth's useable fresh-water resources and plays an important part in maintaining soil moisture, stream flow, and wetlands. Over half the world's population depends on ground water for drinking-water supplies. This invisible resource is vulnerable to pollution and over-exploitation. Contamination of ground-water supplies is often not detected until noxious substances appear in drinking water. Most ground-water contaminants are derived from agricultural, urban, and industrial uses, including fertilizers, pesticides, septic tank systems, street drainage, and air and surface-water pollution. Effective conservation of ground-water supplies requires the integration of land-use and water management. (See HYDROLOGICAL CYCLE.)

H

habitat A place where an organism or a community of organisms live(s), including all BIOTIC and ABIOTIC elements in the surrounding ENVIRONMENT. Microhabitat is the term used for the conditions and organisms in the immediate vicinity of an animal or plant. The smallest unit of a habitat with a relative uniformity of animal and plant species and environmental conditions is called a BIOTOPE. In terms of regions, a habitat may comprise a desert, a tropical forest, a prairies field, or the Arctic Tundra. (See also UN CONFERENCE ON HUMAN SETTLEMENTS.)

habitat diversification See BETA DIVERSITY.

habitat evaluation procedure (HEP) A system devised by the FISH AND WILDLIFE SERVICE, US in 1976 for use in the evaluation of habitats in the vicinity of major federal water projects. It determines the quality of a HABITAT type using functional curves relating habitat quality to quantitative BIOTIC and ABIOTIC characteristics of the habitat. Habitat sizes and quality are combined to assess project impact.

haemoglobin (Hb) A respiratory pigment found in the blood plasma. It combines readily with oxygen to form oxyhaemoglobin, but has a still greater affinity for CARBON MONOXIDE. (See CARBOXYHAEMOGLOBIN.)

Hague Conference on the Environment The summit of world leaders convened by the prime ministers of France, Norway, and The Netherlands, to consider ways to combat global climatic change arising from the GREENHOUSE EFFECT. Twenty-four countries participated in the conference. A Declaration of The Hague Conference on the Environment was released in 1989 and, among other recommendations, the conference urged that a new institutional authority should be created within the framework of the United Nations to devise means of combating climatic change.

half-life The period of time in which half of the nuclei in a given sample of a particular radioactive nuclide will undergo decay; the half-life is a characteristic of a particular nuclide. (See RADIOACTIVITY.)

halocarbon Any chemical compound of the element carbon and one or more of the halogens (bromine, chlorine, fluorine, iodine). Two important categories of halocarbons are the chlorocarbons, containing only carbon atoms and CHLORINE, and the fluorocarbons, containing only carbon atoms and fluorine. Examples of chlorocarbons include carbon tetrachloride and tetrachloroethylene. (See CHLORINATED HYDROCARBONS; CHLOROFLUOROCARBONS.)

halocline A vertical zone in oceanic waters in which salinity changes rapidly with depth, located below the normally well-mixed uniformly saline surface water.

halogen Any of the five highly reactive elements, namely fluorine, chlorine, bromine, iodine, and astatine in group VIIB of the periodic table. Most of the halogens are found in small quantities in the Earth's surface, with the exception of astatine, which consists of short-lived radioactive isotopes and does not occur naturally. Due to their reactivity, however, the halogens occur in compounds; they readily react with metals and many non-metals. The members of the halogen family closely resemble one another in general chemical behaviour.

halogenated organic waste Organic compounds containing chlorine, bromine, iodine, or fluorine; they are sometimes disposed of by high-temperature incineration at temperatures of around 1200°C.

Hamburg Manifesto on Climatic Change See WORLD CONGRESSES ON CLIMATE AND DEVELOPMENT.

Han River Environmental Master Plan A plan prepared by the Seoul Metropolitan Government, Korea, to dramatically improve and develop the Han River in the vicinity of the capital. The Master Plan extends to the year 2000. It encompasses new sewage treatment plants, recreational facilities, riverside roads, industrial effluent treatment and recycling, and augmenting the improvements that were introduced for the Olympic Games.

hardness A characteristic of water representing the total concentration of calcium and magnesium ions. Hard water is difficult to lather, may block safety valves on hot water systems, increase corrosion, and cause taste problems. Where hardness is due mainly to the presence of the bicarbonates of calcium and magnesium it is described as 'temporary'; if due to the sulphates and chlorides of calcium and magnesium, it is described as 'permanent'. However, both kinds of hardness may be reduced substantially by the use of appropriate techniques; for example, the lime-soda process followed, if necessary, by the ion-exchange process. Hard water is unsuitable for some industrial processes but it is preferred for irrigation purposes as it reacts more favourably with soils and readily reaches the root zone.

hardwood The wood of broad-leaved trees from the Angiosperms (flowering plants). They are deciduous, except in the warmest regions, when their leaves are shed at the end of each growing season. Hardwood accounts for about one-fifth of the world's commercial timber and includes oak, beech, American black walnut, Indian rosewood, teak, ebony, primavera, maple, satinwood, greenheart, and mahogany. (See SOFTWOOD.)

Harrisburg nuclear power plant incident An incident in March 1979 involving the release of radioactive material to the atmosphere from the Three-mile Island nuclear power station, near Harrisburg, Pennsylvania, USA. For a few hours a state of emergency was declared. The main anxiety was the formation of a hydrogen bubble above the core which restricted circulation of cooling water. A core melt-down was feared and possibly a massive gas explosion. In the event, the gas bubble subsided and the reactor gradually cooled.

Hawai Volcanoes National Park Established in 1916, in Hawaii, USA, a national park incorporating an active volcanic region on the island of Hawaii including Mauna Loa and Kilauea. The park also has luxuriant vegetation and rare endemic plants and animals.

HAZAN See HAZARD ANALYSIS.

hazard A physical situation with a potential for human injury, damage to property, harm to the environment, or some combination of these as a result of fire, explosion, toxic or corrosive effects, or radiation.

Hazard analysis (HAZAN) Identification of events that could lead to a HAZARD; the analysis of the routes by which a hazard could occur with estimates of the extent, magnitude, and likelihood of harmful effects. Hazard analyses may be carried out at various stages as a design evolves. Early in a project, a preliminary study may be carried out to identify the most serious potential hazards, possibly requiring fundamental design changes.

hazard and operability study (HAZOP) The application of guide words to identify all departures from the original design of an industrial plant that are likely to have undesirable effects on safety or operability. Guide words are lists of words applied to items of equipment or can function as an aid to identifying undesirable deviations. The aim is to facilitate safe and prompt commissioning without extensive last-minute modifications, and to ensure a subsequent trouble-free continuing operation. The study should be carried out by a group of senior representatives from design, project, safety, and operations units.

hazard and risk assessment An essential component of many an ENVIRONMENTAL IMPACT STATEMENT, embracing the potentially adverse effects of a project involving fire, heat, blast, explosion, or flood, arising from a manufacturing plant or transportation system. An assessment reveals hazards to life and limb and property, and is expressed in the form of risk probability. Safety depends both on the location of a plant and the safety precautions and back-up arrangements adopted, together with the degree of training and alertness in the plant. Buffer zones and correct routeing of vehicles are also essential.

hazard indices A checklist method of HAZARD identification which provides a comparative ranking of the degree of hazard posed by particular plant layout and equipment.

hazardous wastes Wastes that by reason of their chemical reactivity or toxic, explosive, corrosive, radioactive, intractable, or other characteristics cause danger, or are likely to cause danger, to HEALTH or the ENVIRONMENT. Examples of such wastes include polychlorinated biphenyls (PCBs), pesticide wastes, inorganic cyanides, asbestos, lead wastes, heavy metals, biological wastes, radioactive wastes, halogenated hydrocarbons, arsenic wastes, dioxins, and wastes which have mutagenic, teratogenic, or carcinogenic effects. (See CONVENTION ON THE CONTROL OF TRANSBOUNDARY MOVEMENT OF HAZARDOUS WASTES AND THEIR DISPOSAL; INTRACTABLE WASTES.)

hazard survey/hazard inventory Undertaken at the conceptual design stage of a new industrial project, the identification of all proposed and existing stocks of hazardous materials or energy resources, with relevant details of the conditions of storage.

hazen units (HU) A unit of measurement for colour in water. A reading of 15 HU would mean a colour just noticeable in a glass.

HAZOP See HAZARD AND OPERABILITY STUDY.

HCB HEXACHLOROBENZENE.

HDR system See HOT DRY ROCK SYSTEM.

health Defined by the WORLD HEALTH ORGANIZATION as 'a state of complete physical, mental, and social wellbeing and not merely the absence of disease or infirmity'. However, most assessments of health still rely upon morbidity and mortality statistics, such as infant- and child-mortality rates and average expectations of life in different countries.

health-related guideline value A concentration or measure of, say, a water-quality characteristic (a physical or chemical property, or the presence of a microorganism or a radionuclide) that does not, in the context of present knowledge, result in any significant risk to the HEALTH of the consumer over a lifetime of consumption.

hearing conservation programme A planned procedure to protect hearing; the components of such a programme may include noise-exposure determinations, engineering noise-reduction surveys, the setting of maximum noise ratings in purchase specifications, noise-control measures, audiometric and medical examinations, the provision and use of personal hearing protective devices, and the provision of adequate instruction to employees,

hearing loss

managers, and employers regarding their roles in the programme.

hearing loss The amount in decibels by which the threshold of audibility for an ear exceeds the normal threshold, the threshold of audibility being the minimum sound-pressure level at which a person can hear a sound of a given frequency. (See ACOUSTIC TRAUMA.)

heat treatment The controlled heating and cooling of metals and alloys after fabrication to relieve internal stresses and improve their physical properties; methods include annealing, quenching, and tempering.

heavy metals (toxic metals) A term loosely applied to a whole range of elements, not all of them metals, which may contaminate the ENVIRONMENT. Heavy metals include antimony, arsenic, beryllium, cadmium, chromium, cobalt, copper, germanium, lead, mercury, molybdenum, nickel, selenium, and zinc. Lead is the most widespread potential hazard; quite small increases in lead consumption can raise blood lead levels to a point posing threats to the nervous system. Cadmium poisoning has occurred in areas in which the soil or irrigation water has been contaminated. Mercury has been a problem in areas where the population eats large amounts of fish taken from contaminated waters or waters with a high natural mercury content. (See ARSENIC; ARSENIC POISONING; BERYLLIUM; CADMIUM; LEAD; MERCURY.)

hectare (ha) The metric area measurement equal to $10\,000$ m^2 or approximately 2.47 acres.

hedonic price technique (property value method) An attempt to assess the value attributed by buyers to the environmental attributes of a dwelling (or other asset). It is a generally accepted fact that, all other things being equal, a house located in a poor environment (broadly considered) will sell for a lower price than a very similar house in a better environment. Thus, the difference in house prices can be used as an estimate of buyers' willingness to pay for a better environment. The differential paid for the superior environment is known as the 'hedonic price'. (See EXTERNALITY.)

helminths Internal parasitic worms of animals.

Helsinki Agreement A declaration agreeing to end by the year 2000 all production and consumption of CHLOROFLUOROCARBONS (CFCs) which damage the OZONE LAYER. The agreement was reached in 1989 by 80 countries at a UN ENVIRONMENT PROGRAM meeting in Helsinki, Finland. (See CONVENTION FOR THE PROTECTION OF THE OZONE LAYER.)

Helsinki convention See CONVENTION ON THE PROTECTION OF THE MARINE ENVIRONMENT OF THE BALTIC SEA.

Helsinki Protocol See CONVENTION ON LONG-RANGE TRANSBOUNDARY AIR POLLUTION.

Henan dams catastrophe In August 1975, the bursting of two dams in southern China (the Banqiao and the Shimantan) located on the Huai River in Henan Province. More than 85 000 people died within two hours of the almost simultaneous failures, and an additional 145 000 perished as a result of epidemics and famine.

HEP HABITAT EVALUATION PROCEDURE.

hepatitis An inflammatory disease of the liver, usually caused by a virus. Hepatitis viruses A and E can be spread in drinking water contaminated with faecal material or sewage effluent.

herbaceous Green or leaflike in appearance or texture; describing those plants which are perennial, living, growing, flowering, and producing seeds for many years. A herbaceous plant is a vascular plant that does not develop persistent woody tissues above ground.

herbicide Any substance or mixture of substances intended for preventing, destroying or controlling any unwanted plants, including algae or aquatic weeds.

herbivore An organism or animal that eats plants or other photosynthetic organisms to obtain its food energy; primary consumers in the FOOD CHAIN. Herbivores form a vital link in the food chain between plants and carnivores, carnivores being animals that eat other animals. (See AUTOTROPH; CARNIVORE; DETRITIVORE.)

heritage Essentially, something passed down from generation to generation, within a family, group, community, race, nation, or simply, collectively.

heritage conservation Measures adopted to restore, preserve, and maintain buildings, townscapes, structures, objects, and landscapes of HERITAGE significance. Legislation may provide for permanent conservation orders to be imposed on heritage items, to prevent their removal or demolition, to ensure careful management and restoration, and to ensure that any works, alterations, and developments are wholly consistent with the heritage qualities being preserved.

heritage criteria Criteria that are used to identify significant buildings and structures in HERITAGE CONSERVATION terms. These include associations with an historical event or person, phenomenon or institution, or an example of an architect's work or of a particular style that is a focal point in a townscape or stretch of countryside, or displays the evolution of technical or planning methods. (See WORLD HERITAGE LIST.)

hertz (Hz) The SI unit of frequency equal to one cycle per second.

heterotroph   An organism that cannot manufacture organic compounds and so must feed on complex organic food materials that have originated in plants or other heterotrophs. All animals are heterotrophic. (See AUTOTROPH; CARNIVORE; DETRITIVORE; HERBIVORE.)

hex   Short for uranium hexafluoride, the gaseous compound of uranium that is used for enrichment by gaseous diffusion and centrifugation. (See RADIOACTIVITY.)

hexachlorobenzene (HCB)   A fungicide for seed grain and vegetable seeds; it is a by-product from the production of CHLORINE gas and CHLORINATED HYDROCARBONS. Residue levels have been found in the fatty tissues of cattle and sheep at the time of slaughter for human consumption.

HFCs   HYDROCHLOROFLUOROCARBONS.

high-level inversion   On occasions a TEMPERATURE INVERSION formed in the ATMOSPHERE well above the Earth's surface. This acts as a lid preventing the ascent and diffusion of chimney plumes and motor vehicle exhaust gases. It is formed by the slow descent of air which becomes warmer by adiabatic compression, since the descent is inevitably into a level of higher pressure. This is also known as a SUBSIDENCE INVERSION. Eventually, thermal convection may bring some of the accumulated pollution to ground level by way of FUMIGATION.

high-level waste   The most highly radioactive waste from nuclear fuel reprocessing containing most of the fission products from spent fuel and typically containing millions of curies per cubic metre. The waste also contains small amounts of unseparated uranium and plutonium.

high-pressure polythene   See POLYTHENE.

high-rise folly   A condemnation of many high-rise residential developments of the 1950s and 1960s in both the inner and outer suburbs of many cities. These developments proved unpopular among families with children, were sometimes structurally unsafe and led to extensive vandalism. High-rise developments took place in Australia, Britain, and Sweden, in state-supported projects around Paris and in public projects in the largest American cities. Cities began pulling down tall housing blocks after less than twenty years' life, or allocated them to students.

high-temperature incineration   Incineration designed essentially to deal with INTRACTABLE WASTES. Incinerators for this purpose operate under stringent conditions aiming at almost complete destruction of the substances introduced. However, public confidence in the performance of such incinerators has yet to be established.

Holford rules   Basic rules proposed by the late Lord Holford, professor of town planning, University College, London, as a guide to minimizing the effects of electricity transmission lines. These rules are set out in Box 15.

Holmepierrepont Inquiry   A public inquiry conducted in Nottingham, England in 1961 into a development application by the Central

---

Box 15   Holford amenity rules for electricity transmission lines

(1) Avoid altogether, if possible, the major areas of highest amenity value, by so planning the general route of the line in the first place, even if the total mileage is somewhat increased in consequence.

(2) Avoid smaller areas of high amenity value or scientific interest by deviation, provided that this can be done without using too many angle towers.

(3) Other things being equal, choose the most direct line, with no sharp changes of direction and thus with fewer angle towers.

(4) Choose tree and hill backgrounds in preference to sky backgrounds whenever possible, and when the line has to cross a ridge, secure this opaque background as long as possible and cross obliquely when a dip in the ridge provides an opportunity.

(5) Prefer moderately open valleys with woods, where the apparent height of the towers will be reduced and views of the line will be broken by trees.

(6) In country which is flat and sparsely planted, keep the high-voltage lines as far as possible independent of smaller lines, converging routes, distribution poles, and other masts, wires, and cables, so as to avoid a concatenation or 'wirescape'.

(7) Approach urban areas through industrial zones, where they exist, and where pleasant residential and recreational land intervenes between the approach line and the substation, go carefully into the comparative cost of undergrounding, for lines other than those of the highest voltage.

Electricity Generating Board to construct and operate a 2000 MW coal-fired power station on a site close to the city. The development application was rejected by the Minister for Power entirely on environmental grounds relating to air pollution, aesthetics, and traffic congestion.

homeostasis A self-regulating process by which biological systems try to maintain equilibrium or stability while adjusting to changing conditions that are optimal for survival. For example, the regulation of population implies a homeostatic mechanism. As a population increases and approaches a point when the physical limitations on nesting sites and shelter will support few more, population growth slows and may even decline; on the other hand, when a HABITAT can support more, the birth rate increases, the death rate declines and the population expands again. Hence, homeostatic processes account in part for long-term fluctuations in animal populations; such fluctuations may also be influenced by weather and climate.

homeothermy In living organisms, the maintenence of a steady body temperature, particularly through the operation of internal physiological mechanisms.

home range The area over which an animal or organism moves in seeking and obtaining its food.

homosphere Generally speaking the ATMOSPHERE, comprising the TROPOSPHERE, STRATOSPHERE, and MESOSPHERE, dominated by a mixture of gaseous NITROGEN and OXYGEN.

horticulture A branch of agriculture concerned with the cultivation of garden crops such as fruit, vegetables, flowers, and ornamental plants.

host An organism or animal on or in which a parasite lives.

hot dry rock systems (HDR systems) Geothermal reservoirs in which rocks are artificially fractured at an accessible depth. Water is injected into the artificial reservoir returning to the surface through a production well as steam or hot water to generate electricity. HDR systems differ from natural GEOTHERMAL ENERGY systems in that there is no natural reservoir of hot water and steam. Successful HDR projects have been completed in the USA, Britain, and France. However, difficulties still remain in creating large artificial heat exchangers at depths of several kilometres having low hydraulic resistance. However, HDR systems offer a route for converting the limitless energy of the Earth's core for human and environmental purposes.

Huascaran National Park Established in 1966, in Peru, a mountainous region of the Cordillera Blanco; the park's special feature is the Nevada Huascaran rising to 7000 metres.

hum, environmental A disturbing sound with a frequency of about 40 Hz heard by individuals sensitive to this range, but not to the rest of the population. It may be caused by industrial or traffic noise, or fast-flowing high-altitude masses of air.

human development index Introduced in 1990 by the UNITED NATIONS DEVELOPMENT PROGRAM (UNDP) as an alternative measure of economic and social progress, GROSS DOMESTIC PRODUCT (GDP) being a poor measure of relative living standards and conditions. The index is a cocktail of life expectancy, adult literacy, years of schooling, and GDP per capita measured at purchasing-power parity. The UNDP calculates the index for most countries in the world. Some countries rank low in GDP per capita, but much higher in terms of the index; others rank much lower on the index than their per capita GDP might suggest. For rich countries, the differences in ranking tend to be smaller, there being little difference in the cases of Canada, Japan, France, the USA, and Britain. The poorer countries reveal much greater diversity.

human ecology The collective interaction of humanity with the ENVIRONMENT; influenced by the work of biologists on the interactions of organisms with their environments.

humus A complex organic component of the SOIL, resulting from the decomposition of plant and animal tissue; it gives soil its characteristically dark colour and is of great importance for plant growth. The term is also used for a dark brown or almost black complex organic material residue left after the completion of biochemical processes in sewage treatment works.

humus tank A final settlement tank through which the effluent from a percolating filter or aeration tank in a sewage treatment plant flows for the purpose of separating settleable solids or humus.

hydrocarbons Chemicals whose molecules contain only carbon and hydrogen atoms. Crude oil is largely a mixture of hydrocarbons. They are subdivided into aliphatic and cyclic hydrocarbons, according to the arrangement of the carbon atoms in the molecule. The aliphatic hydrocarbons are in turn subdivided into: (1) alkanes (paraffins); (2) olefins or alkenes; (3) diolefins and others, according to the number of double bonds in the molecule. The cyclic hydrocarbons are subdivided into: (1) aromatics; (2) naphthenes or cyclo-paraffins. In all types of hydrocarbons, hydrogen atoms may be replaced by other atoms making the formation of a virtually endless number of compounds possible. Hydrocarbons are found commonly in fossil fuels and in the products of the partial COMBUSTION of these substances as in the exhaust gases of petrol-driven vehicles.

hydrochlorofluorocarbons (HCFCs) Chemicals used to replace CHLOROFLUOROCARBONS (CFCs), most notably HCFC-123 instead of CFC-11, as industrial refrigerants and as blowing agents in the manufacture of plastic foam. While HCFC-123 appears less damaging to the ozone layer compared with CFC-11, many authorities continue to investigate its occupational health, public health, and environmental effects. Workers may be exposed to HCFC-123 during manufacturing, installing and maintenance operations. Exposure to high levels of HCFC-123 are known to cause acute effects, such as cardiac sensitization and asphyxiation. Other substitutes for CFCs include hydrofluorocarbons (HFCs), which are also under continuing investigation. (See CONVENTION FOR THE PROTECTION OF THE OZONE LAYER; MONTREAL PROTOCOL ON SUBSTANCES THAT DEPLETE THE OZONE LAYER.)

hydroelectric power The use of water power to drive water turbines, which in turn drive electricity generators. The degree of utilization of water power varies from nil in some countries to over 50% in Switzerland.

hydrogen A colourless odourless gas, the lightest substance in existence, extremely FLAMMABLE, combining with oxygen to form water with the liberation of heat.

hydrogenation A chemical reaction with HYDROGEN to yield a product containing an increased proportion of hydrogen; for example, the addition of hydrogen to coal at elevated temperature and pressure increases the hydrogen/carbon ratio, approaching that of petroleum, yielding a motor spirit.

hydrogen economy A concept for the future, in which HYDROGEN is generated on a vast scale, possibly from the electrolysis of water. It has been proposed that hydrogen be used instead of petrol (gasoline) for internal combustion engines, as a source of energy in fuel cells, in the home for domestic purposes and as a feedstock for hydrogen-based industrial processes. The concept envisages a very cheap production process for hydrogen, not yet realised.

hydrogen sulphide A colourless somewhat dense gas, with a characteristic foul odour of rotten eggs. It arises in the decomposition of organic matter; other sources include oil refineries, sulphur recovery plants, some metallurgical processes, and various chemical industries using sulphur-containing compounds. It is associated with GEOTHERMAL ENERGY in New Zealand. Hydrogen sulphide tends to be a localized problem; in ordinary combustion processes the gas is readily burned to produce SULPHUR DIOXIDE. It may be formed in water, however, by sulphate-reducing microorganisms or hydrolysis of soluble sulphide under anoxic conditions.

hydrography The art and science of compiling and producing charts and maps of water-covered areas of the Earth's surface.

hydrological cycle The continual exchange of water between the Earth and the ATMOSPHERE. Of the many processes involved, the most

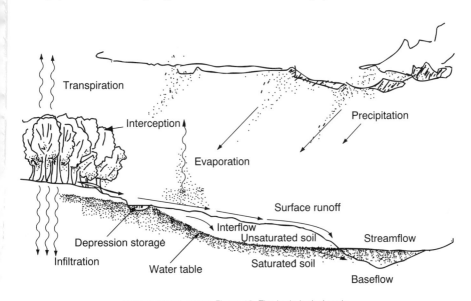

HYDROLOGICAL CYCLE: Figure 13. The hydrological cycle.

hydrology

important are evaporation, transpiration, condensation, precipitation, and runoff. While the total amount of water within the cycle remains essentially constant, its distribution between the various processes is continually changing. (See Figure 13.)

hydrology The scientific discipline concerned with the waters of the Earth, including their occurrence, distribution, and circulation through the HYDROLOGICAL CYCLE and the interactions with living things. It also deals with the physical and chemical properties of water in all its phases. Hydrology has as its primary aim the study of the interrelationship between water and its environment.

hydrolysis The breaking down of a substance into smaller molecules, through a chemical reaction with water or its ions. The hydrolysis of carbohydrates is of major importance in the food and brewing industries. (See HYDROGEN ECONOMY.)

hydrometeorology A branch of METEOROLOGY that deals with problems involving the HYDROLOGICAL CYCLE, particularly rainfall statistics. The results often serve as a basis for the design of flood-control and water-usage products such as dams and reservoirs.

hydrophyte A plant adapted to live in water or in waterlogged soil.

hydroponics The cultivation of plants in nutrient-rich water, with or without mechanical support. A wide variety of vegetables and florist crops can be grown in this way. (See AQUACULTURE.)

hydrosphere That part of the Earth comprising water. It includes oceans, lakes, and rivers, all liquid and frozen surface waters, GROUND WATER held in soil or rock, and atmospheric water vapour. The components of the hydrosphere and the HYDROLOGICAL CYCLE itself have been significantly modified by the activities of contemporary society. For example, the discharge of industrial and domestic wastes have contaminated rivers and streams, while sewage and fertilizer effluents have caused EUTROPHICATION. Global climatic changes are associated with GREENHOUSE GASES and the GREENHOUSE EFFECT.

hydrostatic pressure The pressure exerted by water at rest, at any point within the water body, as, in the case of GROUND WATER, the pressure caused by the weight of water in the same zone of saturation above a given point.

hygiene The science of the preservation of HEALTH and the prevention of disease. It is concerned with personal cleanliness, SANITATION, public health measures, the proper handling of food, anti-vector measures, clean and safe drinking water, efficient garbage collection and disposal, protective clothing, anti-litter measures, anti-infection measures, and necessary protective packaging.

hyperkinesis Excessive motility (activity or movement) of a person, or of muscles.

hypolimnion The bottom stratum of water in a lake, below the THERMOCLINE which, like the EPILIMNION, shows a temperature gradient of less than 1°C per metre of depth.

I

IAIA INTERNATIONAL ASSOCIATION OF IMPACT ASSESSMENT.

IATAFI INTERNATIONAL ASSOCIATION OF TECHNOLOGY ASSESSMENT AND FORECASTING INSTITUTIONS.

IAEA INTERNATIONAL ATOMIC ENERGY AGENCY.

IBP INTERNATIONAL BIOLOGICAL PROGRAM.

IBRD INTERNATIONAL BANK FOR RECONSTRUCTION AND DEVELOPMENT.

ice age or **glacial age** Any geological period during which thick ice sheets have covered vast areas of the Earth; such periods may last several million years. The earliest known ice age took place during the Precambrian time dating back more than 570 million years. The most recent period occurred during the Pleistocene epoch (about 1 600 000 to only 10 000 years ago). A lesser, quite recent, glacial age was known as the 'Little Ice Age' beginning in the 16th century and advancing and receding over three centuries. Its worst phase was reached in about 1750. These phases of cooling and warming over the longer term have complicated the discussion regarding the GREENHOUSE EFFECT.

ichthyology The scientific study of fishes; it has a number of specialized sub-disciplines including taxonomy, anatomy, behavioural science, ecology, physiology, and economics. Much fish research is conducted in government laboratories, who are responsible for the management of FISHERIES and their sustained productivity.

ICOLD INTERNATIONAL COMMISSION ON LARGE DAMS.

ICRP INTERNATIONAL COMMISSION ON RADIOLOGICAL PROTECTION.

ICSU INTERNATIONAL COUNCIL OF SCIENTIFIC UNIONS.

IDA INTERNATIONAL DEVELOPMENT ASSOCIATION.

IEA INTERNATIONAL ENERGY AGENCY.

IFC INTERNATIONAL FINANCE CORPORATION.

igneous rock Crystallized or glassy rock formed by the cooling and solidifying of molten earthy material; one of three principle classes of rocks, the others being metamorphic and sedimentary. The different types of igneous rocks vary considerably in composition.

ignition The beginning of COMBUSTION; ignition is effected by raising the temperature of a fuel to a point at which the rate of burning provides the heat essential for the process to continue. Burning does not occur below this temperature, which varies with different fuels, and also requires the presence of oxygen.

IGY INTERNATIONAL GEOPHYSICAL YEAR.

IIED INTERNATIONAL INSTITUTE FOR ENVIRONMENT AND DEVELOPMENT.

illiteracy rate Generally defined as the percentage of the population aged 15 or older who cannot read and write a short simple statement about everyday life. The OECD countries have the lowest levels of illiteracy, while illiteracy is widespread in parts of Asia, Africa, and South America. Illiteracy is another grave handicap to the emergence of participatory democracy.

ILO INTERNATIONAL LABOUR ORGANIZATION.

IMF INTERNATIONAL MONETARAY FUND.

Imhoff tank A primary sewage treatment process combining sedimentation with ANAEROBIC sludge digestion in one tank; there is no mechanical equipment. The sewage flows through the upper or sedimentation chamber, the deposited sludge passing through slots into a digestion chamber below.

immission A term commonly used in Europe referring to the level of pollution to which a receptor is exposed. It must be clearly distinguished from the word EMISSION, which indicates what is discharged and not the level to which a receptor is exposed.

immunology The study of the body's resistance to invasion by other organisms; that is the study of immunity. Immunology became established around 1900 when it was finally recognized that proliferating microorganisms in the body cause many infectious diseases and that the body has cellular and chemical components that recognize and destroy foreign substances within the body. This emerging science was associated with highly successful techniques of immunization that could mobilize and stimulate the body's natural defences against infectious diseases. Immunology has gained a comprehensive understanding of the formation, mobilization, action, and interaction of antibodies and antigens.

IMO INTERNATIONAL MARITIME ORGANIZATION.

impairment A term used in medicine for anatomical loss, or loss of bodily or intellectual functioning; it is also used in the context of environmental damage and loss of AMENITY.

imperfect competition A market situation in which neither absolute monopoly nor perfect competition prevails; a situation closest to real life in most circumstances. It is characterized by: the ability of sellers to influence demand by product differentiation, branding and advertising; restriction of entry of competitors into various lines of production and service, either because of the initial investment required or because of restrictive practices; the existence of uncertainty with imperfect knowledge of profits earned in similar or other lines of production; and the absence of price competition in varying degrees.

inceptisol A type of SOIL in which the soil horizons are just beginning to develop; in terms of land area, it is ranked third among the ten worldwide soil types. They are found in every climatic zone being especially common in the Arctic tundras of North America, Europe, and Asia; they are also found associated with volcanic ash in the tropics. Inceptisols are widely used in the tropics for the cultivation of coffee and sugar cane.

incidental take The unintentional capture of other animals or plants during commercial harvesting for a particular species. The term often applies to fishing operations where non-usable species are caught in nets and discarded.

incineration A method of treating wastes and rubbish (garbage) by burning, to reduce the volume and weight of the waste and to leave an innocuous residue. Incinerator plant for domestic wastes are usually built in places where suitable sanitary LANDFILL facilities are not available or too distant from the centre of generation of the waste, making transfers and transportation of the waste a more costly proposition. An incinerator may also be installed to deal with intractable and hazardous wastes in relatively isolated locations, Incinerators are commonly used on industrial sites to destroy gaseous, solid, and liquid wastes. Destruction by incineration is essential for hospital wastes.

incipient lethal level That level of a toxic substance beyond which 50% of a population cannot live for an indefinite period of time. Such levels denote threshold concentrations.

incompatible use Land uses, situated in relation to each other in such a way as to be regarded by many as in conflict; for example, reducing the environmental and social quality of residential areas. Incompatible uses may exist as a consequence of EXISTING USE rights, and may be eliminated only slowly over time. Examples of incompatible uses include factories or scrap yards in residential areas, mining or quarrying inside national parks, and heavy trucks passing hospitals and schools.

indenture agreements Contractual arrangements between governments and developers, often called FRANCHISE AGREEMENTS.

India, evolution of environmental policy See Box 16.

indicator A substance or organism used as a measure of air or water quality, or biological or ecological wellbeing. (See AIR POLLUTION; BIOLOGICAL BENCHMARK; WATER POLLUTION.)

indicator species Those SPECIES in a HABITAT which are most sensitive to slight changes in environmental factors. When identified, their decline can serve as an early warning of the endangerment of the HEALTH of the community.

Box 16 India: evolution of environmental policy

1908 Koziranga National Park, Assam, established
1935 Corbett National Park, Uttar Pradesh, established
1972 National committee on environmental planning and coordination formed to examine, *inter alia*, the environmental implications of development projects and to set standards and guidelines to safeguard environmental quality; Silent Valley project abandoned to save rich ecological resources
1972 Chipko Andolan Movement (tree hugging) movement begins in Uttar Pradesh, as a protest against the cutting down of trees
1979 Social forestry programme launched in Uttar Pradesh, the project being World Bank financed
1980 Department of Environment established; Forest Conservation Act
1981 Air Pollution Control Act
1984 Bhopal disaster: a catastrophic leak of methylisocyanate at a pesticide plant results in over 2000 deaths
1986 Introduction of EIA procedures; Environment Protection Act
1987 Air Pollution Control Act
1988 Technology impact statements introduced
1992 Sardar Sarova water projects initiated
1993 India decides not to seek further World Bank loans due to disagreements over benchmarks; the projects would proceed, however. 100 000 people are to be resettled

indifference analysis An important technique in ECONOMICS, analysing consumer behaviour, whereby choices are made between products and services, more being consumed of some and less of others, theoretically to a point where a consumer may be indifferent between purchases, all being equally desirable.

indigenous Literally, native to a particular region or country; not introduced from the outside. A term used, for example, to differentiate the status of Aborigines as the original inhabitants of Australia from the status of people who have settled in Australia during the past two centuries.

individual risk The frequency at which an individual may be expected to sustain a given level of harm from the realization of specific hazards, as distinct from societal risk.

individual risk criteria Criteria relating to the likelihood with which an individual may be expected to sustain a given level of harm from specified hazards. (See HAZARD.)

Indonesia: evolution of environmental policy See Box 17.

indoor air pollution The levels of pollutants to which people indoors are exposed resulting from: the use of fuels for heating, cooking, and lighting; cooking processes; facilities for ventilation; and exposure to radiation. Studies have shown marked levels of air pollution in some circumstances.

industrialization The evolution of forms of production characterized by: increasing capital intensity; an intensive division of labour; increasingly complex forms of transport; increasingly complex forms of industrial organization; increasing interdependence between persons and groups; growth of financial instruments; growth of marketing and shopping centres; growth of complex labour, trade-union, and corporation laws; and, generally speaking, in the longer term, an improvement in the standard of living and QUALITY OF LIFE of most people but presenting contemporary problems relating to SUSTAINABLE DEVELOPMENT, the GREENHOUSE EFFECT, INTERGENERATIONAL EQUITY, INTRAGENERATIONAL EQUITY, the control of air, water, and noise pollution and the management of hazardous wastes.

industrial-waste strategy A strategy that outlines policies and procedures to be adopted for industrial-waste management. It seeks to achieve the following goals: (1) to minimize the creation of waste material by cleaner production methods and alternative techniques; (2) to provide proper facilities for the storage, treatment, and disposal of wastes; (3) to improve the management of hazardous and intractable wastes; (4) to reduce the environmental impact of waste-disposal operations; (5) to minimize health and safety risks; (6) to involve the government and the public in the evolution of waste-management policies and

---

Box 17 Indonesia: evolution of environmental policy

1978 Environmental policy guidelines endorsed by a General Session of the Consultative People's Assembly in order to: (1) safeguard natural resources vital to development; (2) promote EIA development projects, sectorally and regionally; (3) ensure the rehabilitation of impaired natural resources; and (4) secure an improvement in the living environment of low-income people

1982 Environmental Management Act, providing a legal basis for Indonesian environmental policy embracing natural resources, environmental management, conservation, pollution control, and EIA

1986 Regulations for EIA (AMDAL) introduced, providing a framework for comprehensive EIA in Indonesia

1987 EIA procedures established by all Indonesian ministries

1989 National clean waters programme inaugurated, a number of key river systems being targeted

1990 Water pollution control regulations; Environmental Impact Management Agency (BAPEDAL) created to develop and implement policies on pollution control, hazardous waste management and EIA

1993 Legislation requiring all companies to produce environmental audit reports on a five-year cycle, the only system worldwide of compulsory environmental audit

1994 The World Bank finances a programme for the supply of clean and safe water and sanitation services for about 1.5 million low-income people in Indonesia

industrial zones

facilities; (7) to provide a training programme for those involved in the strategy.

industrial zones   Zones allocated for industry within a town-planning scheme or environmental plan. The range of industries accommodated in a plan may include: (1) light industry, being industry that does not cause any interference to the AMENITY of any area and which does not impose a greater-than-normal load on any public utility or road; (2) service industry, being industry that provides transport, repair, and maintenance services for other industry; (3) general industry, not of a hazardous or offensive nature; (4) hazardous, noxious or offensive industry which could constitute a danger to persons, properties, or the ENVIRONMENT; (5) waterfront industry, requiring direct access to a river, stream, or other body of water; (6) extractive industry, industry that takes material from the ground with subsequent processing. Standards are usually defined for industrial zones relating to access and roads, drainage, car parking, aesthetics, landscaping, buffer zones, noise levels, and air and water pollution.

industry   In its widest sense all the manifold activities of a country which provide employment in the production of goods and services. Specifically, an industry comprises all those activities directed to the production of a given class of goods or services, for example minerals, automobiles, agricultural produce, or tourism. An industry may be composed of many business units as in agriculture, or perhaps only one business unit as in the case of steel. An individual company may make a wide range of dissimilar goods and be participating in the activities of several industries.

industry, optimal location of   Industrial location considered the most efficient, having regard to natural resource, labour, industrial, social, and national costs. Location is determined by the complex interaction of factors such as: the availability and costs of skilled labour and raw materials; access to markets for the finished products; transportation costs; the prospects for future expansion and the long-term availability of essential inputs; government policies, incentives, charges, and taxes. Governments may encourage the establishment of industry through subsidies, or through the provision of INFRASTRUCTURE. From time to time, public and private interests differ as when, for example, a mining company finds its most profitable location in the middle of a national park, or when a processing company finds its best location next to a residential district, or where the establishment or expansion of an airport will create or intensify noise problems over urban districts.

inert gases   Gases such as nitrogen and CARBON DIOXIDE that do not support COMBUSTION.

inert gas system   A system that permits oil tankers to replace the inflammable gases in ullage spaces above cargo when loaded and those that remain in empty cargo tanks with gases that are low in oxygen and therefore non-combustible. Frequently, ship's boiler gas is cleaned and piped to cargo tanks. The installation of inert gas systems has greatly reduced the chances of cargo tank explosions, which can injure crew and endanger vessels. (See Figure 14.)

infant industry   A young and growing industry, often considered not quite able to stand on its own feet in the face of foreign competition. Such industries have often been afforded the protection of a tariff or have been granted a subsidy until strong enough to stand alone. Experience suggests, however, that tariffs and subsidies, once introduced, are not easily removed due to less efficient companies coming into being behind the tariff. (See GENERAL AGREEMENT ON TARIFFS AND TRADE.)

infant mortality rate   The annual deaths of infants aged under one year per thousand live births. An early crude indicator of social welfare, it continues to be a valuable indicator of those aspects of the QUALITY OF LIFE impinging on HEALTH and the development of health services generally. The infant mortality rate in advanced countries is about seven deaths per thousand or less and in poorer developing countries about 20 to 85 deaths per thousand.

infill development   See URBAN CONSOLIDATION.

information and documentation system for environmental planning   (UMPLIS)   A highly coordinated systematic procedure developed in Germany to achieve purposeful planning and monitoring of government-sponsored environmental research. It is run by the Federal Environmental Agency.

INFOTERRA   See INTERNATIONAL REFERRAL SYSTEM.

infrared photography   Photography based on exposure to infrared radiation, outside the visible light spectrum. From the air, such photography reveals details of waste-heat distribution, waste discharges, atmospheric conditions and the pattern of damage to vegetation that cannot be recorded by conventional photography.

infrasound   Sound whose frequency is below the range of audio frequency; infrasound has frequencies below 100 Hz and is consequently not catered for on the dB(A) scale. Insulation is rather ineffective in this frequency range, the only way of abating infrasound being to eliminate it at source.

infrastructure   The framework of key facilities which support communities and their industrial and commercial activities, including services from: (1) public utilities providing electrical power, telecommunications,

interaction matrix

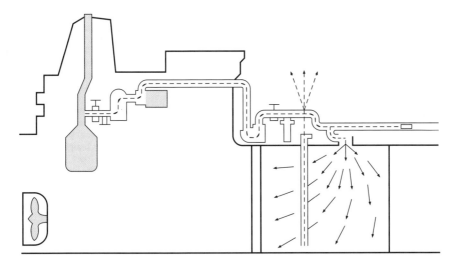

INERT GAS SYSTEM: Figure 14. Inert gas system for oil tankers
The flue gas is conducted from the steam generators through cleaning systems and into the tanks, while air is expelled through an outlet pipe
Source: IMO.

piped water supplies, drainage and sewerage, solid-waste collection and disposal, and piped gas: (2) public works providing roads, urban and intercity railways, other urban transport, ports, waterways, dams, and airports; and (3) schools, colleges, and universities, health services and hospitals, entertainment facilities, and community support services generally. Infrastructure is an umbrella term sometimes referred to as 'social overhead capital'.

injurious affection   A term referring to the depreciation in the value of land caused by the adverse effects of public works through noise, vibration, smell, smoke, fumes, artificial lighting, overshadowing, loss of support, and restriction or loss of access. Injurious affection is usually associated with the carrying out of substantial public undertakings such as the construction of a motorway or an airport. Land around an airport may fall in value due to aircraft noise. In some cases, injurious affection may result in compensation, though usually only when the land is acquired for the project. However, under British legislation compensation is not so limited, being payable outside the actual acquisition area.

inorganic pesticide   A PESTICIDE which does not contain carbon, such as copper sulphate, sodium arsenate, sulphuric acid, and salt.

inorganic solid waste   Generally, waste of a non-biological origin; for example, bottles, tins, and plastics.

insect vector   An insect that transmits a disease-inducing organism or agent such as the mosquito transmitting malaria.

Institute of Terrestrial Ecology   A component body of the NATURAL ENVIRONMENT RESEARCH COUNCIL, based in Cambridgeshire, England. Its work encompasses environmental management with particular reference to ENVIRONMENTAL IMPACT ASSESSMENT, HABITAT restoration, nature conservation, and the development of policy for the natural environment.

intangibles   See COST–BENEFIT ANALYSIS.

integrated logging   The harvesting of both pulpwood and sawlogs in a single operation.

Integrated pollution control (IPC)   Introduced into Britain under the Environmental Protection Act 1990, a procedure whereby all major emissions to land, air, and water are considered simultaneously and not in isolation to avoid situations in which one control measure for one medium adversely affects another. Thus, an air pollution control measure may result in water pollution and excessive noise. The BEST AVAILABLE TECHNIQUES NOT ENTAILING EXCESSIVE COSTS are required to minimize pollution of the environment as a whole. Under the Act, Her Majesty's Inspectorate of Pollution (HMIP) are able to recover the costs incurred in operating the Integrated Pollution Control programme.

intensive animal industry   An industry in which many animals are concentrated in a small area being fed by humans rather than free-ranging; it includes cattle feedlots, piggeries, and poultry farms.

interaction matrix   One of the earlier methodologies used in the USA and elsewhere to examine the environmental impacts and effects of development projects. The matrix consists of a display of project actions or activities along one axis, with appropriate environmental factors being listed along the other axis. When a given action or activity is anticipated to cause a change in an environmental factor, this is noted at the interaction point in the matrix and further described in terms of magnitude and importance. Many variations of the simple interactive matrix have been utilized in environmental impact studies. (See ENVIRONMENTAL IMPACT ASSESSMENT.)

interest   See DISCOUNTED CASH FLOW; PRESENT VALUE OR WORTH.

interest group   See STAKEHOLDER.

intergenerational equity   A concept that those living today should not compromise or restrict the opportunities open to future generations. It envisages a partnership among all the generations that will expect to thrive on the world's resources. It implies, at the very least, that each generation should hand over to the rising generation a world in as good an order as possible and with the full benefits of SUSTAINABLE DEVELOPMENT. The WORLD COMMISSION ON ENVIRONMENT AND DEVELOPMENT in *Our Common Future* (1987) stated that: 'Sustainable development is development that meets the needs of the present without compromising the ability of future generations to meet their own needs'. Problems arise with the application of the concept. Clearly, a treeless barren planet scorched with ultraviolet radiation and littered with radioactive waste, would not meet the criteria. Yet, considering more likely outcomes, it is impracticable to envisage the circumstances of generations as yet unborn. One has only to ask whether anyone living in 1795 could predict the needs and aspirations of those living in 1895; and whether anyone living in 1895 could predict the needs and aspirations of those living in 1995 (recalling in that period alone the development of motor cars, aircraft, telecommunications, computers, satellites, medical and surgical techniques, industrial productivity, and a general advance in the standard of living). It is equally difficult in 1995 to envisage the changes that will occur by 2095, or to identify the needs and aspirations of that generation. Intergenerational equity is therefore a noble concept, which cannot be translated into any practical policies other than the pusuit now of responsible ENVIRONMENTAL MANAGEMENT and SUSTAINABLE DEVELOPMENT.

Intergovernmental Panel of Experts on Radioactive Wastes   A body established by the CONVENTION ON THE PREVENTION OF MARINE POLLUTION BY THE DUMPING OF WASTES AND OTHER MATTER (the London convention). From 1993, there was an indefinite moratorium on all radioactive dumping at sea.

Intergovernmental Panel on Climate Change   A panel created by the WORLD METEROLOGICAL ORGANIZATION and the UN ENVIRONMENT PROGRAM to assess scientific information on climatic change and its environmental and socio-economic consequences and to formulate responses to such change. The panel released its first assessment report in 1990; the report was discussed at the 45th General Assembly of the UN, where the main findings were accepted. The panel predicted a rate of increase of global mean temperature during the next century of 0.3°C per decade, without remedial measures, with sea-level rises of between 3 and 10 cm per decade. By 2030, the global mean sea level might rise by 20 cm and might reach 65 cm by the end of 3000. The panel said its predictions were limited by: a lack of understanding of greenhouse gas sources and sinks; of clouds and oceans and how they influence climate change; and of polar ice sheets and their impact on sea-level rises.

internalization of environmental costs   A central principle of SUSTAINABLE DEVELOPMENT; it implies that as far as practicable the environmental costs of a venture shall be borne by the initiator through pollution-control measures at source, permit or licence fees, contributions to INFRASTRUCTURE, provision of landscaping and buffer zones, acquisition of properties, compensation to members of the public, and environmental levies and taxes. All such measures appear in the accountancy costs of the enterprise, while the adverse environmental effects on the public are minimized or abated. (See EXTERNALITY; POLLUTER-PAYS PRINCIPLE; PRINCIPLES OF SUSTAINABILITY.)

International Association of Impact Assessment (IAIA)   Established in 1980, a voluntary body intended to bring together EIA administrators and researchers. Some 1000 members reside in over 50 countries, representing a whole range of disciplines and professions. The essential goal of IAIA is to advance the science and art of EIA and to foster its application around the world.

International Association of Technology Assessment and Forecasting Institutions (IATAFI)   Formally established in 1993 under the auspices of the UN with headquarters in Bergen, Norway, a body intended to link developing countries to technology assessment capabilities in the developed world, and to promote technology assessment, including environmental impact assessment. The founding institutions include the US Office of Technology Assessment, the China National Research Centre for Science and Development, the Polish State Committee for Scientific Research, the Russian Academy of Sciences, and the Pan-African Union for Science and

Technology. The 1995 workshop was held in Budapest, Hungary and the 1996 conference in Brussels, Belgium.

International Atomic Energy Agency (IAEA) Established in 1957, an agency of the United Nations with headquarters in Vienna, Austria, its primary purpose is to promote the peaceful uses of nuclear energy and to ensure that nuclear material is not diverted to the manufacture of nuclear weapons. The IAEA also promotes research in the application of atomic energy to medicine, agriculture, water location, and industry. The success of IAEA in the international arena depends very much on the cooperation of the countries participating in the safeguards programme with free access for the IAEA nuclear inspectors. Difficulties arise when individual nations attempt to secretly manufacture nuclear weapons.

International Bank for Reconstruction and Development (IBRD)   See WORLD BANK.

International Biological Program (IBP)   See INTERNATIONAL COUNCIL OF SCIENTIFIC UNIONS.

International Charter for the Conservation and Restoration of Monuments and Sites   A charter offering guidelines in relation to items of environmental heritage and their preservation, maintenance, restoration, reconstruction, and adaptation. It outlines the principles and aims of conservation in relation to historic buildings and places, and examines the processes through which conservation may be achieved. The charter was adopted in Vienna in 1966 by UNESCO.

International Commission for the Protection of the Rhine   A commission created by the riparian countries in 1950 for the protection of the River Rhine against pollution. The commission includes representatives from France, Germany, Luxembourg, the Netherlands, and Switzerland. The secretariat is based in Koblenz, Germany. The commission was given legal status by the CONVENTIONS FOR THE PROTECTION OF THE RHINE AGAINST POLLUTION (the Bonn conventions).

International Commission on Large Dams (ICOLD)   An international body concerned with the geophysical and environmental problems which may arise during and after the construction of large dams. A large dam is defined as a dam that is more than 150 m in height, or whose volume is more than 15 million cubic metres, or whose reservoir has a capacity of more than 15 billion cubic metres. Large dams are linked with increased earthquake frequency in seismically active areas, with ecological effects downstream and possible increases in water-borne diseases. ICOLD, which holds regular international congresses, publishes reports on the environmental consequences of dam projects based upon experiences throughout the world.

International Finance Corporation

International Commission on Radiological Protection (ICRP)   A commission founded in 1928 in Stockholm to set international standards for radiation protection and to study their effects. About 50 countries are represented on the commission. ICRP has concluded that there is no wholly safe dose of radiation; it believes that the policy of assuming some risk at low doses is the most reasonable basis for radiation protection.

International Conference on the Economics of Climate Change   A conference convened in June 1993 following the UN CONFERENCE ON ENVIRONMENT AND DEVELOPMENT by the OECD and the International Energy Agency to help ensure that the UN CONVENTION ON CLIMATE CHANGE was based on sound economic reasoning, and to help OECD governments to consider practical options and priorities for action. The conference was attended by about 250 economists and policy makers. The conference received the assessment report of the INTERGOVERNMENTAL PANEL ON CLIMATE CHANGE.

International Conventions   See CONVENTION linked to the appropriate subject.

International Council of Scientific Unions (ICSU)   A non-government body with a membership drawn from universities and research institutes throughout the world; many members are environmental scientists. Since the 1950s, ICSU has planned, promoted and successfully developed a series of research programmes of diverse character, including the International Biological Program (1964–74), the International Geophysical Year (1957–59), the International Geosphere-Biosphere Program (1986–96), and the long-term Global Atmospheric Research Program.

International Court of Justice   The principal judicial organ of the United Nations; the seat of the court is at The Hague. The function of the court is to pass judgement on disputes between nations, hence only nations can be represented before the court.

International Development Association (IDA)   See WORLD BANK.

International Energy Agency (IEA)   An autonomous body established within the framework of OECD to implement an international energy programme, adopted by the participating countries in 1974. IEA has issued a list of recommended energy conservation measures covering pricing and marketing, public education, and technical and engineering measures. The basic aim of the IEA group of countries is to bring about a better structure of energy supply and demand over the longer term, supporting the objective of SUSTAINABLE DEVELOPMENT.

International Finance Corporation (IFC)   See WORLD BANK.

International Fund for Agricultural Development (IFAD) A United Nations agency based in Rome that supports increased food production in the poorer countries; it was created by the World Food Conference in 1974. The richer countries are required to provide revenue for the fund. IFAD seeks, not only to increase food production in developing countries, but also to provide agricultural employment and reduce malnutrition.

International Geophysical Year (IGY) See INTERNATIONAL COUNCIL OF SCIENTIFIC UNIONS.

International Institute for Environment and Development (IIED) Based in London, England, an independent, non-profit-making organization which promotes sustainable patterns of world development through research, services, training, policy studies, consensus building, and public information. Special attention is paid to the economics of SUSTAINABLE DEVELOPMENT such as: (1) the valuation of the costs and benefits of environmental goods and services, particularly in relation to the poor; (2) analysis of the impact of economic policy on natural resource use and environmental management; and (3) the development of appropriate economic incentives linking environmental conservation with poverty alleviation. Attention has also been paid to the economics of sustainable forestry, waste-paper recycling, and the linkages between international trade and the environment. The aim is to influence policy and practice on the ground.

International Labour Organization (ILO) Established in 1919, an intergovernmental agency in which representatives of governments, employers, and employees participate to improve occupational and working conditions. The ILO was recognized by the United Nations in 1946 as a specialized agency. Within the context of the working environment, the ILO has produced codes and guidance in respect of air pollution, noise and vibration, occupational cancer, chemical hazards, and control of risks. It cooperates closely with the WORLD HEALTH ORGANIZATION.

International Maize and Wheat Improvement Center Known by its Spanish acronym CIMMYT, a global leader in scientific research and training on maize and wheat, and in economics and natural resource management research related to these crops. Focusing on the needs of the poor in developing countries, CIMMYT generates improved maize and wheat germplasm, as well as other technologies, that increase agricultural productivity and preserve the natural resource base. CIMMYT is based in Mexico City, USA, with 15 regional offices.

International Maritime Organization (IMO) An arm of the United Nations system, a body with far-reaching responsibilities for maritime safety and regulation. A Marine Environment Protection Committee was established in 1973 to address the issues of pollution discharges from ships, to provide practical measures to deal with pollution when it occurs and to provide the means of compensation for pollution damage. (See CONVENTION FOR THE PREVENTION OF MARINE POLLUTION BY DUMPING FROM SHIPS AND AIRCRAFT; CONVENTION FOR THE PREVENTION OF POLLUTION FROM SHIPS; CONVENTION ON THE PREVENTION OF MARINE POLLUTION BY THE DUMPING OF WASTES AND OTHER MATTER.)

International Monetary Fund (IMF) An international financial institution set up in 1945 as the result of the Bretton Woods Agreement (along with the WORLD BANK). The IMF gives balance-of-payments assistance to countries whose reserves have come under pressure. In 1994, the IMF had over 150 members and a permanent staff of over 2000. In 1986, the IMF and the World Bank began a new multi-billion-dollar lending pool to serve the world's poorest nations.

International Organization for Standardization (ISO) Founded in Geneva, Switzerland, in 1946 a specialized international organization concerned with standardization in most technical and non-technical fields. Beyond the technical, ISO 9000 established international management quality standards. In 1994, the ISO issued ISO 14 000, a set of environmental management guidelines that can be adopted by virtually any organization in the world.

International referral system (INFOTERRA) An environmental information system created by the UN CONFERENCE ON THE HUMAN ENVIRONMENT 1972. The system, which became fully operational in 1977, aims to link those who seek information on environmental issues with those best able to provide such information.

International register of potentially toxic chemicals (IRPTC) A component of the Earthwatch programme maintained by the UN Environment Program (UNEP), a programme initiated by the UN CONFERENCE ON THE HUMAN ENVIRONMENT 1972. The register collates information on the most potentially hazardous and toxic chemicals in use in the world, complementing the services of the INTERNATIONAL REFERRAL SYSTEM (INFOTERRA).

International Seabed Authority An organization based in Kingston, Jamaica, created by the CONVENTION ON THE LAW OF THE SEA (the Montego Bay convention) to oversee and regulate the exploitation of the rich mineral resources of the oceans through a licensing system.

International Tropical Timber Agreement (ITTA) In 1983, a trade agreement, establishing a system of consultation and cooperation between consuming and producing countries. The agreement has the support of some 47 countries, representing about 95% of the international trade in tropical timber. The ITTA recognizes in its charter the importance of conservation and the principles of SUSTAINABLE YIELD. In 1986, the ITTA created an International Tropical Timber Organization (ITTO). The ITTO has attempted to promote the sustainable management of tropical forests; in 1990, the ITTO adopted a set of guidelines for sustainable tropical forest management and set the target of the year 2000 by which the entire tropical timber trade should come from sustainably managed forests. By 1994, perhaps no more than 1% of tropical timber came from sustainably managed forests; further, the parties could not agree on the meaning of sustainable management. Clearly, incentives such as preferential entry to markets are needed to encourage good management. During a renegotiation of the ITTA in 1993, the producer countries endorsed the putting of tropical forests under sustainable management by the year 2000, though with the qualification that the target be also applicable to timber from all forests, tropical, temperate, and boreal.

International Union for Conservation of Nature and Natural Resources (IUCN) See WORLD CONSERVATION UNION.

International Union of Air Pollution Prevention Associations (IUAPPA) An international association founded in 1964, by six national bodies; the union now has 26 member organizations.

International Whaling Commission (IWC) An international body formed in 1946 with the purpose of framing regulations for the 'conservation, development, and optimum utilization of whale resources'. The body operates under the CONVENTION FOR THE REGULATION OF WHALING. It meets each year to review whale stocks. In view of the virtual failure of whale management programmes, the IWC in 1982 voted to discontinue commercial whaling worldwide, commencing in 1986. Some countries still persisted, however, in some commercial whaling. In response, the IWC in 1994 decided to create a vast oceanic sanctuary which would give whales permanent protection from commercial hunts. First proposed by France, the SOUTHERN OCEAN SANCTUARY would cover an area more than five times the size of Australia.

Interparliamentary Conference on the Global Environment Held in Washington DC in May 1990, with the US Congress as host, an international conference with representatives from 42 countries. The outcome was a call for a 'global Marshall Plan' to help newly industrializing countries to reduce their environmental problems, with a sharing by industrialized countries of technologies for environmental improvement. The Marshall Plan, generous in scope, provided aid for countries in western Europe whose economies had suffered as a consequence of the Second World War; it operated from 1948 to 1951. The conference echoed somewhat the earlier recommendations of the BRANDT COMMISSION, while its views found later expression in AGENDA 21 endorsed by the 1992 UN CONFERENCE ON ENVIRONMENT AND DEVELOPMENT.

intractable wastes Potentially hazardous or persistent wastes that are difficult to treat and adequately dispose of, such as the following: polychlorinated biphenyls (PCBs); halogenated hydrocarbons such as vinyl chloride monomer, hexachlorobenzene, or DDT; pesticides and herbicides, both chlorinated and unchlorinated; mercury wastes; arsenic wastes; organometallics such as tetraethyl lead; biological wastes; high boiling point combustible wastes (greater than 300°C); explosives; radioactive wastes; peroxides; and dioxins. (See ARSENIC; ARSENIC POISONING; CHLORINATED HYDROCARBONS; CHLOROFLUOROCARBONS; DDT; DIOXINS; HALOGEN; HALOGENATED ORGANIC WASTE; HEAVY METALS; HERBICIDE; HEXACHLOROBENZENE; INCINERATION; LEAD; MERCURY; PESTICIDE; POLYCHLORINATED BIPHENYLS; RADIOACTIVITY.)

intragenerational equity The concept of fairness between individuals and groups, within society, locally regionally, nationally, and globally. The concept of human rights and dignity was embodied in the UN Universal Declaration of Human Rights in 1948, and in a whole range of conventions and declarations since then. Such rights include: equality before the law; protection against arbitrary arrest; the right to a fair trial; the right to own property; freedom of thought, conscience and religion; freedom of opinion and expression; freedom of peaceful assembly and association; the right to work; the right to equal pay for equal work; the right to form and join trade unions; the right to rest and leisure; the right to an adequate standard of living; and the right to education. These rights are compromised in all countries by reasons of birth, gender, race, property, class, caste, political division, territorial ambition, inequality of income, denial of rights, persecution, genocide, conquest, prejudice, bigotry, arrogance, maladministration, and non-democratic forms of government. These factors work to the detriment of effective ENVIRONMENTAL MANAGEMENT and SUSTAINABLE DEVELOPMENT, and to the concept of INTERGENERATIONAL EQUITY.

inversion A temperature inversion in the ATMOSPHERE in which the temperature, instead

invertebrate

of falling, increases with height above the ground. With the colder and heavier air below, there is no tendency to form upward currents and turbulence is suppressed. Inversions are often formed in the late afternoon when the radiation emitted from the ground exceeds that received from the sinking sun. Inversions are also caused by katabatic winds, that is cold winds flowing down the hillside into a valley, and by anticyclones. In inversion layers, both vertical and horizontal diffusion is inhibited and pollutants become trapped, sometimes for long periods. Low-level discharges of pollutants are more readily trapped by inversions than high-level discharges, hence the case for high stacks. Furthermore, high-level discharges into an inversion tend to remain at a high level because of the absence of vertical mixing. (See KATABATIC WIND; LAPSE RATE; Figure 15.)

invertebrate A species of animal not having a backbone; for example, a butterfly or a lobster.

invisible hand The influence of market forces through which self-interest often, but not always, benefits the community as a whole. A term used by Adam Smith (1723–90) in his *Inquiry into the Nature and Causes of the Wealth of Nations*, stating: 'The individual intends only his own gain and is led by an invisible hand to promote an end which was no part of his intention'. The explanation is that in order to profit, the entrepreneur must sell goods and services, often in a competitive situation, to consumers. Meeting effective demand, gaining and keeping market share, and maximizing profits depends on supplying quality goods and services in a timely manner to consumers at large and to particular segments of the economy.

involuntary resettlement The arbitrary transfer of substantial numbers of people to accommodate the reservoirs of hydroelectric schemes; the circumstances, nature, timing, and financing of resettlement schemes often present grave difficulties in major schemes. (See ASWAN HIGH DAM; THREE-GORGES WATER CONSERVATION AND HYDROELECTRIC PROJECT, CHINA; WORLD BANK.)

ion An electrically charged atom.

ion exchange A process in which ions in solution are exchanged with ions from a solid material, usually a synthetic resin. Ion exchange is a method of softening water which depends on the properties of certain synthetic resins that enables them to give up ions in exchange for ions from the water. Thus, calcium and magnesium cations can be replaced by sodium cations.

ionization The process by which a neutral atom or group of atoms becomes electrically charged through the loss or gain of electrons.

ionizing radiation Radiation which, by reason of its nature and energy, interacts with matter to remove electrons from the atoms of material absorbing it, producing electrically charged atoms called ions.

ionosphere The upper ATMOSPHERE above the OZONE LAYER and the STRATOSPHERE. As with the ozonosphere, the ionosphere is a warm layer, arising from the absorption of ultraviolet radiation from the sun. The ionosphere extends from about 80 km above the Earth, indefinitely upwards. The density of the ionosphere is extremely low.

IPC INTEGRATED POLLUTION CONTROL.

IRPTC INTERNATIONAL REGISTER OF POTENTIALLY TOXIC CHEMICALS.

irradiation Exposure to rays; for example, X-rays and ultraviolet rays.

irrigation Various kinds of techniques to facilitate the artificial application of water to land. Much irrigation water is pumped from lakes, ponds, streams, and wells, while other irrigation water is obtained simply by diverting water from a stream or river, without pumping.

ISO INTERNATIONAL ORGANIZATION FOR STANDARDIZATION.

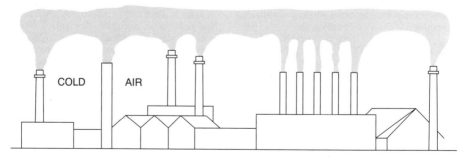

INVERSION: Figure 15. Temperature inversion: a pall of pollution trapped within a stable layer
Source: Warren Spring Laboratory, England.

Box 18 Israel: evolution of environmental policy

1955	Water Metering Act
1955–62	Water Drilling Control Acts
1957–61	Drainage and Flood Control Acts
1959	Water Measurement Act
1959–65	Water Act (Consolidated Version)
1965	Planning and Building Act, establishing a comprehensive physical planning system
1970	Water Pollution Act
1972	Environmental Protection Service established
1977	125 million trees planted; Mount Carmel National Park created
1982	EIA requirements: regulations define the types of projects for which an EIS must be prepared and set out the basic requirements
1992	Regulations set out specific EIA guidelines for quarries, industries, roads, and other categories
1993–94	Review of EIA policy in Israel by the Technion-Israel Institute of Technology

Box 19 Italy: evolution of environmental policy

1865	Water Supply and River Management Act
1902	Cultural Heritage Act
1922	Parks and Nature Reserves Act: First national park
1931	Fauna and Flora Act
1934	Air Pollution Act
1939	Coastal Protection Act; Cultural Heritage Act
1941	Waste Pollution Act
1942	Town and Country Planning Act
1961	Water Pollution Act
1966	Air Pollution Act
1973	Environmental Impact Assessment Act
1974	Toxic and Hazardous Substances Act
1976	Water Pollution Act
1977	Hunting and Wildlife Protection Act
1979	Water Pollution Act
1983	Air Pollution Act; Marine Pollution Act
1984	Parks and Nature Reserves Act; Toxic and Hazardous Substances Act
1985	Water Pollution Act
1986	Ministry of the Environment created
1988	Decree requiring an EIA for certain public and private projects
1989	EIA legislation comes into force; EIA Commission created; introduction of public inquiries
1992	EIA requirements extended; Minister of the Environment reviewing legislation to ensure full implementation of EC Directive 85/337
1994	Investigation of major contraventions relating to pollution of the River Arno, which empties into the Gulf of Naples, involving raw sewage and untreated industrial wastes.

isobar　A line drawn on weather charts linking places with the same atmospheric pressure, usually measured in millibars.

isohaline　A line drawn on a map of an area, indicating points of equal salinity.

isohyet　A line drawn on a map of an area, connecting points receiving equal rainfall during a stated period.

isokinetic　In respect of taking a sample of flue gas or other exhaust product, describing a situation in which the flow rate of gas into the sampling probe has the same rate and direction as the gas being sampled, hence, 'isokinetic sampling'.

isopleth　A line on a map connecting points at which given variables have the same numerical values; for example, topographic contour lines.

isothermal　Of constant temperature.

Israel: evolution of environmental policy　See Box 18.

Italy: evolution of environmental policy　See Box 19.

ITTA　INTERNATIONAL TROPICAL TIMBER AGREEMENT.

IUAPPA　INTERNATIONAL UNION OF AIR POLLUTION PREVENTION ASSOCIATIONS.

IUCN　INTERNATIONAL UNION FOR CONSERVATION OF NATURE AND NATURAL RESOURCES.

IUCN categories　Ten categories of protected natural areas recognized by the WORLD CONSERVATION UNION. The nomenclature was devised in 1978 and intended to facilitate ready identification of types of natural sites and necessary degrees of protection. The categories are scientific reserves/strict nature reserves; national parks/provincial parks; national monuments/natural landmarks; nature conservation reserves/wildlife sanctuaries; protected landscapes; resource reserves; anthropological reserves; multiple-use management areas; biosphere reserves; and world heritage sites.

IWC　INTERNATIONAL WHALING COMMISSION.

Ixtoc I incident　A major blow-out from an oil well off the coast of Yucatan, Mexico, during 1979 in which large quantities of oil escaped into the ocean. The huge oil slick eventually stretched 500 km northwestward from the well across the Gulf of Mexico. This massive incident had implications for offshore oil exploration around the world.

J

Japan: evolution of environmental policy See Box 20.

Jarrah dieback A root-infecting fungus *Phytophthora cinnamomi*, which has occurred on a significant scale in the jarrah forest of Western Australia. Jarrah trees growing in dieback-affected areas which are poorly drained may die following exceptionally heavy winter and summer rainfall.

Jasper National Park Established in 1907 in Alberta, Canada, a national park contiguous with Banff National Park in Alberta and Yoho National Park in British Columbia. Jasper is an area of ridges and peaks in the eastern Canadian Rockies including Mounts Columbia, Alberta, and Crown. It contains the extensive Columbia ice field, the largest sheet of glacial ice in North America outside the Arctic Circle. The flora includes pine, Douglas fir, spruce, alpine larch, aspen, paper birch, balsam poplar, and many flowering plants. Fauna includes elk, deer, moose, mountain goat, bighorn sheep, mountain caribou, cougar, bear, wolverine, wolf, marmot, porcupine, coyote, and many birds.

JET JOINT EUROPEAN TORUS.

jet flame The combustion of liquid or gases emerging with significant momentum from an orifice. (See BOILING LIQUID EXPANDING VAPOUR EXPLOSION; FLASH FIRE.)

jhumming A form of shifting cultivation practised in the northeast of India. A tribe cuts down a patch of trees, cultivates the soil for two or three years, then plants trees and moves on. The same patch would be used only once every 20 or 30 years; now the cycle is down to 5 or 6 years, which does not give the forest time to regenerate. The reasons are that much of the forest has been progressively cleared for other purposes and the population of the tribal peoples has increased.

jobless growth Increases in production and PRODUCTIVITY made possible by more and better use of machinery and innovations, without an increase in employment. (See CAPITAL-INTENSIVE.)

Box 20   Japan: evolution of environmental policy

1918	Wildlife Protection and Hunting Law
1957	Natural Parks Law
1962	Smoke Control Law
1965	Japan Environment Corporation established to finance pollution control projects
1967	Basic Law for Environmental Pollution Control
1970	Water Pollution Control Law; Soil Pollution Control Law; Waste Disposal and Public Cleansing Law
1971	Japan Environment Agency established; Offensive Odour Control Law; Amendment of Agricultural Chemical Regulation Law
1972	Air Pollution Control Law; Nature Conservation Law; Adoption by Cabinet of environmental impact assessment procedures for major public projects
1973	Pollution-related Health-damage Compensation Law; Chemical Substances Control Law; Amendment of Nature Conservation Law and Natural Parks Law
1984	Adoption by Cabinet of environmental impact procedures (EIA), for general application
1986	National survey of the natural environment (the Green Census); Revision of Chemical Substances Control Law
1987	Comprehensive Resort Area Development Law; Amendment of Pollution-related Health-damage Compensation Law
1988	Protection of the Ozone Layer Law
1989	Amendment of Water Pollution Control Law
1990	Revision of Nature Conservation Law and Natural Parks Law
1991	Revision of Wildlife Preservation Law. Goal of seventh Five-Year program (1991–95) to increase the percentage of the population served by sewers to 55%. Recycling Resources Law

Joint European Torus (JET) A major nuclear fusion project located at Culham, England, following a decision of the European Community in 1978 in respect of this joint product. Its aim has been to develop a fusion reactor based on the torus, or doughnut-shaped design.

jojoba bean (Pronounced ho-ho-ba) A bush which grows in arid and semi-arid areas and has a high tolerance for salinity. It can live for more than 150 years, producing an olive-shaped bean. The bean yields half its weight in oil; it is one of the finest lubricants known and chemically almost identical with the oil from sperm whales. American Indians in California and Arizona, USA, have used jojoba oil for cooking for centuries. Plantations have been established in Israel.

joule The unit of energy, including work and quantity of heat; the work done when the point of application of a force of 1 newton is displaced through a distance of 1 m in the direction of the force. One British thermal unit equals 1055.06 joules; or 1.05506 kilojoules. The joule (J) is one of the International System of Units (SI), adopted worldwide in 1954.

judicial review An action in a higher court to review the decisions of lower courts, tribunals, and administrative bodies, following which, various orders may be issued. The process seeks to ensure that the decisions made have conformed with the provisions of the constitution; actions that do not conform are unconstitutional and therefore null and void. The institution of judicial review historically presumes the existence of a written constitution. A number of constitutions throughout the world have incorporated provisions in various forms for judicial reviews, undertaken usually by supreme courts and high courts. Through this procedure, the validity of central-government, state-government, or provincial-government environmental laws might be tested for legal and constitutional validity.

Jurassic Period An interval of geological time from 208 to 144 million years ago; it is one of the three major divisions of the Mesozoic Era, being preceded by the Triassic Period and followed by the Cretaceous Period. During the Jurassic Period the dinosaurs rose to supremacy on land. Climates worldwide were equable with forests of conifers and ferns flourishing. Birds evolved, while deposits of limestones and iron ores were deposited.

jurisprudence The science or theory of law; the study of the principles of law.

K

Kabalega National Park Established in 1952 in Uganda, Africa, a hilly region bisected by the Victoria Nile, containing the Kabalega (formerly Murchison) Falls. The flora is mainly savannah, well-wooded in places, while the fauna includes the hippopotamus, elephant, rhinoceros (black and white), buffalo, giraffe, lion, leopard, and various species of antelope.

Kafue National Park Established in 1950 in Zambia, Africa, a plateau area with Kalahari sand to the south. The flora comprises mixed forest, thicket, woodland, and grass; while the fauna includes the black rhinoceros, hippopotamus, elephant, lion, buffalo, zebra, various species of antelope, and crocodile, with much bird life.

Kakadu National Park Established initially in 1975 in the Northern Territory, Australia, and then considerably enlarged, one of the most magnificent natural reserves in Australia now entirely included in the WORLD HERITAGE LIST. The topography ranges from tidal flats and flood plains to dramatic sandstone plateaus and escarpments. The wetlands are frequented by very large numbers of waterbirds. Kakadu contains more than 1000 known Aboriginal art sites.

Kalahari Gemsbok National Park Established in 1931, in South Africa, a national park located in the Kalahari desert of extreme northern Cape Province, adjoining the Gemsbok National Park of Botswana. The park consists generally of reddish dunes, with acacia growing in the river beds; various shrubs provide forage for large herds of gemsbok (or oryx), wildebeest, springbok, and some wild hartebeest. Other wildlife includes lions, dogs, jackals, ostriches, and numerous other birds.

Kaldor compensation principle See PARETO OPTIMALITY.

kaolin or **china clay** A soft white clay that is an essential ingredient in the manufacture of china and porcelain, and is widely used in the manufacture of refractories, paper, rubber, paints, adhesives, ink, organic plastics, some cosmetics, and many other products.

kaolisol A SOIL that is formed in tropical conditions, with heat and heavy rainfall, with a leaching-out of silica and bases. The soil is highly weathered, silica-poor, poor in weatherable minerals; it is, however, rich in iron, released as a result of weathering. The clay content may increase with depth because of the permeability of the soil and the transportation of clay minerals by water from the upper layers. Once formed, these soils may become subsequently the parent material for new soils.

karst topography A limestone terrain characterized by rocky ground, caves, sinkholes, and underground rivers, resulting from the excavating effects of underground water. Limestone (calcium carbonate) dissolves relatively easily and rapidly in slightly acidic water. Many of the cave areas of the world are areas of karsts.

karyotype The characteristics of the set of chromosomes (in respect of sizes, shapes, and numbers) of a representative somatic cell of a given species or individual. A CYTOTYPE is a member of a population composed of individuals with essentially similar karyotypes.

katabatic wind A wind caused by cold air flowing downhill. When a sloping land surface cools by night-time radiation, the cold air in contact with the ground flows downhill and along valley bottoms. Opposite of an ANABATIC WIND. (See Figure 16.)

Katmai National Park Established in 1918 in Alaska, USA, an area of dying volcanoes on the east coast of the Upper Alaska Peninsula in the Aleutian range; the park includes Katmai Volcano, Mount Denison and the Valley of Ten Thousand Smokes. There are also, by contrast, several glaciers and lakes. The flora comprises pine, spruce and fir, and other sub-arctic plants.

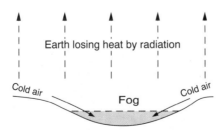

KATABATIC WIND: Figure 16. Katabatic wind: cold air accumulates in a valley to form a persistent temperature inversion
Source: Warren Spring Laboratory, England.

Keep America Beautiful

The fauna includes the world's largest land carnivore, the Alaskan brown bear, and also caribou, moose, wolf, and many birds.

Keep America Beautiful A voluntary US body that created a standardized and systematic approach to picking up litter in urban areas and in smaller communities. Many of the US States have adopted anti-litter laws to discourage littering, encourage recycling, impose refundable deposits on certain types of beverage containers, or to ban certain types of container or retainer. Some states have introduced taxes on businesses whose products may end up as litter. (See Figure 17 for a typical US anti-litter logo.)

Keep Australia Beautiful Council A national voluntary organization which promotes Australia-wide anti-litter programmes stressing litter reduction, beautification, and recycling.

Keep Singapore Clean Campaign An ambitious plan of action launched in the late 1960s to transform Singapore into one of the cleanest cities of the world whereby litter, dirt, and other forms of waste pollution would be banished. The campaign was backed with appropriate publicity and substantial fines; it has been an undoubted success.

Kenting National Park Established in 1984 in southern Taiwan, Republic of China, the park is located on the Hengchun peninsula with the Pacific ocean to the east, Taiwan Strait to the west, and the Bashi Channel to the south. The topography embraces coastal cliffs, coral reefs, sandstone peaks, KARST terrain, limestone caves, shell sands, estuaries, and lakes. The tropical flora ranges from coastal coral-reef plants and tropical coastal forests to grassland, coastal bushes, marsh plants, and low-altitude natural forest communities. Fauna diversity embraces many birds, reptiles, amphibians, and all kinds of insects. (See TAIWAN, EVOLUTION OF ENVIRONMENTAL POLICY.)

Keynesian revolution Appellation for the profound changes in fundamental economic thinking brought about by the challenging contributions of John Maynard Keynes (1883–1946). Keynes challenged the assumption that all labour could find employment if it lowered its price, arguing that this simply reduced total effective demand upon which all productive activity depended. He believed that high unemployment might prevail regardless of the price of labour, favouring government budget deficits during times of recession to boost market demand, and budget surpluses when the economy is booming. In other words, instead of consistently balanced budgets, government budgets should become countercyclical; tax rates should vary accordingly.

keystone The central wedge-shaped stone at the top of an arch that locks the parts together.

keystone species A species that plays a vital role in maintaining the diversity and success of an ECOSYSTEM, perhaps as a PRODUCER that provides food for several consumer species; or as a predator that maintains a balance among several species.

kilowatt (kW) One thousand watts; a unit of electrical power. Relevant to many electrical domestic appliances, power station capacities are measured in terms of megawatts (MW). A modern power station may range in capacity from 2000 to 4000 MW.

kilowatt-hour (kWh) The unit of electrical energy, as distinct from power. A measure of output or consumption of electrical energy. One kWh equals 3.6 megajoules (MJ) or 3600 kilojoules (kJ).

kinship A socially recognized relationship between individuals in a culture who are biologically related or are given the status of relatives by marriage, adoption, or other relationship. A broad-ranging term for all the relationships that people are born into or acquire or create later in life that are considered binding and recognizable in the eyes of their society. Kinship defines rights and responsibilities.

Klebsiella A genus of bacteria that is widespread in the environment and spread by handling. It is sometimes detected in drinking water.

knocking Detonations in engine cylinders, accompanied by a knocking sound and associated with a decrease in power output. Also called 'pinking'.

KEEP AMERICA BEAUTIFUL: Figure 17. 'Do the Right Thing: Put your Litter in the Bin': a popular anti-litter logo.

Kobe earthquake A massive earthquake measuring 7.2 on the RICHTER SCALE which struck 30 km off the west coast port of Kobe, Japan, at 5.46 a.m. on 17 January 1995. It was Japan's worst earthquake for almost 50 years, involving some 6000 deaths. Devastation covered a 100 km radius around Kobe, including Osaka and Kyoto. Nearly 10 000 houses, buildings, and other structures were destroyed, while an elevated highway in Nishinomlya overturned.

Komi pipeline oil disaster In 1994, major leakages from a pipeline in Russia's far north, in the republic of Komi. The republic is centred on the Pechora River. The climate is severe with prolonged cold winters and cool summers. The Arctic Circle crosses the north of the republic. The pipeline system is operated by Komineft. The pipeline systems of northern Russia have suffered acutely from lack of maintenance, the emphasis being on new pipelines. Huge pools of thick black oil from the fractures have threatened the area's fragile ecosystem. Some reports have placed the amount of leakage as many times that from the 1989 EXXON VALDEZ DISASTER off Alaska.

Komoé National Park Established in 1926 in the Ivory Coast, Africa, a series of mountain ranges. The flora is comprised of Guinean savannah, sedge, bulrushes, and bombax, while the fauna includes hippopotamus, elephant, lion, leopard, panther, hyena, and many species of antelope.

Box 21   Korea: evolution of environmental policy

1963   Pollution Prevention Act
1970   Saemol Undong (new community movement) initiated
1973   First 10-year national reforestation plan initiated
1976   First municipal sewage treatment plant constructed
1977   Environment Preservation Act; Marine Pollution Prevention Act
1978   Government institutes charter for nature conservation
1979   Environmental impact assessment (EIA) adopted as a regulatory mechanism
1980   Environmental Administration established; new constitution adopted guaranteeing the right to live in a clean and healthy environment; Korea Resource Recovery and Reutilization Corporation established
1981   Environment Preservation Act amended to introduce an emission charge system to enforce the measurement of emission standards; low-sulphur fuel oil policy for large cities; EIA process introduced
1986   Solid Waste Management Act; six regional offices of the Environment Administration established; Environment Preservation Act amended to extend the application of the EIA process to non-governmental projects
1987   Environmental Management Corporation established new emission standards for cars; introduction of unleaded petrol (gasoline); environmental education in schools strengthened
1990   Ministry of Environment established, replacing the Environment Administration; Basic Environmental Policy Act; Air Environment Preservation Act; Water Environment Preservation Act; Noise and Vibration Control Act; Hazardous Chemical Substance Control Act; Environment Pollution Damage Dispute Coordination Act; public hearings to be held in some cases
1991   Amendment of the Solid Waste Management Act and the Marine Pollution Prevention Act; Natural Environment Preservation Act; Central Environmental Disputes Coordination Commission established; Position of Ambassador for Environmental Affairs established; new deposit system for waste recovery and treatment
1992   Ministry of Foreign Affairs establishes the Science and Environment Office; eco-mark introduced for consumer products which are environmentally friendly; nationwide recycling programme
1994   Environmental Assessment Act

'Kon-Tiki' The legendary Sun King who ruled the country later occupied by the Incas; the name was adopted by the explorer Thor Heyerdahl (b. 1914) for his raft which sailed from Peru to the Pacific Islands in 1947. This and a later expedition were intended to prove the possibility of ancient transoceanic contacts between distant civilizations and cultures.

Korea: evolution of environmental policy See Box 21.

Kosciusko National Park Established in 1944 in New South Wales, Australia, an area of alpine peaks and rolling plateaus on the western face of the Great Dividing Range. A special feature is Mount Kosciusko, Australia's highest peak. The flora is alpine and subalpine with snowgum forests and zones of eucalyptus. The fauna includes the grey forester kangaroo, brush-tailed rock wallaby, wombats, marsupial and pouched mice, koala, duck-billed platypus, spiny anteater, and many birds, including black-backed magpie, currawongs, lyrebird, and emu.

Kruger National Park Established in 1898 in the Transvaal, South Africa, a gently undulating region of hills and doleritic dikes, with plains to the south. The flora includes grassy plains, park savannah, dry deciduous forest and thornbush, many flowering plants, and a wide variety of trees. The fauna includes hippopotamus, elephant, white rhinoceros, lion, leopard, cheetah, hyena, giraffe, African buffalo, numerous species of antelope, crocodiles, and ostrich. It is one of the finest wildlife areas in the world.

L

labelling See ECO-LABELLING.

labour One of the FACTORS OF PRODUCTION; the combination of all exertions by individuals, whether they be manual, physical, mental, or entrepreneurial, directed towards the creation of wellbeing. Labour is quantitative, involving both effort and time; it is also qualitative, both intelligence and skill determining its efficiency. Labour may also be regarded as a flow of services generated by a heterogeneous stock of human capital. It may command a market price as paid employment, or be of a domestic, voluntary, or charitable nature.

labour costs per unit of output The cost of LABOUR in real terms involved in making each unit of output from a factory. High wages associated with high PRODUCTIVITY may mean a lower labour cost per unit than low wages associated with low productivity.

labour, division of The specialization of LABOUR which intensifies with agricultural, commercial, and industrial development. Its advantage lies in that individuals gain greater skill and dexterity by doing one job, rather than several. It allows the exploitation of individual aptitude and encourages the use of CAPITAL. The result is an increase in productivity and an improvement in quality. On the other hand, the individual becomes 'narrowed' to the job in hand, and faces difficulties when that particular skill is no longer required, while the community becomes more vulnerable to the strike threats of minorities holding key positions. Today, multiskilling breaks down some of the barriers of excessive specialization and provides more varied and interesting employment.

labour economics In ECONOMICS, the study of one of the factors of production, namely LABOUR: the distribution of the workforce, the determination of rates of pay, the growth and decline of employment opportunities, industrial relations, and the relationship with the other factors of production, LAND and CAPITAL.

labour-intensive industry Industry in which the ratio of LABOUR input to CAPITAL input is higher than the average for industry as a whole. Examples of labour-intensive industries are the textile, leather, furniture, and service industries and professions.

labour laws That body of law devoted to such matters as employment, remuneration, conditions of work, trade unions, industrial relations, occupational health and safety, minimum wages and conditions, the settlement of disputes, and the functioning of collective and enterprise agreements.

labour mobility The ease of movement of LABOUR between areas and occupations. Occupational mobility may be lateral or vertical. Many factors govern mobility including work opportunities in expanding industries, the costs and disturbances of moving, the opportunities for migration, training and retraining opportunities, perceptions of the future for the family, the prospect of 'bringing home the golden fleece', and political considerations.

La Brea Tar Pits Tar pits located in Hancock Park, Los Angeles, USA, discovered by Gaspar de Portola's expedition in 1769. They were found to contain the fossilized remains of the skulls and bones of prehistoric animals, together with plants and vegetation from the same period. More than one million prehistoric specimens have been exhumed from the pits.

lacustrine ecosystem Any pond, lake, or body of water considered as an ecological unit of the physicochemical ENVIRONMENT and the biotic community within which energy and mass are cyclically exchanged. While the characteristics of water bodies vary according to depth, stillness, size, and seasonal changes, three major zones of habitat are usually present: (1) the LITTORAL (or shallow-water) ZONE, with light penetrating to the bottom, supporting rooted plants and bottom-dwelling animals; (2) the LIMNETIC ZONE, with some light penetration, supporting plant and animal plankton; and (3) the PROFUNDAL ZONE, the deepest part being beyond light penetration, supporting organisms adapted to darkness.

lagoon (1) An area of relatively shallow, quiet water, with access to the sea and probably separated from it by sandbars, barrier islands, or coral reefs. Lagoons may be coastal and found on most land margins, or coral-reef, occurring in areas where warm-water corals thrive. (2) An oxidation pond in which waste water is purified through sedimentation and both AEROBIC and ANAEROBIC biochemical action take place over a period of time. Lagoon treatment or 'ponding' involves storing liquid waste in natural or artificial lakes, although this storage is provided in different circumstances to achieve different ends. The two main types of lagoon are: a stabilization lagoon or oxidation pond, in which crude sewage or industrial waste is treated; and a maturation lagoon, which is a pond for the tertiary treatment of sewage works effluent, a great improvement in the organic quality of an effluent being achieved.

laissez-faire (French for: let (them) act) A policy based on a minimum of government

lake

interference in the economic affairs of individuals and society at large. Belief in laissez-faire was a popular view during the 19th century. Early economists arguing in favour of unregulated individual activity and the INVISIBLE HAND. The practical expression of the philosophy did much to ensure the development of commerce and industry without governmental and ecclesiastical interference. However, the system did not produce all the good results that were expected and the reactions to its failing found expression in factories, housing, public health, food and drug industries, education, consumer protection, land-use planning, pollution control, environmental and national parks legislation, social services, public works programmes, trade unionism, and various shades of socialism and communism. Today state intervention in the economic, social, and environmental spheres is widespread in all countries, with growing international cooperation to deal with the consequences of unregulated activity. (See SUSTAINABLE DEVELOPMENT.)

lake Any relatively large body of water, slow-moving or standing, that occupies an inland basin; a lake may consist of fresh water or salt water. An example of an extremely salty lake is the Great Salt Lake in northern Utah, USA, while the largest freshwater lake is LAKE BAIKAL in Russia.

Lake Baikal Located in eastern Siberia, Russia, a lake which contains one-fifth of the world's surface fresh water; it functions as a huge biofilter, as the water in the rivers that flow into it are much more polluted than the water that flows out from it. Early public concern at the possible adverse effects of development on Lake Baikal led to the popular slogan 'Hands Off Baikal'. A government report issued in 1981 indicated that the condition of the lake was deteriorating. Since then, stricter measures have become necessary with the closure of those plants causing most pollution.

Lake Clark National Park Established initially in 1978 as Lake Clark National Monument, and renamed two years later, a national park and preserve in southern Alaska, USA, southwest of Anchorage. It has great geological diversity with jagged peaks, granite spires, dozens of glaciers, hundreds of waterfalls, tundra plains, and active volcanoes. Lake Clark provides the headwaters for the most important spawning ground for red salmon in North America. Wildlife includes caribou, Dall sheep, brown and grizzly bears, bald eagles, and peregrine falcons.

Lake District National Park Established in 1951, a famous scenic region and national park in the county of Cumbria, northwest England. There are many lakes including Windermere and Thirlmere and the highest English mountain, Scafell Pike, is situated here.

Lake Pedder An inundated isolated lake of unique character situated in the rugged southwestern area of Tasmania, Australia. In 1967, the Hydroelectric Commission of Tasmania proposed a hydroelectric scheme on the Gordon River, a consequence of which was the flooding of Lake Pedder. Conservationists bitterly opposed this development but lost the battle. However, a later proposal intending to flood the Franklin River was soundly defeated. The area subsequently became incorporated in the WORLD HERITAGE LIST.

land One of the FACTORS OF PRODUCTION; in an economic sense it includes all natural resources such as coal, oil, water, and SOIL. Land is virtually fixed in total area, although the supply of useful land may be increased by the use of fertilizers, irrigation, dykes, and embankments. However, it can be reduced on a large scale through neglecting the principles of SOIL CONSERVATION. Unlike the other factors of production, land as a natural asset has no cost of production; the application of LABOUR and CAPITAL incurs costs of various kinds.

Land and Environment Court A court created in 1979 in New South Wales, Australia, to handle all appeals and prosecutions arising out of environmental planning, pollution control, and heritage legislation in that state. It is a superior court of record.

land breeze The movement of air from land to sea. On clear nights, the land cools faster than the water, cooling more quickly the lowest layers of air. The heavier land air tends to spill seaward displacing the warmer sea air.

land capability The uses to which an area of land may be put, taking into account all the physical, economic, social, and environmental constraints. The concept was developed initially by an American landscape architect, Ian McHarg, in his book *Design with Nature* (1969). Long-term as well as short-term considerations are involved in determining the most appropriate use of land. If land is used inappropriately or beyond its CARRYING CAPACITY, it ultimately loses its productive capacity and social usefulness.

land degradation A decline in the PRODUCTIVITY of an area of LAND or in its ability to support natural ecosystems or types of agriculture. Degradation may be caused by a variety of factors including inappropriate land management techniques, soil erosion, salinity, flooding, clearing, pests, pollution, climatic factors, or progressive urbanization.

land drainage The removal of water from wet or waterlogged LAND to make it fit for the cultivation of crops or for urban development. (See WETLANDS.)

landfill The most common form of disposal of houshold, commercial, and industrial refuse; it appears that 80 to 90% of the world's refuse will be disposed of by this method for several years to come. The type of sites used for such disposal include: mineral excavations, abandoned quarries, low-lying land, valleys, areas involving the reclamation of land from water, or flat land to build up a feature. Generally, in a landfill scheme, refuse is tipped in trenches or cells prepared to such a width that the daily input of refuse can be effectively covered, presenting a clean face each day. The refuse can be tipped either at the bottom of the face and bulldozed into the face, or tipped on top of the previous fill and bulldozed over the face. It is essential that the refuse be adequately covered and compacted to allow traffic over the fill. The landfill method is known in the UK as 'controlled tipping' and in the USA by the title of 'sanitary landfill'; the former emphasizes the system by which the waste is deposited, while the latter emphasizes the hygiene aspects. The landfill technique is often used constructively to provide facilities for sport; its use for urban development involves many years of settlement. Landfills need to be carefully located and managed to avoid, for example, leachates reaching streams, the breeding of vectors and rodents, odour, windblown litter, and an appearance of desolation. The method is sometimes used for the disposal of hazardous wastes, using impervious cells and rigorous control methods.

landform Any conspicuous topographic feature on the face of the Earth or other planetary body; familiar examples are mountains, volcanoes, plateaus, valleys, seamounts, mid-oceanic ridges, and submarine canyons.

Land of Ten Thousand Sinks An area of west-central Kentucky, USA, comprising numerous sink holes and caves, including the interconnected caves of Mammoth Cave National Park and the Flint Ridge Cave System. The whole presents a vast network of underworld caverns, rivers, and lakes.

land pollution The discharge to, or deposition on, land of undesirable or inappropriate solid and liquid waste, including domestic refuse, discarded equipment, waste from building sites, rock and soil from excavation activities, hospital waste, industrial waste, commercial waste from offices and shops, discarded vehicles, and hazardous waste.

land reform Fundamental change in the way in which agricultural land is held or owned, the techniques of cultivation that are employed, or in the relationship of agricultures to the rest of the economy. Such fundamental reforms may be instituted by governments, often revolutionary ones.

landscape analysis Factors applied systematically to landscape units or subareas of a total study area, and rated on a scale from zero to 10 linked with a weighting system. Factors include slope, vegetation, visual significance, flooding, mine subsidence, erosion potential, watershed, geology, natural features, fauna and flora, mangroves, wetlands, mineral resources, and wind exposure. One advantage of such an analysis is that it sets out in quantitative terms the degree to which various factors influence a final LAND-USE PLANNING recommendation.

landscape architecture The creation, development, and decorative planting of gardens, grounds, parks, and other outdoor spaces. Landscape gardening is used to enhance nature helping to create a natural setting for individual residences and buildings, and even towns, particularly where special approaches and central settings are required.

landscape character The appearance of broad areas of the ENVIRONMENT in which certain visual elements predominate.

landscape or amenity conservation The safeguarding, for public enjoyment, of scenery or landscape and of opportunities for outdoor recreation, tourism, field sports, and similar activities; the concept includes the preservation and enhancement not only of what has been inherited but the provision of new amenities and facilities.

landscape unit An individual part of the landscape within the visual ENVIRONMENT having certain uniform characteristics or homogeneity which distinguish that part from other areas; it is used as a classification system.

land-use planning Traditionally, a technical or physical approach to the segregation of incompatible activities, such as housing and industry, through systems of land-use, zoning, and development controls; such planning has been effective in preventing the worst aspects of mixed development in new areas, and has served individual as well as community interests. The individual benefits, for example, when investment in residential property is protected against incompatible local developments, while the community benefits from the protection of open spaces and in recreational opportunities. Orderly development may well help reduce INFRASTRUCTURE costs. Land-use planning is an important instrument in serving economic, social, and environmental objectives. (See ENVIRONMENTAL PLANNING; GEOGRAPHIC INFORMATION SYSTEM.)

land-use prediction model A theoretical model that seeks to describe and predict the locations of households, firms, and other types of urban activities; the model may be based upon the principle that locational decisions are dictated largely by the costs of overcoming the distance which separates

Lane, Franklin K. (1864-1921) interrelated and interactive activities. (See GEOGRAPHIC INFORMATION SYSTEM.)

Lane, Franklin K. (1864-1921) US lawyer and politician who, as Secretary of the Interior (1913-20) made important contributions to conservation. At Lane's urging, the US Congress in 1916 created the NATIONAL PARK SERVICE. Lane appointed as its first director the noted conservationist Stephen T. Mather.

lapse rate Changes of temperature of the ATMOSPHERE with distance above the ground; the vertical distribution or profile of temperature. For different layers of the atmosphere and at different times, this distribution may take one of three forms: (1) a temperature lapse, that is, a decrease of temperature with height expressed in degrees Celsius per kilometre; (2) a temperature inversion, in which the temperature increases with height; or (3) an isothermal layer in which temperature does not vary with height, that is, it has a zero lapse rate. The lapse rate greatly influences the dispersion of gases in the atmosphere. (See INVERSION; Figure 18.)

larva (Plural: larvae) An intermediate form of life in the development of many animals, occurring after birth or hatching and before the adult form is reached. Examples include tadpoles (between fertilized egg and frog); caterpillars (between eggs and butterflies); and maggots (between eggs and flies). Larvae are typical of invertebrates, and some have two or more distinct larval stages. Among vertebrates only the amphibians and certain fish have a larval stage.

latent heat Hidden heat; the number of heat units absorbed or released by unit weight of a substance during a change of physical state, for example, from water into steam or from water into ice. The heat transfer involved is not shown by a thermometer, hence the description 'hidden heat'.

laterite A SOIL layer that is rich in iron oxide. It is derived from a wide variety of rocks weathering under strongly oxidizing and leaching conditions. It tends to form in tropical and subtropical regions where the climate is humid. Other minerals may also be present. The aluminium-rich representative of laterite is bauxite.

Law of the Sea See CONVENTION ON THE LAW OF THE SEA; UN CONFERENCE ON THE LAW OF THE SEA (UNCLOS).

leachate Liquids containing dissolved solids from a LANDFILL; liquids escaping from mining-tailings dams; salts and alkalis removed from soils by irrigation and drainage; and other soluble constituents removed from soils by percolating water. The formation and escape of leachates may be minimized: by restricting the movement of air and water within dumps and emplacements; by restricting the movement of surface waters; and by sealing the surfaces of dumps and emplacements. Where leachates cannot be prevented, treatment may be necessary by, for example, evaporation or by neutralization through the addition of alkalis such as lime. For highly saline waters, where disposal by irrigation may not be possible, evaporation in dams or transportation to suitable places

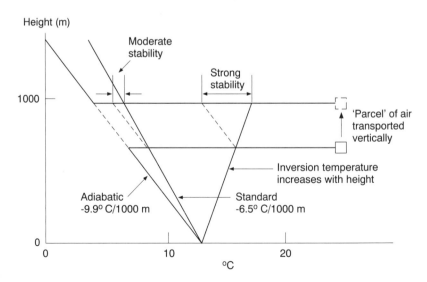

LAPSE RATE: Figure 18. Vertical stability in different atmospheres.

lead One of the HEAVY METALS; a soft metal, historically used in battery accumulators, roofing materials, domestic water pipes, as a basis for paints, and as an additive in the form of tetraethyl and tetramethyl lead in petrol (gasoline). Evidence of lead poisoning in children has resulted in severe restrictions on the use of lead pipes and lead-based paints, and the use of lead in fuels. Standards restrict the presence of lead in water and foodstuffs and the lead levels in the blood of vulnerable workers. Concern continues to be expressed about lead in the blood of children, particularly those living near lead refineries. Children are vulnerable to lead, it being associated with mental retardation, hyperactivity, and nerve impairment.

lean-burn engine A petrol-powered car engine designed to burn a fuel-air mixture containing a higher proportion of air than normal, that is, a lean mixture. The effect is to reduce fuel consumption and the production of unburnt hydrocarbons and nitrogen oxides, without resort to a CATALYTIC CONVERTER.

lee wave A vertical undulation of airstreams on the lee side of a mountain; they may produce clouds and have an important effect on the weather. (See DOWNDRAUGHT; DOWNWASH.)

Legionnaire's disease A form of pneumonia first identified in 1976; the name of the disease and the previously unknown bacillus *Legionella pneumophilia* derives from a 1976 state convention of the American Legion in Philadelphia; 182 Legionnaires contracted the disease, 29 of them fatally. Episodes of this disease have been widespread; it is suspected that contaminated water in central air-conditioning units may serve to disseminate the disease, through droplets into the surrounding atmosphere.

legumes Leguminous plants such as peas, beans, and clover. They are important in the NITROGEN CYCLE as they contain nitrifying bacteria in the root nodules that convert gaseous nitrogen from the ATMOSPHERE into nitrates that can be used by the plants. A crop of legumes enriches the nitrate content of the SOIL.

lethal concentration (LC_{50}) The concentration of air-borne material, the four-hour inhalation of which results in the death of 50% of the test group, within a 14-day observation period. When used in aquatic toxicology, the test period is usually six hours.

lethal dose (LD_{50}) The quantity of material administered orally or by skin absorption or injection, which results in the death of 50% of the test group, within a 14-day observation period.

licence A statutory document, issued by an environmental agency, permitting a person or organization to discharge, emit, or deposit wastes into the ENVIRONMENT, subject to a variety of conditions relating to control measures, monitoring, volume, timing, nature, and composition of the waste. Licences may often be varied or rescinded at any time. Breaches of licensing conditions may result in prosecution.

life cycle In BIOLOGY, the series of changes that the members of a species undergo as they pass from birth to death, or from a given development stage to a similar stage in a subsequent generation.

life cycle EIA The study of the effects of a product or activity on the ENVIRONMENT from inception, manufacture, distribution, use, and final disposal, with all the direct and indirect effects on the environment such as the effects of mining raw materials and the problems of disposal of discarded plant and products. Thus, it is possible to examine the life cycle of the automobile, the power station, the oil rig, the drink container, the household appliance, detergents, fertilizers, or the progress of individual minerals and metals through the fabric of society.

life span The average period of time between the birth and death of an organism, ranging from one day for the mayfly to thousands of years for the bristlecone pine.

life system concept That part of an ECOSYSTEM which determines the existence, abundance, and evolution of a particular population. A life system comprises a subject population and its effective ENVIRONMENT; the latter includes all those BIOTIC and ABIOTIC agencies influencing the population. Population and environment are regarded as interdependent elements which function together as a system.

light Electromagnetic radiation that can be detected by the eye. In terms of wavelength, electromagnetic radiation occurs over an extremely wide range from gamma rays to long radio waves. Wavelengths visible to the human eye occupy a very narrow band, ranging from red to violet light. The adjacent spectral regions are infrared at one end and ultraviolet at the other; neither is visible to the human eye.

light rail transit A technological outgrowth of trams (streetcars); a system of rail transport usually powered by overhead electric wires being used for medium-capacity local transportation. Light rail systems are more segregated from other traffic than street trams, but less so than mainline and suburban heavy rail systems. The principle has been widely adopted to serve outer suburbs.

lignin A complex, oxygen-containing organic substance that, with cellulose, is the chief constituent of wood. Second only to cellulose, it is the most abundant organic material on Earth. It is concentrated in the cell walls of timber

lignotubers

making up about one-third of the oven-dried weight of SOFTWOOD, and about a quarter of the oven-dried weight of HARDWOOD. In the manufacture of paper it is removed from wood pulp usually by treating with sulphur dioxide, sodium sulphide, or sodium hydroxide. Lignin is used industrially as a binder for particle board and similar wood products, as a soil conditioner, as an ingredient of phenolic resins, as an adhesive for linoleum, and domestically, as a fuel.

lignotubers Lumps of dense woody tisse that form at the base of some species of trees; they provide a reservoir of growth hormones and buds that can produce new shoots if the older stems are damaged or destroyed.

limestone A sedimentary rock comprising mainly calcium carbonate; it may contain considerable amounts of magnesium carbonate (dolomite) as well. Limestone has considerable industrial importance in the production of building stone, lime, fertilizer, glass, and in agriculture. In some areas limestone has proved rich in fossil content. It may also be used in the DESULPHURIZATION OF FLUE GASES.

limnetic zone The region of open water beyond the LITTORAL ZONE of a lake, down to the maximum depth at which there is sufficient sunlight for PHOTOSYNTHESIS. This is the depth at which photosynthesis balances respiration, known as the COMPENSATION DEPTH. Rooted plants are absent in this zone, but there is a great abundance of phytoplankton.

limnology A subsystem of HYDROLOGY that deals with the scientific study of particularly, freshwater lakes and ponds; the discipline includes the biological, physical, and chemical aspects of lakes, ponds, and streams.

lindane Used as an insecticide, one of the CHLORINATED HYDROCARBONS; it is made by chlorinating benzene. A white water-insoluble powder with a characteristic musty odour.

liquefied natural gas (LNG) Natural gas stored as a liquid; it may be stored in liquid form at key terminal points in a national or regional transmission system. Special tankers transport LNG from such countries as Algeria, Australia, Borneo, and Indonesia to markets in Europe, Japan, and the USA. Natural gas is primarily METHANE with varying amounts of ethane and inert gases.

liquefied petroleum gas (LPG) Paraffin hydrocarbon gases comprising propane, butane, and pentane, obtained from natural gas wells and in petroleum refining. For efficient transportation, storage, and use, such gases are liquefied under pressure and distributed in cylinders. Apart from domestic use as bottled gas, these gases are used for the manufacture of chemicals and in aviation fuels.

liquid injection incinerator An incinerator commonly used for the destruction of HAZARDOUS WASTES; it consists of a refractory-lined combustion chamber with a burner at one end. Readily combustible liquid wastes are pumped into the incinerator via the burner, which converts the waste into a fine spray. These incinerators operate at very high temperatures and are suited to the processing of stable chlorinated hydrocarbon wastes. (See INCINERATION.

lithosphere The crust enclosing the kernel of the Earth or barysphere. It is commonly considered as extending to a depth of about 80 km from the surface of the Earth; however, only the outer part of it or BIOSPHERE is associated with any form of life. (See HOT DRY ROCK SYSTEMS.)

litigation Legal proceedings. A litigant is a person engaged in a law suit.

litter Any kind of rubbish, refuse, or garbage, and any article or matter that when left, deposited, dropped, or thrown in a public place, public reserve, or on private land leads to the defacement or defilement of that place or land. (See KEEP AMERICA BEAUTIFUL; KEEP AUSTRALIA BEAUTIFUL COUNCIL; KEEP SINGAPORE CLEAN CAMPAIGN.)

littoral Relating to or taking place on or near the shore.

littoral drift Sand moved under the effect of longshore current; it carries sand which has been stirred into suspension by the turbulence of the breaking waves. Onshore and offshore sand movements caused by low swells and steep waves, respectively, coupled with littoral drift, help to explain the major shoreline changes on the open coasts of the world.

littoral rainforest A type of RAINFOREST found next to the sea and influenced by the sea. (See MANGROVE.)

littoral zone The shallower parts of a sea or lake, being areas roughly limited to a depth reached by light and wave action. For the sea, the zone reaches out from the high watermark to perhaps 200 m. The zone sustains a high level of photosynthetic activity and is characterized by rooted vegetation. (See LIMNETIC ZONE; PHOTOSYNTHESIS.)

living resource conservation As defined in the WORLD CONSERVATION STRATEGY, a proposed conservation policy with three specific objectives: (1) the maintenance of essential ecological processes and life-support systems; (2) the preservation of GENETIC DIVERSITY; and (3) to ensure the sustainable utilization of species and ecosystems. (See ECOSYSTEM; SUSTAINABLE DEVELOPMENT; SUSTAINABLE YIELD.)

LNG LIQUEFIED NATURAL GAS.

load characteristics The load pattern for a particular plant, as revealed by variations in demand over time; such variations may be diurnal, seasonal, or secular. Peak loads may often be modified by effective management and peak-load pricing systems.

load factor In respect of an industrial plant, actual output expressed as a percentage of what would have been produced had all the plant been operating continuously throughout a specified period.

load-on-top system A code of practice aimed at minimizing pollution from oil-carrying ships; it involves passing the washings from tank-cleaning operations and residue from the discharge of the original ballast water to an empty cargo tank nominated as the 'slop' tank. Fresh oil cargo is loaded on top of the final residue left after further discharges of water, the resulting mixture being acceptable to refineries despite some additional cost in removing the salt and water. (See CONVENTION FOR THE PREVENTION OF MARINE POLLUTION BY DUMPING FROM SHIPS AND AIRCRAFT; CONVENTION ON THE PREVENTION OF MARINE POLLUTION BY THE DUMPING OF WASTES AND OTHER MATTER.)

loam A type of SOIL that consists of a mixture of clay, sand, and organic matter and is therefore well-draining and rich in nutrients, usually providing a good base for the growth of many plants.

lobbying Any attempt by individuals, public agencies, or private interest groups to influence the decisions of government; lobbying in some form and to some degree appears inevitable in any political system. Formal presentations to members of parliament and government ministers as well as to agencies may be regarded as valuable aspects of democratic life. However, lobbying reinforced by the movement of funds in various ways, or the movement of support used as a threat, or by direct threat and bribery, or by the offer of campaign funds or other forms of 'assistance' are highly undesirable. Even the more innocent kinds of lobbying (a meeting with a minister) may be resented by those who do not have the resources to lobby. Lobbying is therefore something of a grey area utilized by groups and interests for all manner of purposes all the time. In the environmental arena, members of governments may be beset by loggers and woodchippers on the one hand and nature conservation bodies on the other; by objectors to planning schemes and major development programmes and the proponents of those schemes; by people who pursue bloodsports and animal rights activists.

location of industry See INDUSTRY, OPTIMAL LOCATION OF.

location theory In ECONOMICS and geography, theory concerned with the geographic location of industrial and commercial activity. The subject has become an integral part of spatial economics, regional development, and economic geography.

London Conference on Climatic Change A conference convened in 1989 and attended by the representatives of 188 nations to discuss the protection of the OZONE LAYER in the upper atmosphere and the control of the GREENHOUSE EFFECT. The conference strengthened the MONTREAL PROTOCOL ON SUBSTANCES THAT DEPLETE THE OZONE LAYER by banning the production of all CHLOROFLUOROCARBONS by 1995. The conference also recognized the need for a major reduction in the emission of CARBON DIOXIDE from the burning of fossil fuels.

London Dumping Convention See CONVENTION ON THE PREVENTION OF MARINE POLLUTION BY THE DUMPING OF WASTES AND OTHER MATTER.

London smog incidents Acute episodes of heavy pollution associated with natural fog covering the Greater London area; the episode of December 1952, when 4000 people died, led to the BEAVER REPORT and the Clean Air Act of 1956. Since then the smog problem of the London area has abated.

looping The behaviour of a chimney plume in the presence of superadiabatic temperature lapse rates. Large-scale thermal eddies are set up and sporadic puffs of pollutants at high concentration may be brought to the ground at short distances for a few seconds. Looping occurs with light winds. (See LAPSE RATE.)

Lord Howe Island An island located in the south Pacific, 780 km northeast of Sydney, Australia, included in the WORLD HERITAGE LIST by UNESCO in 1982. Lord Howe Island and its surrounding islets are renowned for natural beauty.

Los Angeles smog Smog of a photochemical nature, largely resulting from the interaction of sunlight and pollutants emitted by motor vehicles. Unlike London smog of earlier times, Los Angeles smog contains neither smoke nor fog. Despite stringent exhaust controls the problem persists in the Los Angeles basin.

Los Glaciares National Park Established in 1937, a national park in Santa Cruz province, southwestern Argentina, in the Andes. The park has forests and grassy plains in the east, and needlelike peaks, large glaciers, and snow fields in the west. Wildlife includes guanacos, chinchillas, pudu and guemal, condors and rheas.

loudness In ACOUSTICS, an attribute of sound that determines the intensity of auditory sensation produced. A unit of loudness is the PHON; one phon is equal to a difference in sound intensity of one DECIBEL.

Love Canal In 1977, a crisis when it was revealed that many residents in the Love Canal area of Niagara Falls township were suffering severe chromosome damage from toxic chemical waste buried there. Indeed, an abnormally high proportion of the 3000 members of the community were found to be suffering from various forms of cancer, high rates of birth defect, liver and kidney damage, and respiratory ailments. For several years, two corporations had been using the abandoned Love Canal project for the dumping of toxic wastes; a school was built on part of the landfill site and residences on the remainder. In 1977, a toxic substance began bubbling to the surface of the school playground. Love Canal was declared a disaster area, the state buying almost 1000 homes. Love Canal became a symbol in the US and elsewhere for the problem of finding safe disposal sites for toxic chemical wastes. (See CONTAMINATED SITES.)

low-level waste Part of the waste from various stages of the nuclear fuel cycle typically containing only a few curies per cubic metre; it presents no hazard to the public and is suitable for disposal by burial or dumping at sea.

LPG LIQUEFIED PETROLEUM GAS.

LPG road tanker incident, Spain An incident at Los Alfaques, Spain, in 1978 in which a road tanker loaded with LIQUEFIED PETROLEUM GAS (LPG) exploded in a caravan park, with dozens of fatalities. There were several contributing factors in the condition of the tanker as well as overfilling.

Lucerne Conference A conference of the environment ministers of the EUROPEAN UNION held in 1993 to review the state of the European environment; the conference agreed on an action programme setting out the key policy changes necessary to help central and eastern Europe deal with any adverse environmental consequences of the previous 40 years. A follow-up conference was held in Sofia, Bulgaria, in 1995.

luddites Members of bands of 19th-century English handicraftsmen who attempted to destroy textile machinery that was displacing them from their jobs; they were generally masked and operated at night. The outcome was many hangings and transportations.

L_x noise levels Noise levels in dB(A) which are exceeded for a specific percentage of the measurement period. For example, L_{10} and L_{90} noise levels mean that the noise level in dB(A) was exceeded for 10% and 90% of the measurement period, respectively. The L_{90} index may be adopted to establish a true background noise level during the relatively short period when other sounds do not intrude. The L_{10} index may be adopted for measuring disturbance by traffic noise. For example, an environmental-control agency could recommend that existing residential development should not be subjected to more than 70 dB(A) on the L_{10} index unless some form of remedial or compensatory action is taken.

lysimeter A device used to measure the quantity of water movement through or from a body of soil or other material, such as solid waste, or used to collect percolated water for qualitative examination.

M

Maastricht Treaty A treaty that came into effect in November 1993, creating the EUROPEAN UNION and committing that Union to further steps in economic and monetary integration. The treaty, introduced to the 12 (now 15) members of the European Community in 1991 received uncertain support in some countries, finally being ratified in October 1993. The creation of the European Union in place of the European Community is a further step also in the promotion of uniform environmental policies throughout the Union.

MAC MAXIMUM ALLOWABLE CONCENTRATION.

macroeconomics A branch of ECONOMICS concerned with the analysis of the economy in the large, that is, with such large aggregates as the volume of employment, the GROSS DOMESTIC PRODUCT (GDP), savings and investment, economic growth, inflation, the balance of payments, exchange rates, and the business cycle. In sum, macroeconomics is vitally concerned with the interactions among the goods, labour, and assets markets of the national economy. (See MICROECONOMICS.)

macroenvironmental problems Regional, national, and international environmental problems in the large. (See MICROENVIRONMENTAL PROBLEMS; Box 22.)

macrofauna The larger types of animal life.

macroflora The larger types of plant life.

macromolecule In chemistry, a very large molecule, usually a polymer.

macronutrients Mineral nutrients utilized by organisms in large quantities, for example, carbon, hydrogen, oxygen, nitrogen, phosphorus, sulphur, potassium, and calcium. (See MICRONUTRIENTS.)

macroorganism An organism visible to the unaided human eye.

macrophyte A plant visible to the unaided human eye.

MACT MAXIMUM ACHIEVABLE CONTROL TECHNOLOGY.

Mad Hatters disease A disease suffered by hatters of a century ago who became mentally deranged as a result of absorbing MERCURY used in making felt for hats.

Madrid Protocol See CONVENTION FOR THE REGULATION OF ANTARCTIC MINERAL RESOURCE ACTIVITY.

magma bodies Molten material consisting of silicates and volatiles originating within the lower crust of the mantle of the Earth. Magma that reaches the surface as a liquid is known as lava, while the solidified products are known as IGNEOUS ROCK. The thermal

---

Box 22 Macroenvironmental problems: regional, national and international

(1) Dereliction, slums, and blight in most of the cities and rural districts of the world.
(2) Unsafe water supplies.
(3) Inadequate or non-existent sewerage systems and the pollution of waterways.
(4) Vector breeding.
(5) Regional, transboundary, and global air pollution.
(6) Floods and associated threat to life and property.
(7) Incidence of drought and famine.
(8) Desertification.
(9) Soil degradation.
(10) Population growth in relation to resources.
(11) Mortality and morbidity arising from environmental sources.
(12) Threats to natural resources including ecosystems, forests, woodland, and mangroves.
(13) Threats to endangered fauna and flora.
(14) Increasing noise levels.
(15) The disposal of toxic and nuclear wastes.
(16) Threats to regional seas and marine resources.
(17) The location of hazardous industries.
(18) Visual pollution.
(19) Atmospheric warming and climatic change.
(20) Threats to the ozone layer.

energy in magma systems represents, as in the case of HOT DRY ROCK SYSTEMS (HDR), a huge potential heat resource, while being more accessible. The goal of the US Magma Energy Extraction Program is to determine the engineering feasibility of locating, accessing, and utilizing magma as a viable energy source by the year 2000.

magnetohydrodynamic generation (MHD generation) Direct electricity generation utilizing hot flue gases and a magnetic field. In the simplest form of MHD generation, a hot electrically conducting fluid at about 2500°C is expanded through a nozzle and passed along a duct at high velocity. A magnetic field acts in a direction at right angles to the axis of the duct, the electric currents induced in the flowing gas being collected on electrodes. Seeding of the gas with caesium or potassium enhances electrical conductivity by many orders of magnitude. Such a technique could enhance power station efficiency and conserve resources. Schemes have been successfully developed to generate MHD electricity.

magnetosphere A vast region of the ATMOSPHERE in which charged particles move along the flux lines of the Earth's magnetic field. The lower boundary of the magnetosphere is several hundred kilometres above the Earth's surface while the magnetosphere is some 100 km in thickness.

major accident hazard An imprecise term for situations or conditions of exceptional risk; the term appears in the EU Council Directive 82/501/EEC of 24 June 1982 entitled *The Major Hazards of certain Industrial Activities*. (See BOILING LIQUID EXPANDING VAPOUR EXPLOSION; HAZARD AND RISK ASSESSMENT; UNCONFINED VAPOUR CLOUD EXPLOSION.)

malaria One of the most ancient infections known, transmitted usually by the anopheline mosquito, marked by periodic fever and an enlarged spleen. Malaria occurs throughout the world, though it is most common in the tropics where climatic conditions are favourable for breeding. Control of mosquito populations is of major importance; a global malarial eradication programme was launched in 1955 under the auspices of the WORLD HEALTH ORGANIZATION.

Malaysia, evolution of environmental policy in See Box 23.

Malthusian Theory of Population A theory of population developed by Thomas R. Malthus (1766–1834) and published in *An Essay on the Principle of Population* in 1798. The essence of the theory was: (1) that population would soon outstrip the means of feeding it, if it were not kept down by vice, misery, or self-restraint; (2) that in a state of society where self-restraint does not act at all, or only acts so little that we need not think of it, population will augment until the poorest classes of the community have only just enough to support life; and (3) in a community where self-restraint may act eventually, each class of the community will augment until it reaches the point at which it begins to exercise that restraint. Population, Malthus declared, tended to increase in a geometrical progression, whereas the means of subsistence increased in only arithmetical progression. The world is now characterized by falling birth rates, much higher food production, and much extended life spans while WORLD POPULATION continues to increase. Many countries have passed through a DEMOGRAPHIC TRANSITION in which the dire predictions of Malthus have been dramatically modified by family planning, food production, education, and generally improving living standards.

mammalogy The scientific study of mammals; it is a multidisciplinary field encompassing anatomy, palaeontology, ECOLOGY, behaviour, cytogenetics, biochemistry, and telemetry.

Mammoth Cave National Park Established in 1941, a park containing an extensive system of limestone caverns, located in Kentucky, USA. In 1972, a passage was discovered linking the Mammoth Cave and the Flint Ridge Cave System; the combined underground passages run for nearly 500 km. The caves contain numerous unique geological formations such as the Pillars of Hercules and underground lakes and rivers. Various animals have adapted to the dark ENVIRONMENT including cave crickets, blind fish, and blind crayfish; the flora consists mainly of fungi. Mummified Indian bodies, possibly pre-Columbian, have been found in the caves. Mammoth Cave has been included in the WORLD HERITAGE LIST.

Man and Biosphere Program See UN EDUCATIONAL, SCIENTIFIC AND CULTURAL ORGANIZATION.

management environmental plans Plans that have been developed to guide the present and future management of a geographical area in terms of land use, water, flora and fauna, ecosystems, pollution control, nature conservation, natural resources, and environmental enhancement.

mandatory dedication A measure that local planning bodies may incorporate in subdivision regulations to acquire open land in new residential developments, as a condition for subdivision plan approval. For example, the planning body may require the developer to dedicate a portion of the development site or area for use as a public recreational site, a school, or an amenity site.

mangroves Plant communities and trees that inhabit tidal swamps, muddy silt, and sand banks at the mouths of rivers and other low-lying areas which are regularly inundated by

Box 23	Malaysia: evolution of environmental policy
1920	Waters Enactment
1929	Mining Enactment
1934	Mining Rules
1935	Forest Enactment
1949	Natural Resources Ordinance
1952	Poisons Ordinance; Merchant Shipping Ordinance; Sale of Food and Drugs Ordinance; Dangerous Drugs Ordinance
1953	Federation Port Rules; Irrigation Areas Ordinance
1954	Drainage Works Ordinance
1956	Medicine (Sales and Advertisement) Ordinance
1958	Explosives Ordinance; Road Traffic Ordinance
1960	Land Conservation Act
1965	National Land Code; Housing Development Act
1966	Continental Shelf Act
1968	Radioactive Substances Act
1969	Civil Aviation Act
1971	Malaria Eradication Act
1972	Petroleum Mining Act
1974	Environmental Quality Act; Geological Survey Act; Street, Buildings and Drainage Act; Aboriginal Peoples Act; Factories and Machinery Act; Pesticides Act
1975	Destruction of Disease-Bearing Insects Act; Municipal and Town Boards Act
1976	Protection of Wildlife Act 1976; Antiquities Act; Local Government Act; Town and Country Planning Act
1980	Malaysian Highway Authority Act; Pig Rearing Act
1984	Atomic Energy Licensing Act; Exclusive Economic Zone Act; National Forestry Act
1985	Fisheries Act
1988	EIA becomes mandatory; nineteen categories of industry required to submit EIA reports
1991	331 major projects have now been subject to EIA
1991-95	Sixth Malaysia Plan

the sea, but which are protected from strong waves and currents. Mangroves are the only woody species that will grow where the land is periodically flooded with sea water; individual species have adapted themselves to different tidal levels, to various degrees of salinity, and to the nature of the mud or soil. Mangroves vary in size from substantial trees up to 30 m in height down to miniature forms less than waist high. Mangrove swamps and thickets support hundreds of terrestrial, marine, and amphibian species; have a special role in supporting estuarine fisheries; provide shelter, refuge, and food for many forms of wildlife; are involved in nutrient recycling; help to reduce water pollution; help to prevent bank erosion; and act as visual screens.

Man, Medicine and Environment   Published in 1968, a major work by René Dubos (1901-82), microbiologist and environmentalist, responsible for major advances in antibiotics. Dubos concluded that early environmental influences can have lasting effects on the anatomical, physiological, and behavioural characteristics of adult animals and humanity.

Manovo-Gounda-Saint Floris National Park   Established in 1933, located in the Central African Republic, a national park in the upper basin of the Chari River comprising wooded Sudanian savannah interspersed with marshy area flooded seasonally. Fauna include the elephant, buffalo, antelope, giraffe, lion, hyena, cheetah, and many species of birds including the egret. The park is included in the WORLD HERITAGE LIST.

mantle   The internal region of the Earth, between the core and the outermost crust, about 2900 km thick constituting the bulk of the Earth's volume.

manufacturing   The processing of raw materials to produce a wide range of goods,

manure

with various consequences for the ENVIRONMENT. (See ENVIRONMENTAL IMPACT ASSESSMENT (EIA); LIFE CYCLE EIA; SUSTAINABLE DEVELOPMENT.)

manure   Organic material that is used to fertilize land, usually consisting of barnyard and stable livestock excreta, with or without straw or hay. In many southeast Asian countries, human excrement (nightsoil) is also used. The use of manure as fertilizer dates from the early days of agriculture. (See FERTILIZER.)

marginal cost   The increase in the total cost of producing each successive increment of an output; the cost of producing M plus 1 units, minus the cost of producing M units. As a concept, it includes social and environmental costs, although often construed more narrowly in accountancy terms. The short-run marginal cost is the cost incurred in making marginal or small changes, say in the energy output of a system, within existing capacity. Long-run marginal costs include the marginal costs of changes in the capacity of the system.

marginal land   LAND that is the last to be brought into, or the first to be taken out of, a particular line of production. This will not necessarily be land which is just worth cultivating.

marginal social and private net products   Concepts associated with the name of the economist A. C. Pigou (1877–1959). The marginal social net product is the total net product of physical things or services due to the marginal increment of resources applied in any given use of place, no matter to whom any part of this product may accrue. The marginal private net product is that part of the total net product of physical things or services, due to the marginal increment of resources applied in any given use or place which accrues in the first instance, to the person investing resources there. Pigou's theme was that, in general, industrialists are not interested in the social, but only in the private net product of their operations. Self-interest will tend to bring about equality in the values of the marginal private net product of resources invested in different ways, but it will not tend to bring about equality in the values of the marginal social net products, save where they are incidentally identical. Where there is a divergence between the social (including the environmental) costs and private benefits, self-interest will not close the gap. Only government intervention can achieve this. (See ENVIRONMENTAL IMPACT ASSESSMENT; QUALITY OF LIFE.)

marginal utility   The increase or decrease in the total utility (or satisfaction) of consumption of a good or service that results from increasing or decreasing the quantity of the good or service consumed by one unit or increment.

The law of diminishing marginal utility is a proposition that the marginal utility of a commodity or service to anyone, other things being equal, diminishes after some point with every increase in the supply of it. (See ECONOMICS.)

mariculture   See AQUACULTURE.

marine archaeology   The study of the underwater cultural heritage, notably the investigation and preservation of shipwrecks.

marine biology   The science that deals with FLORA and FAUNA that live in the sea, and life that is airborne or terrestrial that depends on salt water for food and other necessities. Some of its branches are involved with natural history, taxonomy, embryology, morphology, physiology, ecology, and geographical distribution. Marine biology is closely related to the science of OCEANOGRAPHY.

marine geophysics   The scientific discipline that involves the application of geophysical methods to problems of marine geology. Marine geophysics is closely associated with the concepts and problems of seafloor spreading, continental drift, and plate tectonics.

marine park   An area of marine and estuarine habitat set aside as an underwater park and reserve; an extension of the terrestrial NATIONAL PARK idea to the undersea world as a permanent reservation of part of the sea bed. Marine parks exist, for example, in Australia, Britain, Israel, Japan, the Philippines, and the USA. (See GREAT BARRIER REEF.)

marine sediment   Any deposit of insoluble material, such as rock and soil particles, that accumulate on the seafloor from any source.

marine terrace   A rock terrace formed where a sea cliff, with a wave-cut platform before it, is raised above sea level; such terraces are found in California and Oregon, USA, and in Chile, Gibraltar, and New Zealand.

market   A situation, location, or system of communication that provides an opportunity to sell or buy in an orderly way resources, goods, services, stocks and shares, currencies, and contracts. A market may be local, regional, national, or worldwide. Transactions may take time, or be completed within minutes. Procedures may consist simply of buying or selling what is on display or may involve systems of tendering, auction, and open outcry. Markets may be efficient or inefficient. Markets determine values, but only for those things that pass through them. They do not measure the value of voluntary work or environmental assets or the long-term viability of the economic system. (See ALLOCATIVE EFFICIENCY, SOCIAL; COMMON PROPERTY RESOURCES; COMMONS, TRAGEDY OF; ECONOMIC SYSTEM, FUNCTIONS OF; ECONOMIC WELFARE; EXTERNALITY; FREE GOODS; GROSS DOMESTIC PRODUCT; MARKET FAILURE; PRINCIPLES OF SUSTAINABILITY; QUALITY OF LIFE; SUSTAINABLE DEVELOPMENT.)

marketable pollution rights See BUBBLE CONCEPT; EMISSION TRADING PROGRAM.

market economy An economy in which many crucial economic decisions are made by numerous private individuals, companies, and government agencies, catering for competitive markets; the role of government itself tends to be confined to setting the rules of the game, providing a framework of legislation coupled with monetary and fiscal policies.

market efficiency (1) The degree to which a MARKET ECONOMY meets the needs of society and contributes to the QUALITY OF LIFE, encourages efficient production, and responds over time to changing requirements and tastes. (2) In the shorter term, it is a measure of the extent to which the market clears, the quantity of goods and services balancing those demanded, matching supply and demand, particularly in respect of LABOUR.

market failure (1) A situation in which the MARKET ECONOMY does not achieve PARETO OPTIMALITY. (2) A situation in which the market in responding to effective demand only, fails to meet the needs of the less fortunate for housing and other basic needs. (3) A situation in which the market, in responding to effective demand, only fails to harness the available resources of society, particularly with respect to LABOUR. (4) The inability of the market to cater for the PRINCIPLES OF SUSTAINABILITY.

market prices The prices actually paid in the exchange of goods and services.

Marpol convention CONVENTION FOR THE PREVENTION OF POLLUTION FROM SHIPS.

maturation pond See LAGOON.

masking The raising of the threshold of a sound by the presence of another sound; the effect is most marked when the masked sound is of higher frequency than the masking sound.

mass A basic property of matter characterized by inertia and momentum and within the influence of gravity and weight. Although mass is strictly defined in terms of inertia, it is conventionally expressed as weight, the standard unit of mass being one kilogram. However, while a satellite launched into space weighs increasingly less, its mass remains the same, displaying inertia and momentum. For the nuclear physicist, the terms mass and energy are interchangeable.

mass balance A tabulation of all material entering and leaving a system; under steady conditions the total MASS entering the system in unit time must equal the total mass leaving.

mass flow In BOTANY, the most widely accepted explanation for the movement of sugars and other nutrient solutes through the phloem, which is the food-conducting tissue of vascular plants consisting of sieve tubes and other cellular material. Mass flow is also called pressure flow. According to the mass flow hypothesis, dissolved nutrients move from the source into the sieve tubes by 'active transport' and hence to sink areas where they are metabolized or stored. The 'active transport' mechanism is quite complex.

mass movement (mass wasting) The bulk movement of soil and rock debris down slopes, responding to the pull of gravity or the sinking of the Earth's ground surface.

mass number The total number of protons and neutrons in a nucleus, equal to the integer nearest in value to the atomic mass when the latter is expressed in atomic mass units. In the symbol for the nuclide, the mass number is indicated by a number following the element symbol; for example, U-238.

mass production The manufacture of goods on a large scale aimed at low unit costs coupled with large sales; it involves a high degree of specialization of labour and sophisticated automation.

mass spectrometry A technique for obtaining the mass spectrum of a beam of ions by means of suitably disposed magnetic and electric fields. The deflection of any individual ion in these fields depends on the ratio of its MASS to its electric charge. Such a spectrum will appear as a number of lines on a photographic plate, each corresponding to a definite ratio value. Mass spectrometry is widely used to measure the masses and relative abundance of different isotopes and to determine their relative abundance in various natural or enriched samples. Mass spectrometry is also used in gas analysis, for example, in relation to hydrocarbon gases where with automatic recording process control becomes possible. The technique can also be used to measure the geological age of minerals.

mass transfer (1) In respect of a fluid consisting of two or more components in which a concentration gradient exists, the movement or flow of the components so as to reduce the concentration gradient. Mass transfer may be effected by molecular diffusion, a slow process, while in conditions of turbulence molecular diffusion is supplemented by eddy diffusion, a much more rapid mixing process. (2) A process in which matter passes in the molecular state across the boundary between two phases, as, for example, from a gas into a liquid.

mass transit Any transportation system designed to move large numbers of the general public throughout metropolitan areas. Peak periods for mass transit are the morning and evening rush hours, when commuters journey to and from workplaces. Mass transit systems encompass railways, both underground and overground, trams and light-rail systems, trolley buses, helicopters, boats and ferries, hovercraft, buses, streetcars,

mass wasting

minibuses, taxis, people movers, monorails, and moving walkways.

mass wasting   See MASS MOVEMENT.

maximum achievable control technology (MACT)   Emission control requirements that are more exacting than those imposed by BEST AVAILABLE CONTROL TECHNOLOGY; this higher standard may involve the application of new, original, or innovative control technology to emission sources, and almost regardless of cost.

maximum allowable concentration (MAC)   The concentration of a pollutant considered to be harmless to healthy adults during normal working hours of eight hours and a working week of five or six days; these standards are usually incorporated into factory regulations. Sometimes known as a 'maximum exposure limit'.

maximum permissible body burden   The total body content of a radioisotope that, if maintained, will not expose to any critical organ a dose greater than the maximum permissible by regulation.

maximum permissible concentration   The concentration of a radioisotope in air, water, milk, or other products that will deliver not more than the maximum permissible dose to a critical organ, when breathed or consumed at a normal rate.

maximum permissible dose   The dose of ionizing radiation accumulated in a specified time, of such magnitude that no bodily injury may be expected to result in the lifetime of the person exposed and no intolerable burden is likely to accrue to society through genetic damage to descendants.

maximum permissible level   A term used loosely to refer to the maximum permissible dose-rate of radiation, the MAXIMUM PERMISSIBLE CONCENTRATION of a radioisotope, or the maximum permissible degree of contamination of a surface.

maximum sustainable yield   The maximum possible yield or catch that can be removed repeatedly from a population without, in the long term, reducing its size. At this level, the population can sustain the loss of a certain proportion of its members that are continuously replaced. The loss of whale stocks has been a classical example of overharvesting; but examples of overgrazing occur in farming, overharvesting in fishing and forestry, and in other industries. (See SUSTAINABLE DEVELOPMENT; SUSTAINABLE YIELD.)

mediation   An alternative form of resolving disputes, avoiding the costs of a full court hearing. The aim is to bring the parties to a dispute together with a mediator, in a more informal atmosphere, to resolve the problems by mutual agreement. The principle of mediation of environmental disputes has gathered strength, most noticeably in the USA.

Mediterranean   See CONVENTION FOR THE PROTECTION OF THE MEDITERRANEAN SEA AGAINST POLLUTION.

Meeting of the Parties to the Framework Convention for Climate Change   (See BERLIN CLIMATE CONFERENCE.)

megalopolis   A number of coalescing metropolitan areas to form a huge massing of people and economic activities. The term was introduced by the French geographer Jean Gottman (b. 1915) who referred to the northeastern area of the USA as perhaps 'the cradle of a new order in the organization of inhabited space'. Gottman was referring to the area extending from Boston to Washington DC; but the term may also be applied to such areas as metropolitan Los Angeles and San Francisco, metropolitan Dallas and Houston, metropolitan Chicago and Milwaukee. In Europe, the term megalopolis may be applied to the area around London and the Black Country, England; to the area of Greater Paris; to the Ruhr in Germany. In Asia, the term may be applied to the Tokyo–Osaka complex.

Meiofauna   See MESOFAUNA.

Mekong River Commission   A commission established by an Agreement on Cooperation for the SUSTAINABLE DEVELOPMENT of the Mekong River Basin that came into force in 1995; the parties to the agreement are the Kingdom of Cambodia, Lao People's Democratic Republic, Kingdom of Thailand and the Socialist Republic of Vietnam. The longest river in southeast Asia, the Mekong has its source in China's Qinghai province. It then passes through Laos, touching the frontier of Thailand, and then through Cambodia; in Vietnam the Mekong separates into four main branches and enters the South China Sea through a delta. The agreement between the four governments covers the lower Mekong River Basin. The Mekong River Commission consists of three permanent bodies: a council, a joint-committee, and a secretariat. The secretariat is located in Bangkok, Thailand. The function of the commission is to implement the principles of sustainable development enunciated at by the UN CONFERENCE ON ENVIRONMENT AND DEVELOPMENT and the Rio Declaration on Environment and Development. (See COMMISSION FOR SUSTAINABLE DEVELOPMENT, UN.)

melanism   In moths, descriptive of increased dark pigmentation of the wings and other parts of the insect; usually melanism is controlled genetically, but industrial melanism, involving adaptation to a dark, dirty background, is brought about through natural selection. Melanism protects moths from predatory birds.

meltdown   A description of the possible effects of a major breakdown in a nuclear reactor when the reactor becomes uncontrolled, the fuel elements fuse and the molten mass burns through the base of the reactor and foundations and into the ground. (See CHINA SYNDROME.)

membrane   (1) In BIOLOGY, the thin layer that forms the outer boundary of a living cell or of an internal cell compartment. (2) In industrial processes, thin films usually of sophisticated plastics that separate different components for liquids or gases. For example, a reverse osmosis membrane can separate dissolved salts from water by forcing the water through a thin membrane, holding the salts back. Membrane technology can be used to treat sewage and other waste effluents.

mercaptans   Evil-smelling organic compounds containing sulphur; also called thio alcohols. The simpler mercaptans have strong, repulsive garlic-like odours, the latter becoming less pronounced with increasing molecular weight and higher boiling points. Mercaptans may be produced in oil-refinery feed preparation units as a result of incipient cracking; the offensive gases are burnt in plant heaters. Mercaptans arising in cracking units are removed by scrubbing with caustic soda, the mercaptans being removed from the spent caustic soda by steam stripping and subsequently burnt in a process furnace. Mercaptans are also produced by decaying organic matter and emitted at sewage works, food-processing plants, and brick-making works. The most offensive is ethyl mercaptan, although this is readily oxidized to ethyl disulphide.

mercury   A heavy liquid metal obtained generally from the roasting of mercuric sulphide ore (cinnabar). Mercury has been used in older chlor-alkali plants, mercurial catalysts, the pulp and paper industry, and in seed treatment. It is widely used in measuring instruments, in the preparation of pharmaceuticals, in the manufacture of agricultural and industrial fungicides, and in mercury amalgams in dentistry. It is released into the ENVIRONMENT in the burning of fossil fuels and in mining, refinery, and industrial processes. Mercury compounds may function as a cumulative poison affecting the nervous system, particularly in the form of methyl mercury. Mercury may become concentrated in food chains, particularly in freshwater and marine organisms. An epidemic outbreak of methyl mercury poisoning occurred in Japan in the early 1950s, following the ingestion of polluted shellfish and fish, resulting in some fatal cases and congenital disease. This outbreak was closely linked with heavy industrial pollution in Minamata Bay. The source of the mercury was traced to the effluent from a factory. Chronic mercury poisoning may result from the occupational inhalation of mercury vapours and dusts, or from absorption through the skin. The presence of mercury in the environment does not necessarily indicate industrial pollution; mercury in the ocean, for example, may be largely of natural origin. (See CHEMICAL POLLUTANTS; HEAVY METALS; MAD HATTER'S DISEASE.)

merit goods and services   Goods and services distributed by government free, or almost free, to its citizens on the basis of merit or need; for example, hospital services, education, social services, employment services, and the arts. Merit goods differ from public goods in that they provide specific personal benefits for the which the individual would be expected to pay if supplied through the market but might well be unable to do so. (See INTRAGENERATIONAL EQUITY.)

Mesa Verde National Park   Established in 1906, a national park in southwestern Colorado, USA, preserving notable prehistoric cliff dewllings; it contains hundreds of pueblo (Indian village) ruins up to 13 centuries old. The most striking remains are multistorey apartments built under overhanging cliffs. Cliff Palace, excavated in 1909, contains hundreds of rooms, including circular ceremonial chambers. The park also has some very striking and rugged scenery.

mesoclimate   A local climatic effect due to terrain, extending in influence over a distance of several kilometres and one or two hundred metres vertically; this is in contrast to the regional or macroclimate.

mesofauna (meiofauna)   In soil science, animals of intermediate size classed as those greater than $40\,\mu$ in length. Mites, nematodes, and springtails are typical examples. Most mesofauna feed on decaying plant material; in removing roots, they open drainage and aeration channels in the soil.

mesopause   See MESOSPHERE.

mesophyte   A plant that requires an average amount of water; in consequence most trees, shrubs, and herbaceous plants, living in climates of moderate rainfall and temperature are mesophytes.

mesosphere   The region of the upper ATMOSPHERE between about 50 and 80 km above the surface of the Earth. The base of the mesosphere merges with the STRATOSPHERE, with the boundary between the two regions being called the mesopause. The mesosphere extends upwards to meet the base of the THERMOSPHERE. The composition of the air of the mesosphere is similar to that near the Earth's surface, but there is little water vapour and somewhat higher levels of OZONE.

metabolism The sum of the physical and chemical processes that take place within the living cells of an organism, providing energy for vital processes and synthesizing new organic material. Metabolic reactions are of two types: (1) ANABOLIC reactions that use energy to build complex molecules from simpler molecules; and (2) catabolic reactions that break down complex molecules into simpler molecules, releasing energy. Photosynthetic plants use solar energy to synthesize organic compounds from inorganic constituents such as CARBON DIOXIDE, AMMONIA, and water. Non-photosynthetic organisms through digesting proteins, carbohydrates and lipids acquire an indirect source of solar energy as well as materials to synthesize their own organic constituents.

metamorphic rock Rock that has changed character in response to pronounced changes in temperature, pressure, and chemical ENVIRONMENT; the pre-existing rocks may be igneous, sedimentary, or other metamorphic rocks.

metamorphism The process by which consolidated rocks are altered in composition, texture, or internal structure by conditions and forces differing from those under which the rocks originally formed.

metamorphosis In BIOLOGY, a striking change in form or structure in an individual organism after birth or hatching. (See LARVA.)

metastasis The transfer of disease from one organ to another not directly connected with it.

meteorological influences The effect of a range of atmospheric characteristics on the dispersal of pollutants; in respect of AIR POLLUTION, the concentration of pollutants at or near ground level is a balance between the amount emitted and the degree to which it is diluted and dispersed by the ATMOSPHERE, or removed from the atmosphere by gravitation or precipitation. In a modern industrial society, it is only through the enormous dilution produced by the atmosphere that renders the situation at all tolerable. Without turbulence in the atmosphere humanity could not live in large communities. Only in exceptional circumstances will high pollution levels be found when the wind is strong and possessing considerable turbulence. The most severe pollution occurs in non-turbulent conditions during an INVERSION. (See also LAPSE RATE.)

meteorology The scientific study of the ATMOSPHERE; its structure, physical and chemical characteristics, and phenomena. The atmosphere is regarded as consisting of several more or less concentric layers, the layer adjacent to the Earth's surface being the TROPOSPHERE, followed by the STRATOSPHERE, MESOSPHERE, HOMOSPHERE, IONOSPHERE, THERMOSPHERE and MAGNETOSPHERE.

methaemoglobinaemia The presence of methaemoglobin (a compound of HAEMOGLOBIN and oxygen) in the blood. It is produced by the action of oxidizing agents on the blood and gives rise, in infants, to the condition described as 'blue baby disease'. It is associated with drinking water containing nitrogen in the form of nitrates. The stomachs of infants reduce nitrate to nitrite, which gives rise to this condition.

methane A non-toxic gas, though contributing to the GREENHOUSE EFFECT. It is a principal constituent of natural gas, produced during the process of digestion in sludge digestion tanks at sewage works, it occurs in coal-mines as 'firedamp', and is generated at the bottom of lakes and marshes as 'marsh gas'. More generally, it emanates from rice paddy fields, livestock, landfill waste dumps, the inefficient burning of biomass and coal, and increasing agricultural activity. The increase in methane in the atmosphere is closely linked to the world's growing population and its growing needs for food, for example, as expressed in increasing numbers of ruminant animals (cattle and sheep) and increasing hectares of rice paddies. An average cow emits about 280 litres of methane each day. When burnt, methane yields carbon dioxide and water vapour. Atmospheric methane has been increasing at a rate of about 1% each year, its concentration being now about 1.7 ppm. (See SEWAGE GAS.)

methylene-blue stability test A test regarded as a means of assessing the stability of an effluent, stability being defined as the ability of an effluent to remain in an oxidized condition when out of contact with air. The test lasts for 5 days at 20°C; if the blue colour persists after 5 days, the sample is considered to be stable. (See BIOCHEMICAL OXYGEN DEMAND.)

methyl mercury See MERCURY.

metropolis A major city, often the result of the coalescence of a number of smaller towns and villages which become suburbs of the new entity. A major school of thought, following the pioneer British city planner Sir Patrick Geddes (1854-1932) and such architects as the American Frank Lloyd Wright (1869-1959), maintains that the huge metropolis, let alone the MEGALOPOLIS, is an inhuman institution. The evils of noise, congestion, traffic, tension, and impersonality suggest policies, it is argued, to check metropolitan growth with the construction of new towns. Such alternative policies around the world have had limited success, however, while metropolitan areas are considered by many leaving country districts to be of magnetic appeal.

Meuse Valley incident An air pollution incident in December 1930 involving respiratory illness and death; it occurred in the Meuse

Valley, Belgium, in the steep-sided valley between Huy and Liège, a distance of some 24 km. The illnesses and deaths over five days involving several hundred people were caused by poisonous waste gas from many factories, including sulphur dioxide and fluorides, in conjunction with unusual climatic conditions.

Mexico City industrial disaster   In 1984, a fire and explosion at a natural gas storage and distribution facility at San Juan Ixhuatepec, Mexico City, which resulted in several hundred deaths, many more injured, and 10 000 homeless; it was the worst industrial disaster in Mexican history. The plant was situated in the very heart of a poor crowded district, ignoring all environmental protection principles.

MHD generation   See MAGNETOHYDRODYNAMIC GENERATION.

Michigan episode, US   A tragic incident, occurring in June 1973, when at the premises of the Michigan Chemical Corporation a manufactured fire-retarding substance known as PBB or Firemaster was accidentally substituted for a magnesium oxide product called Nutrimaster to add to stock feed to make it more palatable. The two products were similar in appearance. A number of bags of Firemaster were despatched with Nutrimaster stencils, and for six months stock all over Michigan were fed on contaminated feed. From the milk, people suffered a wide range of disabilities. In 1976, tests on over a thousand people revealed that one-third had serious symptoms of PBB poisoning. By then over 30 000 cattle and large numbers of other animals had been destroyed. The effects of this error were felt for many years.

microbiology   A scientific study of microorganisms including protozoans, algae, moulds, bacteria, and viruses. Microbiology is concerned with the structure, function, and classification of such organisms and with ways of managing their activities.

microburst   Intense wind that descends from rain clouds for short periods affecting a limited area; however, by causing a sudden divergence in wind direction speed (known as wind shear) microbursts create a hazard for aircraft at take-off and landing, the pilot being confronted with a rapid and unexpected shift from headwind to tailwind.

microclimate   A climatic condition confined to a relatively small area within a few metres above and below the surface of the Earth and within canopies of vegetation. Vegetation, soil conditions, small-scale topography, structures, and industrial activities may create pronounced microclimate differences.

microeconomic reform   Reform and restructuring of branches of industry, financial institutions, industrial relations, and tariff structures to improve the overall performance of segments of the national economy.

microeconomics   Sometimes termed ECONOMICS in the small; a branch of economics concerned with the analysis of the behaviour of individual consumers and producers, particularly with the optimizing behaviour of individual units such as households and firms. It examines the determination or prices in particular markets and the effects of monopoly and oligopoly in such markets. (See MACROECONOMICS.)

microenvironmental problems   Environmental problems immediately affecting the lives of citizens. (See Box 24.)

microfauna   Small, often microscopic, animals, such as single-celled protozoans, small nematodes, unsegmented worms, and eight-legged arthropods. Many inhabit water or soil feeding on smaller microorganisms.

microflora   Tiny plants which embrace algae, fungi, bacteria, yeasts, and related microscopic organisms.

micromerograph   A fast accurate instrument for determining the particle-size distribution of powdered materials in the subsieve size range. It uses the principle of sedimentation in still air at atmospheric pressure, with a cloud of particles being introduced at the top of a sedimentation column. The particles fall at their terminal velocities on to the pan of a recording balance.

micrometeorology   The detailed study of physical phenomena in the lowest layers of the ATMOSPHERE, perhaps over a restricted area such as the site of a new town.

micronutrients   Mineral nutrients utilized by organisms only in minute amounts such as iron, boron, copper, manganese, zinc, molybdenum, and chlorine.

microorganisms   Microscopic plants (bacteria, and certain fungi and algae) or animals (protozoa, rotifers).

middleground   Landscape perceived by an observer at a distance of 4 to 6 km; this forms a linkage between the FOREGROUND and BACKGROUND parts of the landscape. Within this range, the observer experiences overall shapes and patterns and discerns the relationship between landscape units.

MIGA   MULTILATERAL INVESTMENT GUARANTEE AGENCY.

migration   In ETHOLOGY, the regular, usually seasonal movement of all or part of a faunal population to and from a geographical area; regular migrants include many birds; fish such as salmon, whales, and porpoises; hoofed animals especially in East Africa and in the Arctic tundra; bats; plankton and crustaceans; and turtles. Northern hemisphere birds tend to go south for the winter, while tropical birds migrate according to the cycle of wet and dry

Minamata disease

> **Box 24 Microenvironmental problems immediately affecting lives of citizens**
> (1) Dereliction, slums and blight in certain localities.
> (2) Unsafe water supplies.
> (3) Inadequate or non-existent sewerage systems.
> (4) Vector breeding.
> (5) Air, water, and noise pollution.
> (6) Fumes and vibration from industrial processes.
> (7) Hazards from traffic in the street.
> (8) Poorly located industrial plant.
> (9) Loss of light and over-shadowing from other buildings and overhead roads.
> (10) Severance of communities and neighbourhoods by highways, railways, traffic management schemes, or large-scale developments.
> (11) Lack of space for play or recreation.
> (12) Visual squalour due to litter, garbage, abandoned vehicles and equipment, and overhead wirescape.
> (13) Dereliction arising from abandoned and closed dwellings, business premises, and factories.
> (14) Inadequate street maintenance and drainage.
> (15) Loss of heritage buildings and the special character of areas.
> (16) Inequitable, obtrusive, and antisocial developments.
> (17) Loss of privacy.
> (18) Loss of views and vistas.
> (19) Loss or deterioration of natural assets in the immediate neighbourhood.
> (20) Loss of existence value, that is, the loss of things elsewhere whose existence has been appreciated such as rainforest or brown bears.

seasons. Migration has considerable ecological significance, resulting in the optimal use of food surpluses, though it is not yet fully understood how migrating animals find their way over long and complex routes. (See ANADROMOUS FISH; CATADROMOUS FISH.)

Minamata disease (See MERCURY.)

mine bonding The posting or placing of a bond (sum of money) with a planning or consent-giving body before the commencement of earth-moving at a surface mine, as a guarantee for the restoration or rehabilitation of the area to be mined. If a miner goes into bankruptcy or refuses to reclaim the site, the bond is forfeit to the government which then undertakes the reclamation of the mine site.

mine drainage The liquid effluent discharges from mines either by natural flow or by pumping. Mine drainage often contributes to acid pollution, hardness, metallic compounds, and sediments to waterways. These pollutants often have an adverse effect on aquatic life, water supplies, and the recreational use of waters. Pollution arises from both working and abandoned mines, their refuse heaps, tailing ponds, and washings from preparation processes. (See Figure 19.)

mineral A naturally occurring inorganic substance in the form of an element or compound that has a distinctive chemical composition and physical properties. Examples include bauxite, diamond, gold, tin, and salt. The term mineral is also applied to certain organic substances such as coal, graphite, oil, natural gas, and sulphur.

mineral fraction That portion of SOIL which comprises fragments of rock and particles of sand, containing many of the elements necessary for plant growth. The exact composition of the mineral fraction is influenced by the parent rock.

mineralization The microbial breakdown of HUMUS and other organic material in SOIL to inorganic substances.

mineralogy A scientific discipline that is concerned with all aspects of minerals, including their physical properties, chemical composition, crystal structure, occurrence and distribution, and their origins.

mineral sands Sands that have a high content of commercially important minerals such as rutile (titanium dioxide), ilmenite (a compound of titanium, iron, and oxygen), and zircon (zirconium silicate). If commercial ore-bodies

mining wastes

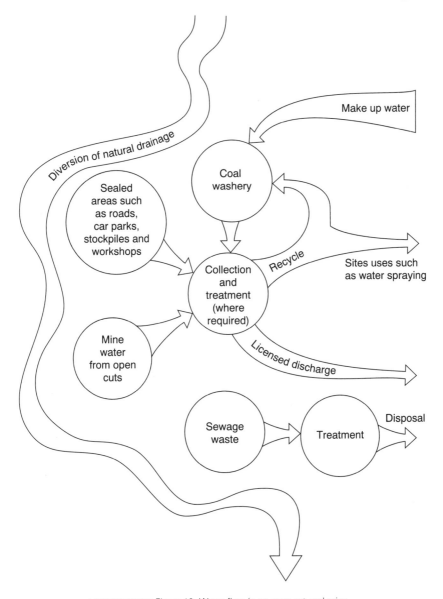

MINE DRAINAGE: Figure 19. Water flow in an open-cut coal mine.

are found in a scenic and sensitive coastal zone, for example, serious land-use conflicts arise. Rehabilitation of mined areas is another important issue.

mineral water Water that contains a large quantity of dissolved minerals or gases; mineral water from natural springs often has a high content of calcium carbonate, magnesium sulphate, potassium, and sodium sulphate.

minimum tillage See CONSERVATION TILLAGE.

mining wastes Unwanted material arising from all classes of mining operations in two principal forms: (1) rock waste, that may occupy valuable land and disfigure the landscape; and (2) tailings from mills, that as silt can impede the natural flow of streams or may contain chemicals hazardous to vegetation and animal life. Modern mining procedures utilize abandoned workings such as shafts, tunnels, and adits to store unwanted waste as backfill, thus conserving space. Some methods of treatment and disposal for coal-washery waste reject material are shown in Figure 4.

minimum-maximum value approach   See EXTERNALITY.

Ministerial Conference on Pollution of the North Sea   A conference held in London, England in November 1987 attended by representatives of the Nordic countries, the Netherlands, Germany and Britain to review growing concerns about pollution of the North Sea and the Skaggerak by nutrient salts. The Netherlands, Germany, and Britain were responsible for the largest proportion of the discharges. The discharges of nutrient salts to the North Sea had doubled over the previous 30 years and the southern parts of the sea were now seriously threatened. The following aims were agreed at the conference: a 50% reduction in the discharges of nutrients and persistent toxic substances by 1995; the dumping of hazardous waste material to end by 1989; the incineration of industrial waste to cease in 1994; more stringent standards to be introduced for discharges of waste by shipping and discharges of oily wastes from oil platforms; the use of the best available technology to minimize discharges of radioactive substances; and the promotion of more comprehensive research. Several of these steps agreed on for the North Sea and Skaggerak have since been extended to include the entire northeast Atlantic. (See CONVENTION FOR THE PREVENTION OF MARINE POLLUTION BY DUMPING FROM SHIPS AND AIRCRAFT; CONVENTION ON THE PREVENTION OF MARINE POLLUTION FROM LAND-BASED SOURCES.)

mitigating measures   Physical actions taken to prevent, avoid, or minimize the actual or potential adverse effects or a project or activity. Measures may include: avoiding an impact by abandoning or modifying a project or by not taking a certain action; minimizing an impact by limiting the magnitude of an action; rectifying an impact by restoring, repairing, or rehabilitating the affected environment; reducing or eliminating an impact over time through adequate maintainance and efficient operation; providing compensation for an impact through relocation, new facilities, sound proofing, airconditioning; providing adequate treatment of all effluents; recycling material; employing waste minimization techniques; landscaping; and undertaking public consultation procedures.

mixed economy   An economy in which resources are allocated partly through the decisions of private individuals and business enterprises and partly through the decisions of government, government agencies, and state-owned enterprises. The two sectors are referred to as the private and public sectors. The history of the 20th century has been marked by continuous controversy over the contributions of the two sectors, some parties favouring small government in a market economy and others favouring a large, even comprehensive, public sector. The issues are often summed up as 'capitalism' or 'socialism', while many countries at the close of this century favour a middle course.

mixed forest   A vegetational transition between coniferous forest and broad-leaved deciduous forest; or a forest with two or more dominant tree species.

mixing height   The depth of ATMOSPHERE within which pollutants are dispersed and mixed; the depth available is often determined by the height of the base of an INVERSION layer.

mixing ratio   In METEOROLOGY, the mass of water vapour per unit mass of dry air in a sample of moist air. The relative humidity is the ratio between the actual weight of water vapour in a given volume of air and the amount which would be present if the air was saturated at the same temperature, expressed as a percentage.

mobile industry   See FOOTLOOSE INDUSTRY.

mobile source   A source of wastes and discharges that is in motion during its normal operating mode; the term includes all vehicles, mobile machinery, ships and aircraft, rockets and space vehicles

Model Cities Program   During the 1960s, an experiment conducted in several dozen US cities in attacking the problem of major blighted areas with massive federal financial aid. It included programmes of physical improvement coordinated with social and economic upgrading through job training, school improvements, encouragement of economic enterprise, and a whole range of measures aimed at reducing poverty and all its adverse effects. The programme was introduced by the Department of Housing and Community Development.

molecular biology   The study of the chemical structures and processes of biological phenomena at the molecular level; the discipline is particularly concerned with the study of proteins, nucleic acids, and enzymes, the macromolecules essential to life processes. It seeks to understand the molecular basis of genetic processes. Techniques used include X-ray diffraction and electron microscopy.

molecule   The smallest particle of any substance that can exist independently and still exhibit the chemical properties of a substance. Molecules are composed of atoms, and bonds which hold the atoms together.

mollisol   A dark-coloured SOIL type being the basis of the world's most productive agricultural soils. Mollisols are found in the wetter area that border desert regions; the ground cover in these areas yields abundant organic matter that decomposes within the soil, giving the soil a well-aggregated structure. These soils compose the Great Plains of North America, the pampas of Argentina, and the steppes of Eurasia, where the farming of wheat and maize are the main forms of agriculture. (See ALFISOL SOIL.)

monetary policy Measures employed by governments to influence the level of economic activity, such as the varying of the supply of money and credit and the manipulation of the rates of interest; such measures are often used in tandem with FISCAL POLICY. The usual goals of monetary (and fiscal) policy are to achieve and maintain full employment, to achieve and maintain a high rate of economic growth, and to avoid inflation by the stabilization of prices and wages. Monetary policy is usually the domain of the central bank and the national government.

monitoring The systematic deployment of monitoring equipment for the purpose of detecting or measuring quantitatively or qualitatively the presence, effect, or level of any polluting substance in air or water, noise and blasting, radiation, transport movements, land subsidence, or change in the character of vegetation.

monoculture The cultivation of a single crop, commonly on a large area of land for commercial purposes.

monorail A light railway using a single rail; the rail may be located either above or below the railway cars. They are usually confined to short journeys around shopping centres and amusement parks. Critics maintain that monorails involve greater cost, unsightliness, inconvenience, low speed, and hazards for those below.

monotremes Mammals that lay eggs instead of having pouches, or having young that develop inside the mother attached to a placenta. There are only three species of monotremes in the world: the platypus, the short-beaked echnida (both endemic to Australia), and the long-beaked echidna (endemic to Papua New Guinea).

Montego Bay convention See CONVENTION ON THE LAW OF THE SEA.

Montreal Protocol on Substances that Deplete the Ozone Layer, 1988 An international agreement reached in 1988 by over 30 countries, aimed at protecting the OZONE LAYER by controlling the emission of CHLOROFLUOROCARBONS (CFCs) and halons. Under the agreement, initiated by UNEP, CFC consumption is to be progressively reduced worldwide, being phased out completely by 1996. The consumption of halons was not to be allowed to increase above the levels of 1988. The protocol followed the CONVENTION FOR THE PROTECTION OF THE OZONE LAYER (the Vienna convention) adopted in 1985. By the mid-1990s, the Third World continued to provide a substantial market for ozone-damaging chemicals made, but not sold, elsewhere.

morbidity The incidence of sickness in a population. The morbidity rate for a particular disease is the number of individuals who contract that disease in a year, usually quoted per thousand of the population, or in a particular age group.

morphogenetic region A theoretical area devised by geomorphologists to relate climate, landforms, and geomorphic processes.

morphology In BIOLOGY, the study of the size, shape, and structure of animals, plants, and microorganisms and the relationships of their internal parts; it is concerned primarily with explaining the shapes and arrangement of parts of organisms in terms of evolutionary relations, function, and development.

mortality The frequency of death in a population or community. The mortality rate is the number of individuals who die in a year, usually quoted per thousand of the population, or by age group.

motile Living organisms capable of spontaneous movement.

motor vehicle pollution The emission of undesirable particles and gases from motor vehicle exhausts and engines contributing to visible and invisible pollution, haze, and PHOTOCHEMICAL SMOG; pollutants include airborne LEAD, CARBON MONOXIDE, oxides of nitrogen, hydrocarbons, and noise. Motor vehicle pollution has been progressively reduced in many countries by the imposition of increasingly stringent standards relating to these pollutants, though better performance per vehicle may be offset in some cases by an increasing number of vehicles. Generally speaking, however, lead in the atmosphere has been much reduced through the widespread introduction of lead-free petrol (gasoline).

mound springs The natural outlets of the Great Artesian Basin, Australia, the largest known ground-water basin in the world. Hundreds of these springs occur between Maree and Oodnadatta in South Australia.

Mound State Monument Established in 1933, near Moundville, western Alabama, USA, a monument containing 40 flat-topped earth mounds that were used as foundations for dwellings and buildings occupied between AD 1200 and 1500 by Indian farmers and pottery makers. The archaeological excavations began in 1906.

Mount Apo National Park Established in 1936, in the Philippines a rugged region surrounding Mount Apo, a volcano, the highest peak in the nation; there are many waterfalls and medicinal hot springs. With diverse fauna and flora, the park is the only habitat of the monkey-eating eagle.

Mount Aspiring National Park Established in 1964, a national park in the South Island of New Zealand, embracing a substantial area of the Southern Alps; the landscape is varied and complex being the source of headwaters for seven rivers, and featuring glaciers, rocky mountains, waterfalls, and passes. Its forests are protected to control soil erosion. The park has scientific, wilderness, natural environment, and development areas.

Mount Carmel National Park Established after 1948 in Israel, a national park embracing the Mount Carmel region overlooking the Mediterranean, with diverse fauna and flora.

Mount Cook National Park Established in 1953 in the South Island of New Zealand; Mount Cook, the highest point in New Zealand is located within the park dominating the valleys, glaciers, and numerous surrounding peaks. More than a third of the park is covered by permanent snow and glacial ice. Trees include beech, tussock, ribbonwood, alpine scrub, pine, and others. Birdlife abounds.

Mount Rainier National Park Established in 1899 in Washington, USA, to preserve the Cascade Range including Mount Rainier. The lower areas have dense forests while during the warmer months, the park and mountain are covered with wild flowers that bloom progressively higher up the slopes. Fauna include deer, elk, bears, mountain goats, raccoons, and squirrels.

Mount Revelstoke National Park Established in 1914 in British Columbia, Canada, a national park occupying the western slope of the Selkirk Mountains near the city of Revelstoke. The park affords a spectacular view of three mountain ranges, Monashee, Selkirk, and Purcell.

mountain wind See ANABATIC WIND.

Mountain Zebra National Park Established in 1937, a national park in southwestern Cape Province, South Africa; it is situated in the semiarid Great Karroo region, west of Cradock. It was established to protect the diminishing mountain zebra as well as other animals.

Muir, John (1838-1914) A naturalist and advocate of US forest conservation, who was largely responsible for the creation of the Sequoia and Yosemite national parks in California, USA.

mulch A layer of organic material applied to the surface of the ground to retain moisture; mulching is the spreading of leaves, straw, or other loose material on the ground to prevent erosion, evaporation, or freezing of plant roots. (See SOIL.)

multiculturalism A concept of society in which numerous ethnic groups retain many of their cultural features in an atmosphere of tolerance and mutual respect, yet whose ultimate loyalties lie with the nation of their adoption.

Multilateral Investment Guarantee Agency (MIGA) Created in 1988 as an arm of the WORLD BANK, the MIGA promotes sustainable economic growth in developing and transitional countries requiring the stimulation of private enterprise and foreign direct investment. It provides investment risk insurance and technical and marketing services.

multiple land use A sharing of land for various land uses including reserves, national parks, heritage areas, recreational uses, agriculture, urban developments, forestry, mining, special-use corridors, and the management of various surface and subsurface resources so that they are utilized in the combination that best meets the present and future needs of the community. (See LAND; LAND-USE PLANNING.)

multiplier A ratio indicating the effect on total employment or on total income of a specified amount of capital investment or expenditure. Investment or expenditure represents income for the factors of production; after tax, some of that income will be saved and the rest spent. This expenditure will in turn become the income of others, along an almost endless chain. If the factors were previously unemployed the benefits are considerable; employing one additional person may well create through the multiplier the equivalent of work for another. Whatever the factor of production, there is a chain reaction.

Murray-Darling Basin Commission A commission created in 1987 by the Australian and several state governments to combat salinity, soil erosion, and environmental problems generally throughout the vast Murray-Darling river system, as well as the management of water resources. Many of the activities in the basin have been unsustainable, leading to the present crisis.

Mururoa Atoll An atoll near Tahiti used by France since the early 1960s as a test site for nuclear devices. Australia, New Zealand, and Papua New Guinea have constantly protested at the radioactive risks to their populations. In 1988, Jacques Cousteau revealed that the coral structure at the test site was deeply cracked.

mutagen An agent capable of modifying genes, the material of heredity. Many forms of electromagnetic radiation such as cosmic rays, X-rays, and ultraviolet light, together with a variety of chemical compounds, are mutagenic.

mutualism An interaction between individuals of different species that benefits each individual. The terms symbiosis and mutualism are often used to mean the same thing.

mycology The study of fungi, a group that includes mushrooms, moulds, and yeasts;

many fungi are useful in medicine and industry. Mycological research led to the development of penicillin, streptomycin, and tetracycline; it also has important applications in the dairy, wine, baking, and dyes and inks industries.

mycorrhizae The structures formed by an association of certain fungi and the roots of certain plants.

myxomatosis A disease of rabbits, caused by a virus and marked by fever, inflammation, and often death. It destroyed immense numbers in Australia from 1950 onwards; the rabbit population today is only a fraction of its pre-myxomatosis level. The disease was artificially introduced to the rabbit population.

myxovirus Any of a group of viruses, members of which can cause the common cold, mumps, and measles in humans; canine distemper and rinderpest in animals; and Newcastle disease in fowl.

N

Naegleria fowleri A free-living protozoon; bathing or swimming in water contaminated with this organism can cause rare but fatal meningoencephalitis.

NAFTA NORTH AMERICAN FREE TRADE AGREEMENT.

Nahuel Huapi National Park Established in 1934, as Argentina's first national park; the park and adjacent nature reserve include a region of dense forests, numerous lakes, rapid rivers, waterfalls, snow-clad peaks, and glaciers. Its peaks include El Tronador and Mount Catedral. It encompasses Lake Nahuel Huapi in the Andes.

Nairobi Conference An international conference held in 1982, 10 years after the UN CONFERENCE ON THE HUMAN ENVIRONMENT.

Nairobi National Park Established in 1948, a national park 8 km south of Nairobi; it consists partly of thick woods near the city outskirts, partly of rolling plains and valleys, and partly of a wooded confluence of several rivers. Its flora is of the dry transitional savannah type. Fauna include the lion, gazelle, black rhinoceros, giraffe, antelope, and zebra, together with hundreds of species of birds.

Namarda River Controversy See INDIA, EVOLUTION OF ENVIRONMENTAL POLICY; SARDAR SAROVA DAM PROJECTS.

Nam Choan Hydroelectric Project A proposed project in Thailand to help meet the country's rising energy demands at the expense of its finest natural forest. A public inquiry into the proposal was conducted in 1988. After consideration of all the relevant factors including the degradation of Thailand's largest natural forest with rare fauna and flora, the Cabinet later in 1988 decided to postpone construction of the Nam Choan Dam for an indefinite period.

Namid Desert/Naukluft National Park Established in 1907 in Namibia, Africa, a park offering a wide range of landscapes including sand dunes, canyons, gorges, pools, and waterfalls and the Naukluft mountains. Flora includes true desert vegetation. Fauna include the elephant shrew, the desert golden mole, gerbil, black-backed jackal, bat-eared fox, hyena, leopard, cheetah, zebra, gemsbok, springbok, and ostrich.

Nasser, Lake A reservoir on the Nile River in Upper Egypt and northern Sudan, created by the impounding of the Nile's waters by the ASWAN HIGH DAM.

natality rate The rate of addition of new individuals to a population by birth or hatching.

national accounts Statistics prepared by nations on an annual basis, or more frequent intervals, on the value of income, expenditure and production within the nation, both at current and constant prices. In addition to providing an overview of total economic activity as measured by prices, the national accounts provide information on the relationships between different parts of the economy and on changes in individual components over time. The figures reveal such important aggregates as GROSS DOMESTIC PRODUCT, gross fixed capital expenditure, gross national expenditure, overseas trade, and national and overseas debt. They should also reveal the nation's subsidies and welfare payments. A uniform system of national accounts has been evolved by the UN in conjunction with the European Union, the IMF, the OECD, and the WORLD BANK. The presentation of accounts is constantly being reviewed, revised, and updated to more effectively meet national and global economic policy-making into the 21st century. National accounts in most countries do not embrace NATURAL RESOURCE ACCOUNTING, accounting for the maintenance of national wealth, assessment of environmental costs and benefits or environmental protection expenditures, or physical, as distinct from financial, accounting.

National Biological Service, US (NBS, US) A service created in 1993 to bring together the scientific parts of seven bureaus within the US Department of the Interior. The objective is to collect and analyse biological data about America in a coherent way, paving the way towards rational management of the national assets identified. The aim also is that the information will become part of a National Biological Information Infrastructure, reachable over the Internet.

national conservation strategy A conservation strategy evolved by many countries stemming from the WORLD CONSERVATION STRATEGY released in 1980.

National Environmental Policy Act, US (NEPA, US) An Act of the US Congress declaring a national policy: to encourage productive and enjoyable harmony between humans and their ENVIRONMENT; to promote efforts which would prevent or eliminate damage to the environment and BIOSPHERE; to enrich the understanding of ecological systems and natural resources; to require the preparation of EISs for major federal projects; and to establish a COUNCIL ON ENVIRONMENTAL QUALITY.

National Environmental Technology Centre A centre established in Britain in 1994, merging the work of the principal government air-pollution laboratory at Warren Spring with that of the Atomic Energy Authority Laboratory. The aim was to bring the best environmental research under one roof.

national estate   Components of the natural or cultural ENVIRONMENT of a nation that have special value for future generations, as well as for the present community; these include wilderness areas and landscapes, unique natural features and geological monuments, special aspects of the built environment, and places of importance in cultural history. (See CONVENTION FOR THE PROTECTION OF THE WORLD CULTURAL AND NATURAL HERITAGE; WORLD HERITAGE LIST.

national forest   In the USA, any of the forest areas set aside nationally under federal supervision for the purposes of conserving water, timber, wildlife, fish, and other renewable resources, while providing recreational areas for the public. The national forests are administered by the US Forest Service. (See NATIONAL PARK SERVICE, US.)

National Landcare Program (NLP)   An Australian programme, the object of which is to enhance the efficient, sustainable and equitable management of the nation's natural resources for the benefit of the community at large; it aims to encourage with appropriate funding community groups and landholders to identify and solve the soil, water, vegetation management, and nature conservation problems which they share. Community groups and local government authorities are able to apply for NLP funding through a 'one stop shop' process. The NLP involves the federal, state and territorial governments and such bodies as the Australian Nature Conservation Agency and the Murray-Darling Basin Commission. In 1995, there were over 2200 landcare groups.

national monument   In the USA, numerous areas reserved by act of the US Congress or presidential proclamation for the protection of places or objects of historic, prehistoric, or scientific interest, including natural physical features, or relics of Indian cultures. In 1906, President Theodore Roosevelt created the first national monument at Devils Tower in Wyoming. The jurisdiction of national monuments was unified in 1933 under the NATIONAL PARK SERVICE, US.

national park   A relatively large area of land set aside by legislation for its features of predominantly unspoiled natural landscape, and its flora and fauna, permanently dedicated for public enjoyment, education, and inspiration, being protected from all interference, other than essential management measures, so that its natural attributes are preserved. In the USA, the Yellowstone national park was created in 1872; while the Yosemite and Sequoia national parks were created in 1890. The idea of protecting outstanding scenic and scientific resources, wildlife, and vegetation took root and developed eventually into a national policy. Similar policies have developed in most countries of the world, embracing both terrestrial and marine parks. (See GREAT BARRIER REEF.)

National Park Service, US (NPS, US)   An agency created by statute in 1916 to manage national parks, monuments and reservations and to conserve scenery and natural assets, together with the wildlife. In 1970, the US Congress created the National Park System bringing together many existing parks and reserves. The greatest expansion took place in 1980 with the creation of the ALASKAN NATIONAL PARKS.

national physical planning   National planning concerned with physical location. Its aims are: (1) to chart the claims which various activities make on land, water resources, and ENVIRONMENT; (2) to review the assets available to meet those demands; and (3) to draw up guidelines for the management of such natural resources. National physical planning has received some prominence in Sweden and Denmark.

National Principles for Environmental Impact Assessment   Principles established by most countries in North America, Europe and Asia for the administration of ENVIRONMENTAL IMPACT ASSESSMENT systems, whether embodied in legislation or not. The national principles adopted for Australia are reproduced in Box 9.

national settlement   The distribution of population within a country, geographically.

national trails system   A system established in the USA in 1968 through US Congress legislation; a national system of trails has now been established. The Appalachian Trial is one of America's best-known and most popular recreational trails; some 3200 km long, it winds through 14 states and is readily accessible to half of the population of the USA.

National Trust   A British organization founded in 1895 and incorporated by the National Trust Act 1907 for the purpose of promoting the preservation of, and access to, buildings of historic or architectural interest and land of natural beauty. The society was established partly through the efforts of Octavia Hill (1838-1912), the housing reformer. It serves England, Wales, and Northern Ireland; the National Trust for Scotland was founded in 1931. By 1995, the National Trust owned 207 stately homes and about 270 000 hectares of rolling hills and graceful gardens. Its income comes largely from membership fees and gifts, with some support from the Treasury.

natural environment   A concept that includes: national parks, nature reserves, and other places for the protection of flora and fauna;

natural environment modification

the coastline and islands; inland water bodies, rivers, lakes and other wetlands; special landforms, geological features, caves; forests, woodlands, and grasslands; areas of scientific interest, variety or exceptional quality as an ECOSYSTEM; evidence of botanical, geological, or geomorphological evolution; the HABITAT of an endangered species; or aesthetic qualities, all being largely unmodified by the activities of humanity. (See NATURAL RESOURCES INVENTORY; WORLD HERITAGE LIST.)

natural environment modification   Changes that occur over long periods of time as a result of climate modification, fire, clearing, human settlement, agriculture, plant disease, variations in fauna and flora, and industrial activity.

Natural Environment Research Council   Established by Royal Charter in 1965, a British organization to undertake and support research in the Earth sciences, to give advice on the exploitation of natural resources and on the protection of the ENVIRONMENT, and to promote education and training in these fields of study. Research areas have included industrial pollution, waste disposal, biotechnology, geothermal energy, atmospheric circulation, climatic change, and satellite surveying.

natural forest (old-growth forest)   With the dominant trees being older than a certain age, depending on the species involved, native or old-growth forest is important as it provides wildlife habitats in many areas, offers variety in the trees and vegetation, and presents nature least affected by human activity.

natural gas   A hydrocarbon gas obtained from underground sources, often in association with petroleum deposits. It generally contains a high percentage of METHANE, with varying amounts of ethane and inert gases such as CARBON DIOXIDE, nitrogen, and helium. Commercially, natural gas is transported by pipeline, or in a refrigerated liquefied form by ship. (See LIQUEFIED NATURAL GAS; LIQUEFIED PETROLEUM GAS.)

natural increase   The rate of population growth, determined by subtracting the MORTALITY rate from the NATALITY RATE.

natural pollutants   Substances of natural origin present in the Earth's ATMOSPHERE which may be, when present in excess, regarded as air pollutants. These include: (1) OZONE, formed by electrical discharge or photochemically; (2) sodium chloride, or sea salt; (3) nitrogen dioxide, formed by electrical discharge in the atmosphere; (4) dust and gases of volcanic origin; (5) SOIL dust from dust storms; (6) bacteria, spores, and pollens; and (7) products of forest fires. (See AIR POLLUTION.)

natural resource   Any portion of the NATURAL ENVIRONMENT such as the ATMOSPHERE, water, SOIL, forest, wildlife, land, minerals, and environmental assets generally. Natural resources may be renewable or non-renewable. (See FACTORS OF PRODUCTION; MINERAL; NON-RENEWABLE RESOURCE; RENEWABLE RESOURCE.)

natural resource accounting   Accounting procedures applied to the natural resources of a nation. Natural resources provide materials and energy for productive economic processes as well as providing the immediate environment for humans, animals, and plants. Generally, national income accounts and estimates of GROSS DOMESTIC PRODUCT (GDP) do not recognize the existence, use, or depreciation of natural resource assets, only the flow of income from them. Hence, the flow of income from oil resources is measured, but not the depreciation or augmentation of the oil resource itself. Consequently, national income accounts present a narrow and inadequate statement of the economic health of the nation. The concept of natural resource accounting has been endorsed by the United Nations, while overseas it has been the subject of extensive investigation particularly in North America and Europe.

natural resource conservation   The management of living and non-living resources in such a way as to sustain the maximum benefit for present and future generations.

natural resources inventory   A study of the natural resources of an area; the product of such a study should be a series of maps showing the distribution of environmental features coupled with a written explanation of their importance for the community and the area's future land uses. An inventory should reveal hazards such as subsidence or landslide-prone areas; physical limitations of various kinds; and opportunities for multiple use.

natural selection   The agent of evolutionary change by which organisms possessing advantageous adaptations in a given ENVIRONMENT produce more surviving offspring than those lacking such adaptation.

Nature Conservancy Council (NCC)   Established in 1973 by Act of Parliament, the official British body responsible for the conservation of flora and fauna and the geological and physiographical features of importance throughout the nation. It creates nature reserves, gives advice and education on nature conservation, undertakes research, and advises the Secretary of State for the Environment on the designation of sites of special scientific interest. The NCC replaced the Nature Conservancy, created by Royal Charter in 1949. (See NATURE RESERVE.)

nature reserve   An area set aside by a government for the purpose of protecting certain fauna and flora, or both. A nature reserve differs from a NATIONAL PARK in so far as the latter seeks to protect land, fauna, and flora for the enjoyment of the public, whereas a nature reserve is essentially for the protection of fauna and flora for their own sake. Reserves are often used to protect endangered species. In the USA, many wildlife refuges are serving this purpose, especially with respect to birds. Nature reserves are also numerous in Europe and some African and Asian countries such as India and Indonesia. The programme of nature reserves of the NATURE CONSERVANCY COUNCIL (NCC) embrace arctic-alpine vegetation, bogs, woodlands, ferns, grasslands, inland waters, dunes, salt marshes, expanses of shingle, sea cliffs, and islands. A broad purpose of the NCC reserves is to preserve for future generations the best possible range of examples of natural vegetation and animal life. Reserves are also held by the Society for the Promotion of Nature Reserves, the Royal Society for the Protection of Birds, the NATIONAL TRUST, the National Trust for Scotland, and various regional or country trusts.

Nauru An island republic in the southwestern Pacific ocean, with an area of 21 km^2 and a population of a little more than 9000. Nauru's economy has been almost exclusively based on the mining, processing, and export of phosphate, the island being covered with beds of phosphate rock that are derived from rich deposits of guano, the excrement of sea birds. The country's gross domestic product (GDP) has been the highest in the Pacific and among the highest worldwide. The phosphate deposits by the mid-1990s have been virtually exhausted, while the reclamation of land for agricultural purposes has lagged behind. Nauru has been left as a moonscape after decades of mining. In 1993, Australia, Britain, and New Zealand agreed on a compensation package, payment being made over 20 years. The Nauru Government has committed itself to a rehabilitation programme to 'recreate the Garden of Eden that was once Nauru'. Nauru had launched a law suit against Australia in the INTERNATIONAL COURT OF JUSTICE, seeking massive compensation. Australia settled out of court in 1993.

NBS, US NATIONAL BIOLOGICAL SERVICE, US.
NCC NATURE CONSERVANCY COUNCIL.
NEF NOISE EXPOSURE FORECAST.
negative price A concept used in ECONOMICS; an imputed price for, for example, unwanted waste products. There is a strong underlying similarity between the concepts of negative prices and effluent taxes; the latter are a possible alternative to subsidies and standard setting as a means of controlling air and water pollution. (See EFFLUENT CHARGE.)
neighbourhood noise A great variety of sources of NOISE that may, and frequently does, cause disturbance and annoyance to the general public in their homes and general activities. They include, for example: (1) factory noise; (2) noise from road works, demolition work, and constructions activities; (3) noise from ventilation and air-conditioning plant; (4) noise from sports, entertainment and advertising; and (5) human noise.
nekton PELAGIC animals that swim freely, regardless of water motion or wind. Examples include bony fish, sharks, turtles, saltwater crocodiles, whales, porpoises and seals, squids, and octopuses. Nektonic species are limited in their distribution by the barriers of temperature, salinity, supply of nutrients, and the character of the sea bottom.
NEPA See NATIONAL ENVIRONMENTAL POLICY ACT, US.
neritic zone The relatively warm, nutrient-rich shallow-water zone overlying the CONTINENTAL SHELF; the marine counterpart of the LITTORAL ZONE of a lake. Terminating at the edge of the continental shelf, sunlight normally penetrates to the ocean bottom, permitting photosynthetic activity and promoting the growth of a vast population of floating and anchored plants. The total amount of BIOMASS supported by the neritic zone is greater per unit volume of water than any other part of the ocean.
net present value (NPV) A value calculated in the following way:

$$\text{NPV} = \text{total discounted benefits} - \text{total discounted costs}$$

If the net present value is positive, the benefits exceed the costs and the project may be economically acceptable, depending on the magnitude of the benefits.
net reproduction rate The average number of female babies that will be born to a representative newly born female in her lifetime, if existing reproduction and mortality rates continue. If, for example, 1000 girls born in 1995 ultimately produce 1600 baby girls, then the net reproduction rate is 1.6. A net reproduction rate permanently greater than unity, means an ultimate growth in population.
Netherlands, The: evolution of environmental policy See Box 25.
New Deal The policy inaugurated by President Franklin D. Roosevelt in the USA in 1933, as a response to the economic crisis that had erupted in 1929. The New Deal consisted of a series of far-reaching economic and social measures, involving a large programme of public works and an ample supply of cheap credit. The New Deal certainly contributed to a reduction in the number of unemployed. It was, however, controversial in character, being opposed to the traditional American philosophy of LAISSEZ-FAIRE. One aspect of the

> **Box 25   The Netherlands: evolution of environmental policy**
> 1930   Hoge Veluwe National Park established
> 1966   Prevention of Pollution of Surface Water Act
> 1984   The Netherlands Government released a plan to reduce industrial sulphur dioxide emissions by 70%; nitrogen oxide emissions from motor vehicles by 30%; and ammonia released from agricultural fertilizers by 50%, by the year 2000
> 1986   Environment Protection (General Provisions) Act, stipulates EIA procedures, scheduling works, and activities requiring EIA; post-project analysis also required
> 1990   Report of the Evaluation Commission on the Environmental Protection Act concluded that the EIA procedures were working reasonably well
> 1991   European Union informs the Netherlands Government that the Dutch EIA Regulations did not fully comply with EU Directive 85/337/EEC; changes made to Dutch regulations
> 1992   The Netherlands Government adopts results of a review of EU Directive 85/337/EEC; Environmental Protection Act amendments; concept of compensation introduced in respect of unavoidable effects on the environment; EIA to promote sustainable development
> 1993   The Netherlands Government to take the potential consequences for the environment and sustainable development into consideration in national decision-making on policy and programme proposals
> 1994   General Administration Law Act

programme was the creation in 1933 of the TENNESSEE VALLEY AUTHORITY (TVA) to cover a seven-state area supplying cheap electricity, preventing floods, improving navigation, and producing nitrates.

new towns   A concept of achieving a better distribution of population in a new ENVIRONMENT, while relieving pressure on existing cities and their facilities. In the USA, the systematic development of new towns was known as the 'New Communities Program'; over 30 new towns have been constructed under this programme since the First World War. On the whole, however, the towns have remained largely residential in character, unable to attract industry or approach self-sufficiency. The new towns programme in Britain began in 1946, the first generation being around London; under this programme some 30 new towns were also constructed. However, life in the new towns has for some been disappointing. A central problem has been the provision of basic INFRASTRUCTURE such as hospitals, shopping centres, and institutions of higher learning. Some of these could not be provided while populations remained small, while life has not proved better than in large cities. France's programme of new towns was launched in the 1950s; initially five new cluster communities were created around Paris. Subsequently, some 16 cities were created or expanded, some to take more industry diverted from Paris. In Sweden, some 18 communities have been created around Stockholm, each with its own shopping and cultural facilities. New cities have found favour in many countries, including Canberra, the capital of Australia, and Brasilia, the capital of Brazil.

New Zealand: evolution of environmental policy   See Box 26.

NGOs   NON-GOVERNMENT ORGANIZATIONS.

niche   In ECOLOGY, a place in the system that provides all the living needs of a SPECIES, that species being better adapted to occupy that niche than any other species. (See HABITAT.)

niche diversification   See ALPHA DIVERSITY.

night-soil   Human excrement or faecal matter and human urine. (See MANURE.)

NIMBY syndrome   An individual and often collective reaction to new industrial projects: 'Not in My Back Yard'.

Ningaloo Marine Park   Established in 1987, Australia's largest marine park after the GREAT BARRIER REEF; the park is a joint venture between the Australian Government and the West Australian State Government, extending some 260 km between the North West Cape and Amherst Point. It embraces an area of 4300 km^2.

nitrate   A salt of nitric acid formed naturally in the SOIL by microorganisms from nitrites and protein; in this form the salts are available as a plant nutrient. Nitrates are also produced industrially for use as fertilizers; some may

Box 26 New Zealand: evolution of environmental policy

- 1854 Wellington provincial council passes a law to control thistles, an introduced weed
- 1874 Forest Act
- 1894 Tongariro national park (the first) established
- 1907 Animals Protection Act to protect indigenous animal species and native birds
- 1941 Soil Conservation and Rivers Control Act; soil conservation and rivers control council established
- 1967 Water and Soil Conservation Act 1967
- 1972 Clean Air Act
- 1974 Commission for the environment established; environmental protection and enhancement procedures introduced by cabinet (EIA procedures)
- 1977 Town and Country Planning Act
- 1981 Public Works Act
- 1986 Environment Act; Ministry for the Environment established; Parliamentary Commissioner for the Environment appointed; Forest Act amended
- 1987 Conservation Act; Department of Conservation set up; creation of clean air zone in Christchurch; unleaded petrol available
- 1989 Tasman Conservation Accord, to safeguard 52 areas of native forest throughout New Zealand
- 1990 Ozone Layer Protection Act
- 1991 Resource Management Act (EIA); New Zealand signs Antarctica protocol and ratifies the convention for the prohibition of fishing with long drift nets in the South Pacific; additional tax on lead in petrol
- 1994 New Zealand industry required to reduce carbon dioxide emission over three years to meet international commitments, or face a carbon tax in 1997

consequently be leached into aquifers and fresh water. (See NITRATE POLLUTION.)

nitrate bacteria See NITRIFICATION.

nitrate pollution The contamination of fresh water by NITRATE(S), frequently caused by the leaching of nitrogen fertilizer from agricultural land. Such contamination can occur suddenly, following heavy rain. Should such contamination enter the public water supply, there is a danger of METHAEMOGLOBINAEMIA in infants.

nitric acid A colourless, fuming, and highly corrosive liquid; an important industrial chemical in the manufacture of fertilizers and explosives. It is toxic and can cause severe burns.

nitric oxide A colourless toxic gas formed from NITROGEN and oxygen. (See NITROGEN OXIDES.)

nitrification (1) The conversion by certain bacteria of organic compounds of nitrogen to nitrates; the conversion may be completed in several stages by different types of bacteria. (2) In sewage plants, the oxidation of nitrite, ammonia, and other nitrogenous compounds, to nitrates. Bacteria capable of converting ammonia to nitrite and then to nitrate are called 'nitrifying bacteria'.

nitrifying bacteria See NITRIFICATION.

nitrogen An odourless tasteless gas, the most plentiful in the ATMOSPHERE of the Earth, and a constituent of all living matter. It occurs in the atmosphere to the extent of 78% by volume (or about 75% by weight); in living organisms it occurs in complex organic compounds such as proteins. Fauna obtain their nitrogen from vegetable or animal protein, while flora synthesize their proteins from inorganic nitrogen compounds in SOIL and to some extent, from uncombined nitrogen in the air. A bacterium living in the roots of LEGUMES assimilates atmospheric nitrogen. The nitrogen content of cultivated soil is often enriched by fertilizers containing nitrates and ammonium salts. MANURE and the decay of fauna and flora return nitrogen compounds to the soil and air. (See NITRATE; NITRATE POLLUTION; NITRIFICATION; NITROGEN CYCLE.)

nitrogen cycle The cycling of the element NITROGEN from non-living surroundings through organisms and back again; nitrogen is an important part of the molecules of proteins and other compounds that make up much of the bodies of living things. After oxygen, carbon, and hydrogen, nitrogen is the most abundant element in living tissues. Most of the nitrogen taken up by plants comes from the SOIL as NITRATE ions; a smaller proportion as ammonium ions. Atmospheric nitrogen cannot generally be used by green plants as a source of nitrogen. However, some plants do take nitrogen from the air to form nitrogen compounds, a process called NITROGEN FIXATION. (See NITRIFICATION; NITROGEN OXIDES; NITROUS OXIDE.)

nitrogen fixation A natural or industrial process that converts NITROGEN into reactive compounds such as AMMONIA, nitrates or nitrites; thus, nitrogen becomes available as a plant nutrient either directly or after further reactions. Some nitrogen-fixing bacteria are free-living in the SOIL, or are found in the root nodules of LEGUMES.

nitrogen oxides Oxides formed and released in all common types of COMBUSTION; they are formed by the oxidation of atmospheric NITROGEN at high temperatures. Introduced into the ATMOSPHERE from car exhausts, furnace stacks, incinerators, power stations and similar sources, the oxides include NITROUS OXIDE, nitric oxide, nitrogen dioxide, nitrogen pentoxide, and NITRIC ACID. Nitrogen dioxide can be injurious to health when inhaled, combining with moisture in the lungs to form nitric acid. Nitric oxide can combine with HAEMOGLOBIN in the blood. The oxides of nitrogen undergo many reactions in the ATMOSPHERE to form PHOTOCHEMICAL SMOG.

nitrous oxide One of the gases contributing to the GREENHOUSE EFFECT. Sources are fossil-fuel combustion, biomass burning, and the use of fertilizers. Other compounds of nitrogen formed by the oxidation of atmospheric nitrogen at high temperatures include nitric oxide, nitrogen dioxide, nitrogen pentoxide, and nitric acid. They are discharged into the ATMOSPHERE from motor vehicle exhausts, furnace stacks, incinerators, and other sources. Nitrogen oxides are all potentially harmful; they also undergo many reactions in the atmosphere. Emissions from motor vehicle exhausts, both nitrogen oxides and hydrocarbons, catalyse in bright sunlight to form PHOTOCHEMICAL SMOG.

NLP NATIONAL LANDCARE PROGRAM.

NNI NOISE AND NUMBER INDEX.

noise SOUND that is not wanted by the recipient. It may have various disturbing effects depending upon its intensity, frequency, duration, and other subtle characteristics. Much depends on the sensitivity of the listener to various kinds of noise. Further, what may be a pleasant and enjoyable sound to one person may be unpleasant and jarring to another. Noise can arise from road traffic, air traffic, railways, ships, emergency vehicles, building sites, factories, fairgrounds, racing tracks, shooting ranges, pneumatic drills, musical instruments and recordings, dog pounds and pets. Noise can cause minor disorders such as uneasiness and irritation; it may lead eventually to insomnia, mental stress, and ear complaints. The investigation of noise often involved the use of meters equipped with A-filters. (See DECIBEL (A-SCALE).)

noise abatement zone A concept introduced into local government planning in Britain in 1974; the declaration of noise abatement zones within their districts enabled local governments to impose restraints on noise emissions from specified premises. The procedure for establishing such zones was similar to that for establishing smoke-control areas.

noise and number index (NNI) An index for the measurement of disturbance from aircraft noise; developed in Britain in the 1960s and since used extensively. The index was based on a social survey carried out around Heathrow Airport, London, for the Wilson Committee, and takes into account the average peak noise level at the ground due to passing aircraft and also the number of aircraft involved. Local planning authorities have been asked to take aircraft noise into account in considering planning applications.

noise control legislation Legislation introduced by many governments to prevent or restrict the emission of NOISE from industrial, commercial, and domestic premises; from motor vehicles and aircraft; and from consumer appliances and equipment. Such legislation may involve the prior approval of potentially noisy equipment and processes, the licensing of premises with specific conditions, noise limitations for various kinds of equipment and transport, prescribed methods of MONITORING, and procedures for enforcement. (See NOISE LABELLING.)

noise exposure forecast (NEF) Developed in the USA in the late 1960s, a technique for predicting the subjective effect of aircraft noise on the average person, exposure levels being expressed in NEF units. Factors that are taken into consideration are: the frequency of aircraft movements and their distribution by night and day; the magnitude and duration of aircraft noise as determined by type, weight, and flight profile; and the distribution of noise energy over the spectrum of audible frequencies. In applying the NEF technique, a pattern of contour lines is drawn on a map of the area surrounding the airport. A similar system was adopted by Australia in 1982 (the ANEF).

noise labelling The labelling of consumer goods with information on noise levels. The advantages of labelling are that consumers can use information given on the label when comparing possible purchases; demonstrated demand for less noisy equipment provides incentives for the design of less noisy products; and community awareness of noise problems is increasesd. (See ECO-LABELLING.)

noise rating A system whereby manufacturers indicate on their products the noise level emitted at a fixed distance from the appliance. (See NOISE LABELLING.)

noise-rating curves or numbers Sets of curves relating levels of sound in octave bands to acceptability for particular applications, ranging from domestic noise to noise in factories.

noise-reduction techniques Measures to reduce noise at the source, to encourage quieter technologies or equipment, or to prevent or reduce the propagation of sound. Sound is radiated both as air-borne and as structure-borne; most sources produce both, thus various noise attentuation principles must be employed. Measures may include: the isolation and damping of vibration sources; the replacement of components with quieter parts and material; the enclosure of particularly noisy components; the provision of exhaust silencers; the selection of quieter types of fan; the fitting of sound attenuation equipment in ventilation ducts; the replacement of noisy compressed-air nozzles with quieter types; the choice of quieter transmission and cooling systems; the provision of noise bunds, barriers, and buffer zones; and the regulation of times of operation. (See NOISE CONTROL LEGISLATION.)

nomadism A way of life of peoples who do not live continuously in the same place, but move about periodically or cyclically; it should be distinguished from migration, which tends to be non-cyclic and involves normally a complete change of HABITAT. Historically, the term 'nomad' has embraced nomadic hunters and gatherers, pastoral nomads, and trader nomads.

nomenclature In BIOLOGY, the classifying and naming of organisms. In general, the species to which an organism belongs is indicated by the two words, the genus (kind) and species (member) names, called the Linnaean system after Carolus Linnaeus (1707–78), the Swedish botanist and explorer.

non-government organizations (NGOs) Voluntary organizations worldwide that participate actively in the promotion of environmentally sound policies and SUSTAINABLE DEVELOPMENT, lobbying governments, educating the public, and marshalling political support for appropriate environmental protection legislation. The NGOs have held separate forums at all UN Conferences on the human environment. Examples of NGOs include: the AUDUBON SOCIETY, FRIENDS OF THE EARTH INTERNATIONAL, GREENPEACE, SIERRA CLUB, WORLD CONSERVATION UNION (WCU), WORLDWATCH INSTITUTE, WORLD WIDE FUND FOR NATURE.

non-pecuniary values The many values held by individuals and society that cannot be measured readily or at all in money terms. (See COST–BENEFIT ANALYSIS; QUALITY OF LIFE.)

non-point source A source of pollution from diffuse or diverse sources such as agricultural land or urban impervious surfaces as distinct from stacks or pipes.

Non-proliferation of Nuclear Weapons, Treaty on the (Nuclear Non-proliferation Treaty) An agreement signed in 1968 by the USA, Britain, the former Soviet Union, and 59 other countries, under which the three major signatories agreed not to assist countries not possessing nuclear explosives in obtaining or producing them. The treaty became effective in March 1970.

non-renewable resource Natural resources which, once consumed cannot be replaced; for example, a tonne of coal once consumed is gone forever in that form. Mineral resources generally are regarded as depletable assets of this kind. However, it is difficult to predict the consequences of exhausting particular resources. The exhaustion process is gradual and probably accompanied by a steady increase in price. A rising price intensifies exploration and ensures treatment of lower grades of ore, recycling, and reclamation of scrap and residue. Meanwhile, research into substitute materials and processes, accompanied by changes in the pattern of demand, could mean that an 'indispensable' mineral becomes totally redundant. The virtual collapse of the coal industry in Britain illustrates the point. (See INTERGENERATIONAL EQUITY; INTRAGENERATIONAL EQUITY; RENEWABLE RESOURCES; SUSTAINABLE DEVELOPMENT.)

Nordic convention See CONVENTION ON THE PROTECTION OF THE ENVIRONMENT BETWEEN DENMARK, FINLAND, NORWAY, AND SWEDEN.

Nordic Council A Council of Ministers established in 1971 to facilitate consultation and cooperation between the Nordic countries of Denmark, Finland, Iceland, Norway, and Sweden; its formal decisions are usually binding upon the members. In 1974, the Nordic countries signed a convention on the protection of the environment. Under this convention, the environmental impact assessment of a project must place equal weight on potential damage to a neighbouring Nordic state as within its own country. In 1990, EIA procedures were reviewed and improved. (See CONVENTION ON THE PROTECTION OF THE ENVIRONMENT BETWEEN DENMARK, FINLAND, NORWAY, AND SWEDEN.)

normal distribution See GAUSSIAN DISTRIBUTION.

North American Commission for Environmental Cooperation A body created following the NORTH AMERICAN FREE TRADE AGREEMENT (NAFTA), based in Montreal,

Canada. The commission seeks to address some glaring environmental problems such as the appalling environmental conditions that exist on the border between the USA and Mexico. Many of the Mexican border cities, such as Ciudad Juarez, have no sewerage systems, while illegal rubbish dumps are common. Many smaller towns on both sides of the border lack municipal water systems with fresh water being delivered by lorry. There is evidence that untreated sewage seeps into wells, with many cases of hepatitis occurring. The commission has conducted a research programme into these problems, with a view to a programme of improvement.

North American Free Trade Agreement (NAFTA) An agreement concluded in 1994 between the USA, Canada, and Mexico to form a free trade area; its aim is the phased abolition within a decade of tariffs on most goods traded between the three countries, with implications for the uniformity of environmental policies. NAFTA builds upon the pre-existing US–Canada Free Trade Agreement. (See ASSOCIATION OF SOUTH-EAST ASIAN NATIONS; EUROPEAN UNION; GENERAL AGREEMENT ON TARIFFS AND TRADE.)

Norwalk virus A virus that causes gastroenteritis; it can be spread in drinking and swimming water and also in food (especially shellfish) contaminated with sewage or faecal material.

Norway: evolution of environmental policy See Box 27.

Noumia convention See CONVENTION FOR THE PROTECTION OF THE NATURAL RESOURCES AND ENVIRONMENT OF THE SOUTH PACIFIC REGION.

noxious Injurious, prejudicial or harmful to HEALTH.

noy A unit of noisiness related to the perceived noise level in PNdB (perceived noise decibels).

NPK Nitrogen (N), phosphorus (P), and potassium (K); the three principal constituents of most manufactured fertilizers; the initials are sometimes used as a synonym for artificial FERTILIZER.

NPS NATIONAL PARK SERVICE.

NPV NET PRESENT VALUE.

nuclear energy ENERGY from the inner core or nucleus of the atom, in comparison with energy released in chemical processes, including COMBUSTION. Such energy may be released by NUCLEAR FISSION or NUCLEAR FUSION.

nuclear fission The splitting of the nucleus of an atom into two approximately equal fragments; the process is accompanied by the emission of neutrons and the release of energy. Nuclear fission is the basis of nuclear power generation and is used in some nuclear weapons.

nuclear fuel Substances capable of producing heat as the result of NUCLEAR FISSION or NUCLEAR FUSION, and not through chemical

---

Box 27   Norway: evolution of environmental policy

1957	Open-air Recreation Act
1964	Salmon and Freshwater Fishing Act
1970	Nature Conservation Act
1972	Establishment of Ministry of Environment
1976	Product Control Act
1977	Motorized Traffic in Marginal Land and Watercourses Act
1978	Cultural Heritage Act
1981	Pollution Control Act; Wildlife Act
1983	Regulations concerning conservation of the natural environment in Svalbard
1985	Planning and Building Act (EIA provisions)
1986	Regulations concerning conservation of the natural environment of Jan Mayen and its surrounding territorial waters
1987	Gro Harlem Brundtland, prime minister of Norway, presents the report of the World Commission on Environment and Development *Our Common Future* to the General Assembly of the United Nations
1989	Amendment of Pollution Control Act and the Planning and Building Act (EIA provisions); Sale of detergents containing phosphates prohibited
1990	ECE conference on 'Action for a Common Future' held in Bergen
1992	Amendment of Salmon and Freshwater Fishing Act and Cultural Heritage Act; Plan for national parks presented to Parliament

reactions including COMBUSTION. Fissionable isotopes include uranium-235 (from ore), uranium-233 (bred in reactor), and plutonium-239 (bred in reactor).

nuclear fuel cycle A cycle comprising all of the operations from the mining and milling of uranium, through conversion, enrichment and fuel fabrication, to spent fuel storage and ultimate disposal, or reprocessing and recycling of valuable materials and disposal of the radioactive wastes.

nuclear fusion (thermonuclear fusion) A type of nuclear reaction in which two light nuclei come together to form one heavier nucleus, the reverse of NUCLEAR FISSION. However, as in fission, very large amounts of energy are given off. Fusion reactions only take place at very great temperatures; such reactions are the source of energy given off by the sun or when a hydrogen bomb explodes. A controlled fusion reaction has yet to be achieved.

Nuclear Non-proliferation Treaty See NON-PROLIFERATION OF NUCLEAR WEAPONS, TREATY ON THE.

nuclear power programmes The gradual introduction of nuclear power stations into electricity supply systems in many countries. By 1995, there were some 450 nuclear power stations operational worldwide providing some 20% of the world's electricity. Objection on environmental grounds have slowed down, and even halted in some instances, the development of programmes. (See CHERNOBYL NUCLEAR CATASTROPHE; THREE-MILE ISLAND NUCLEAR INCIDENT.)

nuclear reactor A device in which NUCLEAR FISSION takes place as a self-supporting chain reaction. A typical reactor comprises: (1) fissile material such as uranium or plutonium; (2) a moderator; (3) a reflector; (4) control elements; (5) provision for the removal of heat by means of a coolant; and (6) provision for the storage and disposal of spent fuel elements.

nuclear winter A phenomenon likely to be associated with the massive use of nuclear weapons: large areas of the Earth could be subject to long periods of densely overcast skies, hard freezes, killing frosts, toxic snowfalls, and gale-force winds; to a heavily polluted atmosphere and to dangerous levels of radioactivity. Vast areas of territory could be contaminated beyond the possibilities of habitation, food and water being tainted by dangerous radionuclides.

nucleic acid A complex biological molecule that encodes instructions for the operation of each CELL, and for the creation of new cells. The main types of nucleic acid are deoxyribonucleic acid (DNA) and ribonucleic acid (RNA). The genes of all organisms are made of nucleic acids and it is the genes that determine the characteristics of individual organisms.

nucleonics The practical application of nuclear science, and the techniques associated with these applications. The term is also commonly used to mean the electronic techniques used in nuclear instrumentation and measurements.

nuisance (1) Anything that is unpleasant, annoying, or inconvenient; (2) In English law a 'private nuisance' affects a particular occupier of land, such as exposure to noise from a neighbour, while a 'public nuisance' affects an indefinite number of members of the public, such as the obstruction of a highway. In the former, the aggrieved person can apply to a court for an injunction and claim damages, while the latter is a criminal offence; (3) a statutory nuisance may be a particular nuisance, defined in law as such, for example, a deposit of refuse being 'prejudicial to health or a nuisance'; and (4) nuisance microorganisms that occasionally occur in large numbers in drinking water and which, though they may not prejudice health, may cause problems of taste and odour, colour, and staining of laundry.

nutrient A substance that an organism must extract from its surroundings for sustaining life, growth, and procreation. While essential nutrients must be present in the food, non-essential nutrients are those that can be synthesized by the CELL if they are absent from the food.

nutrient budget An estimate for a particular living system setting out the amounts of essential mineral nutrients that are taken up or lost.

nutrient stripping A tertiary treatment of waste waters, either to reduce the rate of EUTROPHICATION of the receiving waters or to permit the reuse of water for domestic purposes. Compounds of phosphorus and NITROGEN are mainly responsible for stimulating excessive growth of ALGAE. Methods of stripping range from chemical coagulation and biological methods to advanced treatment processes such as those developed for the desalination of sea water and brackish waters. The disposal of the concentrates arising in all advanced waste-water treatment processes presents an additional problem.

nutrition The science of food and nutrients and their role in the birth, life and death of humans, FAUNA, and FLORA.

nymph In ENTOMOLOGY, the immature form of insects that do not have a pupal stage; for example, grasshoppers and dragonflies. Nymphs usually resemble the adult but without fully formed wings or reproductive organs.

O

occupational health and safety An area of statutory duty imposed on employers and employees in most countries, for the protection of the workforce from occupationl diseases and stresses and physical hazards through adequate planning, ventilation, lighting, safeguards, safety and emergency procedures, routine inspections, monitoring, personal protection, correct clothing, controlled atmospheres, medical and psychological supervision, and trade-union and staff-association consultation, with adequate reflection in enterprise and individual agreements.

ocean A continuous body of salt water covering nearly 71% of the terrestrial surface of the Earth; customarily divided into three major oceans, the Pacific, the Atlantic, and the Indian. All are in open connection with the Southern Ocean, a stretch of water encircling the Antarctic continent. The average depth of the oceans is 3.7 km. Substantial areas of the sea bed are covered by loose sediments overlying consolidated sediments of crustal igneous rocks. (See CONVENTION ON THE LAW OF THE SEA.)

ocean current The horizontal and vertical circulation of ocean waters created by gravity, wind friction, and variations in water density in various parts of the oceans; the direction and form of oceanic currents is a product, therefore, of natural forces. Ocean currents and atmospheric circulation profoundly influence each other. For example, when at intervals the warm, humid wind circulation above the western Pacific is displaced eastwards, the eastern Pacific waters are warmed, creating the EL-NIÑO EFFECT, that may bring drought to Australia, storms to California, and a warm winter to central North America; it may also upset the fishing industries of Peru and Chile.

ocean dumping Historically, the use of the oceans as a dumping ground for industrial and radioactive wastes, sewage effluents, and sewage sludge. Progressively national and international action has sought to restrict indiscriminate dumping of hazardous and objectionable wastes both immediately offshore and at greater distances to protect bathing beaches, recreational facilities, fishing, and the environmental qualities of the oceans. (See CONVENTION FOR THE PREVENTION OF MARINE POLLUTION BY DUMPING FROM SHIPS AND AIRCRAFT; CONVENTION FOR THE PREVENTION OF POLLUTION FROM SHIPS; CONVENTION FOR THE PROTECTION OF THE MEDITERRANEAN SEA AGAINST POLLUTION; CONVENTION ON THE PREVENTION OF MARINE POLLUTION BY THE DUMPING OF WASTES AND OTHER MATTER; CONVENTION ON THE PROTECTION OF THE MARINE ENVIRONMENT OF THE BALTIC SEA; CONVENTION ON THE LAW OF THE SEA.)

oceanography The scientific discipline concerned with all aspects of the world's seas and oceans, including their physical and chemical characteristics, their original and geological framework, and all the FAUNA and FLORA that inhabit the marine environment. Oceanography is vital to an understanding of the effects of pollutants on ocean waters and to safeguarding the quality of ocean waters.

ocean thermal energy conversion (OTEC) The concept of utilizing the temperature differences of 20°C or more that occur between the surface of an OCEAN and its depths to achieve a continuous supply of power; this temperature difference may be found in the tropical regions of the world. Various small plants have been constructed to demonstrate the principle. One scheme employs AMMONIA as a working fluid. This is boiled in a heat exchanger by the relatively warm ocean surface water, the resulting vapour being used to drive a turbine generator. The ammonia is returned to a liquid state by condensing it in another heat exchanger cooled with cold water from a depth of about 700 m, where the temperature difference is some 22°C lower.

octave-band noise levels Band levels providing a broad frequency analysis of a SOUND. An octave is an interval between two sounds, one of which has a frequency twice that of the other; the interval between two frequencies has a ratio of two. Octave-band sound levels cover the bands centred on 31.5, 63, 125, 250, 500, 1000, 4000, and 8000 Hz. (See NOISE.)

odour control Technically, one of the most difficult areas in AIR POLLUTION control. Many of the more noxious and objectionable odours come from such processes as oil refining, leather tanning, the processing of animal by-products, metallurgical processes, fertilizer manufacture, brick-making, pulp and paper manufacture, and abattoirs, piggeries, and agriculture. Common methods of control include: (1) ABSORPTION in an oxidizing liquid; (2) ADSORPTION by ACTIVATED CARBON; (3) INCINERATION; (4) after-burning; and (5) ozonization. (See MERCAPTANS.)

Odzala National Park Established in 1940 in the Congo, Africa, a national park comprised of a high plateau that is difficult to access. Flora includes humid tropical forest, changing to savannah in the south. Fauna include the elephant, buffalo, antelope (bongo), anthropoid apes, leopard, golden cat, and bushpig. A few groups of Pygmies inhabit the park.

Olympic National Park

OECD ORGANIZATION FOR ECONOMIC COOPERATION AND DEVELOPMENT.

offensive or hazardous industry An industry which, by reason of the manufacturing method or the nature of the materials used or produced, requires isolation from other buildings and residences. Such an industry interferes with the AMENITY of a neighbourhood by reason of noise, vibration, risk of explosion or fire, odour, fumes, smoke, vapour, steam, soot ash, dust, waste water, waste products, grit, or oil. (See ODOUR CONTROL.)

offroad vehicles (ORVs) Vehicles designed essentially for recreational and sporting use in offroad situations such as various types of motor-cycle, four-wheel drive vehicles, and vehicles specifically designed for use in sand or snow. Such vehicles if unregulated can cause considerable environmental damage to sensitive ecological areas.

oil pollution The dumping and accidental spillage of oil and other hazardous substances into waterways from ships and land-based or offshore installations. The use of chemicals for dispersing and sinking oil spills may sometimes be as detrimental as the oil itself. Oil pollution may destroy or damage aquatic life and wildlife such as birds; limit or destroy the recreational value of rivers, coasts, and offshore waters; contaminate water supplies; and create fire hazards. (See BALLAST WATER; KOMI PIPELINE OIL DISASTER; SANTA BARBARA BLOW-OUT.)

oil spill The leakage or discharge of petroleum over the surface of a large body of water. Oceanic oil spills became a major problem from the 1960s, partly as a result of increased oil exploration on the continental shelves of the world and partly because of the increased size of oil tankers. Such oil spills often have dramatic short-term effects, while the long-term adverse effects are more difficult to assess. (See 'AEGIAN SEA' INCIDENT; 'BRAER' INCIDENT; CONVENTION FOR THE PREVENTION OF MARINE POLLUTION BY DUMPING FROM SHIPS AND AIRCRAFT; CONVENTION FOR THE PREVENTION OF POLLUTION FROM SHIPS; CONVENTION ON THE PREVENTION OF MARINE POLLUTION BY THE DUMPING OF WASTES AND OTHER MATTER; 'EXXON VALDEZ' DISASTER; 'TORREY CANYON' DISASTER; Figure 20.)

old-growth forest See NATURAL FOREST.

oligopoly In ECONOMICS, a situation in which a MARKET, or a large part of a market, is supplied by a small number of firms, each of which exercises a significant degree of economic, social, and environmental influence.

oligotrophic Describing an aquatic ENVIRONMENT which, because of the low concentrations of basic nutrients in the water, is relatively unproductive in terms of FLORA and FAUNA. (See EUTROPHIC; EUTROPHICATION.)

Olympic National Park Established in 1938 in northwestern Washington USA, a national park to preserve the Olympic Mountains and their forests and wildlife. There are numerous

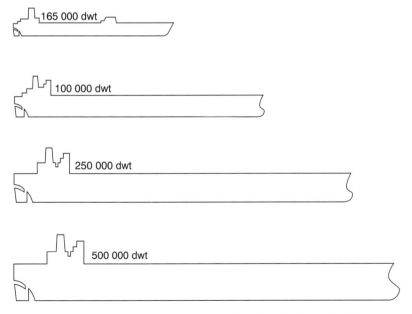

OIL SPILL: Figure 20. Growth in the size of oil tankers since the Second World War.

Olympus National Park

glaciers and extensive rainforest. The ocean-shore portion contains scenic beaches, islets, and points, with three Indian reservations. The fauna include deer, bears, cougars, and rare Roosevelt elks. The park has been included in the WORLD HERITAGE LIST.

Olympus National Park Established in 1938 in Greece, a national park encompassing Mount Olympus. The flora is sparse with beech, pine and broad-leaved evergreens. The fauna includes the wild mountain goat, hare, wolf, fox, roe deer, golden eagle, partridge, and hooded crow.

onchocerciasis A disease also known in its most common form, as river blindness. The black flies that transmit the disease breed on rivers and mostly affect riverine populations. River blindness is common in the savannah area of Africa, in Guatemala and Mexico. In 1987, the WORLD HEALTH ORGANIZATION began to distribute a drug known as ivermectin, found to be safe and effective.

one-hundred year floodplain The land area adjacent to a body of water that is flooded with an average frequency of once in a hundred years. Besides their function as natural drainage channels, floodplains have important water-storage and ground-water recharge capabilities. (See FLOOD PLAIN.)

Only One Earth A definitive report on the state of the human environment commissioned for the UN CONFERENCE ON THE HUMAN ENVIRONMENT held in Stockholm in 1972. The authors were Barbara Ward (Lady Jackson) and René Dubos (1901–82).

open-cut mining A technique of mining employed when the coal is not far below the surface. The overlying earth and rock are mechanically stripped to expose the coal, which is then removed with or without blasting. It includes strip mining and multibench mining. (See MINING WASTES.)

open space Land or water that is not pre-empted by intensive urban uses and which will be kept free of human-made structures on a permanent basis. Examples of open spaces are parks and recreation areas, water and wetland areas, forests, and buffer zones. An open-space system is the network of a city's parks and open spaces, including parkways, that are complementary in scale, function, and design. Open spaces are often described as 'the lungs' of a city.

opportunity cost The cost of satisfying an objective, measured by the value those resources would have had in another attractive alternative use. For example, if capital funds committed to productive equipment could have earned a rate of interest of 20% per annum in another use of similar risk, then that is the opportunity cost of those funds in present use. There is a real cost if the funds perform less well, even though this cost is not revealed in any balance sheet. The opportunity cost of being self-employed is the salary that might be earned doing something else. In turn, the opportunity cost of leisure may be measured by the money that could have been earned instead. A government with a relatively fixed budget may choose to build better highways; such a decision will be made at the cost of other programmes such as schools. A consideration of opportunity costs assists in ensuring that resources are put to the best use and that all costs, explicit and implicit, are taken into account.

optimization Finding the best or most favourable condition, degree, or amount for a particular situation. Optimization exercises, involving applied mathematics, are to be found in many disciplines such as BIOLOGY, ECONOMICS, engineering, physics, and many others. For example, in economics, the question arises as to what size of firm is optimal in any given market situation. In society at large, a continual question is the optimal mix between economic, social, and environmental considerations to maximize the QUALITY OF LIFE. (See ALLOCATIVE EFFICIENCY, SOCIAL.)

optimum population An economic, social, and environmental challenge as to what size of community a country can reasonably support to achieve a good, yet sustainable, standard of living. It is often argued that the world at large, and certainly many countries within it, are overpopulated; or alternatively, that the rate of growth of population is unsustainable. On the other hand, some countries, such as Australia, appear to be underpopulated, though some argue otherwise. (See UN CONFERENCE ON POPULATION AND DEVELOPMENT; UN WORLD POPULATION CONFERENCE; WORLD POPULATION.)

organic farming An agricultural system that avoids the use of synthetic fertilizers, pesticides, growth regulators, and livestock feed additives. As far as practicable, organic farming relies upon crop rotation, crop residues, animal manure, legumes, green manures, off-farm organic wastes, mechanical cultivation, mineral-bearing rocks, and biological pest control to sustain soil productivity and tilth, to supply plant nutrients, and to control insects, weeds, and other pests.

organic fertilizers Examples include animal manure, human excreta, compost, blood and bone, and sewage sludge. Organic fertilizers can provide all the nutrients plants need, improve soil structure and water-holding capacity, help control erosion, and assist in the biological control of soil-borne diseases of plants.

organic matter The humus or decomposed remains of plants and animals in the SOIL.

Organic matter is the food source of fungi and bacteria in the soil. Soils high in organic matter support large populations of microorganisms, plants, and invertebrates.

organic pesticides   Pesticides that contain carbon. (See PESTICIDE.)

organic sludges, industrial   A mixture of organic liquids and solids resulting from manufacturing or processing operations involving animal material, plants, vegetables, or fruit. These wastes, which can be pumped to where they are needed, are generally associated with the food and beverage industries, animal- and fish-processing plants, and tannery operations. Rich organic wastes have a total carbon content greater than 5%; those with less than 5% carbon are known as lean wastes.

organic soil   SOIL that is extremely rich in organic matter; it may have a peaty appearance, peat containing a large proportion of decaying plant matter.

organic wastes   Wastes derived from living and dead organisms. When released into waterways, organic wastes support the growth of aerobic microorganisms. These microorganisms consume oxygen and as the oxygen dissolved in the water decreases, fishlife is affected and gradually disappears. When all of the dissolved oxygen is used up, anaerobic microorganisms convert sulphur compounds into hydrogen sulphide and the water becomes foul-smelling, rendering it useless for any productive and recreational uses. Organic wastes created by humans and animals may also contain pathogenic bacteria, which may contaminate shellfish and also turn swimming areas into health hazards.

Organization for Economic Cooperation and Development (OECD)   An international body which came into being in 1961, succeeding the Organization for European Economic Cooperation (OEEC), which had allocated aid received under the Marshall Plan. The members of OECD include Australia, Austria, Belgium, Britain, Canada, Denmark, Finland, France, Germany, Greece, Iceland, Ireland, Italy, Japan, Luxembourg, The Netherlands, New Zealand, Norway, Portugal, Spain, Sweden, Switzerland, Turkey, and the USA, being some 24 democratic countries with market economies. In 1970, the OECD created an environment committee to advise on patterns of growth and development which would be in harmony with protection of the environment. As a result, the OECD took a lead in promoting the POLLUTER-PAYS PRINCIPLE, in restricting the manufacture and use of POLYCHLORINATED BIPHENYLS (PCBs), in establishing a code in respect of transfrontier pollution, in promoting the environmental assessment of new chemicals, in encouraging energy conservation and the development of alternative energy sources, and in the control of the disposal of radioactive wastes. In 1974, a Declaration on Environmental Policy was adopted. This declaration emphasized the importance of environmental impact assessment (EIA), both in respect of domestic developments and foreign aid. Procedures were set out. (See INTERPARLIAMENTARY CONFERENCE ON THE GLOBAL ENVIRONMENT.)

organochlorines   See CHLORINATED HYDROCARBONS.

Organ Pipe Cactus National Monument   Established in 1937, a national monument in southwestern Arizona, USA at the Mexican border. The monument is named after the organ-pipe cactus, which has branches resembling organ pipes and is found only in this location in the USA. Other desert fauna and flora are found there. Portions of the Devil's Highway are to be seen, along which hundreds of miners and pioneers lost their lives.

ornithology   The scientific study of birds, field research being conducted by both professionals and amateurs. The study of bird movements has been greatly aided by bird banding (or ringing) and also by the use of sensitive radar and radio transmitters (telemeters).

orographic precipitation   The precipitation of rain, snow, or hail produced when moist air rises as it moves over a mountain range. As the air rises and cools, orographic clouds form, becoming the source of the precipitation. Most of the precipitation falls upwind of the mountain ridge, although some, known as spillover, falls on the other side. Normally, the leeside is said to be in the rainshadow.

ORVs   OFF-ROAD VEHICLES.

Oslo Conference on Sustainable Development   A conference held in July 1988 in Oslo, Norway, following the report of the WORLD COMMISSION ON ENVIRONMENT AND DEVELOPMENT of 1987. It was attended by the representatives of 22 UN agencies, the regional economic commissions, the INTERNATIONAL MONETARY FUND and the WORLD BANK. The conference expressed the view that SUSTAINABLE DEVELOPMENT was a joint objective for the entire UN system and emphasized the need for stronger coordination and integration both at the national level, within each international organization, and between international organizations. It was decided to incorporate the recommendations of the World Commission in the various working programmes and budgets. (See BERGEN MINISTERIAL DECLARATION ON SUSTAINABLE DEVELOPMENT.)

Oslo convention   See CONVENTION FOR THE PREVENTION OF MARINE POLLUTION BY DUMPING FROM SHIPS AND AIRCRAFT.

osmosis   The movement of water between two solutions which are separated by a membrane that allows the free passage of water but prevents or slows down the passage of the dissolved substance. The water moves more rapidly from the less concentrated solution to the solution of higher concentration than in the reverse direction; osmosis thus tends to achieve a state of equilibrium between the two solutions. Many cell membranes, being semipermeable, utilize osmosis as an important mechanism in the movement of fluids in living organisms. An example is the movement of moisture from the roots up the stems of plants. Reverse osmosis is a process in which a liquid is forced to flow through a membrane in a direction opposite to its natural tendency, through the application of pressure.

OTEC   OCEAN THERMAL ENERGY CONVERSION.

outage   The amount by which plant availability differs from the total capacity of the system through its being out of service owing to breakdown, essential maintenance, or slackening of demand for its product or service.

outfall sewer   A sewer, pipe, or conduit that transports sewage, or sewage effluent, and other liquid wastes to a final point of discharge. This may be a watercourse, river, or point at a considerable distance offshore.

overburden   The material such as SOIL and rock lying above a mineral deposit that must be removed in order to work the deposit. (See OPEN-CUT MINING.)

overgrazing   The overstocking of territory with animals for the purpose of short-term gain; the problem occurs in many countries. A consequence of overgrazing is that the most palatable grasses are grazed very short with the result that root systems degenerate. Further trampling by cattle hardens the soil, preventing regeneration and causing soil erosion. Examples of overgrazing are to be found in South Africa, India, and Australia.

Overseas Development Administration (ODA) In Britain, the official body that deals with development assistance to overseas countries. Assistance includes financial aid on concessionary terms, together with technical assistance. The ODA also helps developing countries to tackle their national environmental problems and global environmental obligations. Communities receive assistance to achieve SUSTAINABLE DEVELOPMENT and to conserve their ENVIRONMENT. Aid is in any event planned, appraised, implemented, and evaluated in ways that are sensitive to environmental issues.

overspill   A surplus of people greater than a city or town can reasonably accommodate. NEW TOWNS and designated growth centres may be established to deal with overspill.

oxbow lake   A small lake located in an abandoned loop of a meandering river channel. It is usually formed when a river or watercourse cuts through the neck of a meander to shorten its course. The old channel is blocked off and becomes isolated from the river, eventually silting up to form a marsh.

OXFAM (Oxford Committee for Famine Relief) A privately funded British-based agency that provides emergency relief and development assistance to disaster-stricken or impoverished communities worldwide. Originally founded in Oxford, England in 1942, it extended aid to countless refugees in the years following the Second World War. In the 1960s, its assistance was further extended to the improvement of agriculture and food production in impoverished countries. During the 1970s, OXFAM organizations were formed in the USA, Canada, Belgium, and Australia, continuing to provide emergency aid for areas stricken by earthquakes, floods, droughts, and other natural catastrophes.

Oxford Committee for Famine Relief   See OXFAM.

oxidation pond   A basin used for the retention of waste water before final disposal. The basin induces the biological oxidation or organic material through exposure to air, either naturally or artificially. (See LAGOON.)

oxide   Any of a large class of chemical compounds in which OXYGEN is combined with another element. (See CARBON DIOXIDE; CARBON MONOXIDE; NITROGEN OXIDES; SULPHUR DIOXIDE; SULPHUR TRIOXIDE.)

oxisol   An intensely weathered type of SOIL, found in the humid tropics and subtropics. When exposed to repeated wetting and drying, the iron oxides in this kind of soil cement soil particles to form ironstone 'hardpans' (also known as LATERITE). The world's largest areas of oxisols are in the Amazon Basin of South America and the Congo Basin of Africa; they are also found in Hawaii and the Caribbean.

oxygen   A colourless odourless tasteless gas, the most plentiful element in the Earth's crust while comprising 21% of the ATMOSPHERE by volume. During respiration, humans, animals, and lower forms of plant life take in oxygen from the atmosphere and return CARBON DIOXIDE. In contrast, by PHOTOSYNTHESIS higher green plants assimilate carbon dioxide in the presence of sunlight evolving free oxygen. Dissolved oxygen is essential for the respiration of fish and other marine life.

oxygen cycle   The circulation of OXYGEN in various forms through the natural ENVIRONMENT. Plants, animals, and humans use oxygen for respiration returning it to the air and water in the form of CARBON DIOXIDE. Carbon dioxide, on the other hand, is taken up by ALGAE and terrestrial green plants and converted into carbohydrate during the process of PHOTOSYNTHESIS, oxygen being released. Despite the COMBUSTION of fossil fuels and the progressive reduction of natural vegetation, the level

ozonosphere

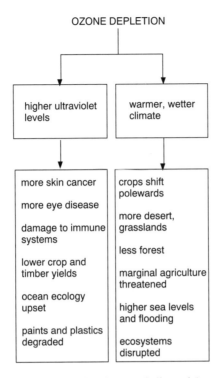

OZONOSPHERE: Figure 21. The social and environmental effects of damaging the ozonosphere.

of oxygen in the atmosphere appears to be relatively stable. (See GREENHOUSE EFFECT; GREENHOUSE GASES; OXIDE.)

oxygen demand The demand of liquid waste effluents for oxygen when discharged to watercourses. (See BIOCHEMICAL OXYGEN DEMAND; CHEMICAL OXYGEN DEMAND.)

oxygen sag The decline and subsequent recovery of the percentage of DISSOLVED OXYGEN in water downstream from a discharge of effluent containing biodegradable material. Oxygen-sag curves relate the dissolved oxygen content of water against time flow, within the context of the process of self-purification.

ozone A triatomic molecule of oxygen; a natural constituent of the atmosphere, with the highest concentrations in the OZONE LAYER (OZONOSPHERE) or STRATOSPHERE. Elevated concentrations of ozone may also occur in the TROPOSPHERE above urban areas, when photochemical reactions occur between oxides of nitrogen and hydrocarbons emitted by motor vehicles in the presence of bright sunlight. Since the 1940s, PHOTOCHEMICAL SMOG has been a characteristic of the Los Angeles basin and has since been identified in other major cities such as Tokyo and Sydney. (See CONVENTION FOR THE PROTECTION OF THE OZONE LAYER; MONTREAL PROTOCOL ON SUBSTANCES THAT DEPLETE THE OZONE LAYER; OZONE LAYER.)

ozone layer (ozonosphere) A layer or strata of the atmosphere about 20 km and 50 km above the surface of the Earth. In this layer, oxygen molecules are split by the sun's ultraviolet radiation, the resulting atomic oxygen recombining with unaffected molecules to produce OZONE. The concentrations of ozone around the globe in this layer vary throughout the year and thinning may occur, for example, over the Antarctic; this is often referred to as a 'hole' in the ozone layer. The preservation of the ozone layer is most important for the survival of humanity, as it is a protective belt moderating the effect of incoming ultraviolet radiation from the sun. Its impairment will undoubtedly promote an increase in the incidence of skin cancer throughout the world. The ozone layer is thought to be threatened by the use of CHLOROFLUOROCARBONS (CFCs) and efforts are being made to restrict the production and use of these substances. (See CONVENTION FOR THE PROTECTION OF THE OZONE LAYER; HYDROCHLOROFLUOROCARBONS; MONTREAL PROTOCOL ON SUBSTANCES THAT DEPLETE THE OZONE LAYER.)

ozonosphere See OZONE LAYER; Figure 21.

P

package treatment plant A compact transportable factory-made sewage-treatment unit capable of achieving a specific effluent quality. It may be deployed to serve a housing estate for a temporary period prior to the installation of a trunk sewer.

packaging The technology and art of protecting a commodity and permitting convenient transport, storage, distribution, display, sale, and delivery. Packaging on the contemporary market aims to protect goods from the hazards of handling and exposure to environmental conditions; to provide a manageable unit for the producer, distributor, and consumer; and to identify the product in a way that appeals to the consumer and meets all legal requirements, displaying all desirable warnings and required information. Materials used include cardboard, metals, timber, plastics, and glass. Packaging has become highly sophisticated in recent years, meeting many desirable ends for the producer, distributor, and consumer. However, many conservation bodies stress the tendencies towards 'overpackaging' and hence overuse of natural resources in the process. The main response has not been to reduce packaging, which has much consumer support, but to encourage RECYCLING, which ensures not only the return of much packaging material, but also the recapture of newspapers and other printed material.

Padjelanta National Park Established in 1962, located in northwestern Sweden; the largest of the Swedish national parks and one of the largest parks in Europe. It contains several lakes, mountains, valleys, and glaciers with a characteristic alpine flora. Among its fauna are the wolverine, Arctic fox, and brown bear; its birdlife includes the golden eagle and merlin.

PAH POLYCYCLIC AROMATIC HYDROCARBONS.

Pak Mun Hydropower Project A hydroelectric power project proposed for Thailand in the early 1990s; the initial plans would have displaced 3300 households. However, after reducing the maximum retention level of the water and moving the location of the dam from the Kaeng Tana rapids to Ban Hua Heo, the number of households requiring resettlement was reduced to 241. These modifications reduced the generating capacity somewhat, but the economics remained favourable. The opportunities for public scrutiny and the commitment of the Thailand Government to sound environmental as well as economic policies ensured, by 1994, a much more satisfactory outcome.

palaeoanthropology An interdisciplinary branch of ANTHROPOLOGY concerned with the origins and development of early humans.

palaeoclimatology The scientific study of the extended climatic conditions of past geological ages. It seeks to explain climatic variations for all parts of the Earth during any given geological period, beginning with the formation of the Earth. (See ICE AGE.)

palaeogeography The scientific study of selected portions of the Earth's surface at specific times in the geological past. Palaeogeology and palaeohydrology are concerned with the GEOLOGY and HYDROLOGY of such periods.

palaeontology The scientific study of life in the geological past that involves the analysis of plant and animal fossils found in rocks, including those of microscopic size. The science is concerned with all aspects of the BIOLOGY of ancient life forms.

PAN PEROXYACETYL NITRATE.

parasitism A relationship between two species of plants or animals, in which one benefits at the expense of the other, without the host being killed. Parasites include ticks, fleas, leeches, lice, tapeworms, flukes, bacteria, and viruses

parasitology The scientific study of animal and plant parasites, which occur in almost all major animal groups and in many plant groups. The science has a number of branches such as veterinary, medical, and agricultural parasitology.

parathion A highly toxic material used as an insecticide; it is effective as a contact poison against many species, acting mainly on the nervous system. It is an organophosphorus insecticide, along with fenthion and malathion. (See PESTICIDE.)

Pareto optimality An economic situation from which it is impossible to deviate so as to make one person or group better off without making some other person or group worse off. This Paretian criterion, as usually conceived, does not concern itself with the distribution of income. Hence, a result may be Pareto-optimal even if the distribution it represents is totally unacceptable on other grounds. The rule does not identify the best or optimal social state. It is named after Vilfredo Pareto (1848-1923). In an attempt to deal with cases which cannot be assessed by the Pareto criterion, that is, cases in which there are losers as well as gainers, Nicholas Kaldor (b. 1908) developed a compensation principle. This states that if those who gain from a policy could fully compensate those who lose and still remain better off, then the policy should be implemented. However, the compensation is hypothetical and not actually paid. A public policy resulting in the well-to-do becoming better off, while the poor become worse off, even if overall the gains exceeded the losses, could have unfortunate repercussions both of a social and political kind.

Paris conventions See CONVENTION FOR THE PROTECTION OF THE WORLD CULTURAL AND NATURAL HERITAGE; CONVENTION ON THE PREVENTION OF MARINE POLLUTION FROM LAND-BASED SOURCES.

Parker Windscale Inquiry Conducted in 1978, a British judicial inquiry reporting on a proposed major expansion of the Windscale nuclear-waste reprocessing plant that would convert uranium oxide waste into reusable uranium and plutonium. The inquiry, conducted by Mr. Justice Parker, heard all sides of the uranium debate concluding that there was no sufficient reason to halt the proposed expansion.

parkway A road passing along a linear landscaped park that is restricted to use by automobiles.

particulates Discrete aggregations of matter, either solid or liquid, found in large numbers in the general ATMOSPHERE. The smallest particles of about 1 μ in size are respirable (can readily enter the human lungs); larger particles of about 100 μ in size may be a source of nuisance. There are various categories of particles: (1) dust particles of mechanical or biological origin such as rock dust, spores, and pollen; (2) fumes such as metallic oxides from metallurgical operations or distillation processes; (3) smoke, being the carbonaceous residues from incomplete combustion; (4) mists or fogs, arising from condensation or by mechanically operated atomizers; and (5) secondary particles such as the visibility-reducing fine particles found in PHOTOCHEMICAL SMOG. (See AIR POLLUTION; SMOG.)

passive dispersion A dispersion or dilution process dependent only on atmospheric conditions.

passive smoking The inhaling of fumes from other people's cigarettes thought to be dangerous to health; for this reason smoking is increasingly forbidden in public places and workplaces.

pasteurization A heat-treatment process that destroys pathogenic microorganisms and fermentative bacteria in food, food products, and beverages. Pasteurization involves raising the temperature of the product to a specified level, maintaining it at that level for a specified period of time, and then cooling the heated products, also within a specified time. Pasteurization has been applied widely to milk, cream, ice cream, and many solid foods.

patent A grant from a government to a person or persons conferring for a specified time the exclusive privilege of making, using, or selling a new invention. Thus, an inventor gains the right to exclude all others from duplicating that invention for use or sale. Patents seek to encourage more inventive activity, protecting the inventor against theft, but nevertheless restrict competition, furthering monopoly in various ways.

pathogenic organisms Organisms responsible for the transmission of communicable diseases such as cholera, bacillary dysentery, typhoid fever, typhus fever, bacterial food poisoning, malaria, yellow fever, amoebic dysentery, and infective hepatitis. Wide varieties of pathogenic bacteria, viruses, and parasites responsible for food, vector-borne, and waterborne diseases are involved.

pathological waste Infectious or potentially infectious waste, the disposal of which could be a public health hazard or nuisance; it includes hospital wastes, some laboratory wastes, and any other material infected with bacteria, viruses, and contaminated blood.

pay-back method A method of comparing the profitability of alternative projects, the object being to determine over what period the net cash generated by an investment will repay the cost of the project.

PCB POLYCHLORINATED BIPHENYLS.

peak concentration The highest concentration of a gas or component predicted at a point by a dispersion model, or measured in the laboratory or in the field. A peak concentration may be transitory or may be maintained over a period.

peak-hour charging The raising of prices for certain services at times of peak demand or conversely, the lowering of prices at off-peak times.

peak load A transient maximum demand on a source of supply such as a steam generator, or gas, or electricity undertaking. Peak loads tend to occur regularly at certain hours of each day. They also occur at certain times of the week and characterize certain seasons.

peak-load pricing (time-of-day pricing) Pricing that endeavours to recover the capital costs of a facility, while at off-peak periods, prices recover only operating costs or short-run marginal costs. In electricity supply, for example, the highest charges correspond with periods of maximum demand, the effect of which is both to recover capacity charges and to discourage excessive use, reducing the amount of generating plant required. At off-peak times, charges are much lower, encouraging demand while making more effective use of plant.

peat The youngest member of the coal series, a fuel consisting of layers of dead vegetation in varying degrees of decomposition, occurring in swampy hollows in cold and intemperate regions. Fresh plant growth at the surface adds material to the decomposing debris. Peat may be found in layers several metres thick. Light in colour near the surface, at deeper levels it is brown and even black.

ped An individual aggregate of natural SOIL, in contrast to a clod produced by disturbance such as ploughing. Peds can readily be seen when most soils are crushed gently in the hand. They give a soil its characteristic structure.

pedal power Synonymous with the use of the smog-free vehicle, the bicycle. Many organizations now promote the use of the bicycle and the establishment of separate bicycle paths through city suburbs. Canberra, the capital of Australia, can be traversed in a number of directions along bicycle paths that are completely separate from roads, the routes utilizing special underpasses. Separate cycle paths began to appear in Britain in the 1930s.

pedestrian malls or precincts A concept associated with the closure of streets to traffic, with some exceptions for business delivery vehicles, with the areas being paved and landscaped. An ENVIRONMENT is created for the pedestrian and shopper that is safe, relaxed, and pleasant; exposure to noise and pollution is much reduced.

pediment (1) In architecture, the triangular part crowning the fronts of buildings in the Greek architectural style; (2) in GEOLOGY, any relatively flat surface of bedrock that occurs at the base of a mountain, being typically found at the base of hills in arid regions where rainfall is spasmodic but intense for brief periods of time.

pedology The scientific discipline concerned with all aspects of soils, including their physical and chemical properties, the role of organisms in SOIL production, the description and mapping of soil characteristics, and the origin and formation of soils.

pedway A pedestrian walkway, including moving pavements, that connects major activities and is separated from vehicular traffic.

pelagic Relating to communities of marine organisms that belong to the open sea, living free from direct dependence on the sea bottom or shore. Pelagic life may be classified as: (1) PHYTOPLANKTON, the food base of all marine animals; (2) ZOOPLANKTON, subsisting on the phytoplankton and relying on water motion for transport; and (3) NEKTON, the free swimmers, dominated by the bony and cartilagenous fishes, molluscs, and decapods.

pelagic zone The entire ocean both in area and depth; PELAGIC life is found throughout, although the numbers of individuals and species decrease with increasing depth. This vertical (and regional) distribution is governed by the abundance of nutrients, dissolved OXYGEN, the presence or absence of sunlight, water temperature, salinity, pressure, and the presence of continental or submarine topographic barriers. (See ABYSSAL ZONE; BATHYAL ZONE; EUPHOTIC ZONE; LITTORAL ZONE.)

Pembrokeshire Coast National Park Established in 1952, a Welsh national park with limestone and sandstone cliffs, sandy beaches, and deep river valleys, comprised of four separate areas with Norman castles and many prehistoric relics. Flora include beech and maple forests, while fauna include badger, otter, grey seal, polecat, voles, squirrels, trout, and salmon.

penicillin One of the first and still widely used antibiotics, derived from the penicillium mould; it is capable of killing many common bacteria that infect humanity, though some strains have developed a specific resistance.

people movers Monorails, moving pavements, escalators and lifts, constructed to connect major central city nodes with each other, with rapid transit stations and peripheral parking facilities.

per capita income The real purchasing power of individuals, calculated as an average for the population. Countries such as Switzerland, the USA, Japan, Germany, and Sweden tend to head the list. At the opposite end of the scale, more than one billion people, some one-fifth of the world's population, live at a level reached by developed countries some two centuries ago. The disparity between nations is analysed in the annual *World Development Report* published by the WORLD BANK. (See BRANDT COMMISSION.)

percolating filter An artificial bed of inert material over which sewage is distributed and through which it percolates to underdrains, promoting the formation of biological slimes that bring about oxidation and clarification of the sewage. It is sometimes referred to as a trickling filter or a bacteria bed.

performance bond An economic instrument for the promotion of environment protection; the placing with an environmental agency of sufficient funds in the form of a bond or security to cover the cost of rehabilitation in the event of the failure of the enterprise concerned. Thus, a guarantee is provided to government against the risk of default by the developer in respect of the rehabilitation conditions laid down. Bonds may also be used for other environmental protection measures, such as security against environmental damage, or for the use of a wide range of natural resources where proper rehabilitation or restoration is required. (See MINE BONDING.)

performance standards Standards employed in ENVIRONMENTAL PLANNING that specify desired results and do not in themselves specify the methods by which the performance criteria should be met. For example, it may be specified that windows shall provide 'ample light and air' rather than, as in the past, that window area should be a specified proportion of the floor area. Similar standards may be laid down in respect of noise, stormwater, erosion, water pollution, overshadowing, access, parking, and public space. However, codes of practice may be issued by local councils and planning agencies giving practical guidance on how performance standards may be achieved.

peri-urban agriculture Areas of land under agriculture and forestry subjected to pressure from expanding urban areas. Such agricultural land may extend as far as 20 or 30 km from a large urban area. The effects on such agriculture may include a considerable rise in land prices, insecurity of tenure, damage due to the passing of people across farming land, and pollution.

permaculture A sustainable agricultural system or 'permanent agriculture' in which many different kinds of perennial trees, shrubs, and vegetables are maintained together in a single area for the purpose of achieving household or community agricultural self-sufficiency. (See MONOCULTURE.)

permafrost Permanently frozen earth, underlying some 20% of the Earth's land surface and reaching depths of over 1500 m in northern Siberia. It embraces 85% of Alaska, and 50% of Russia and Canada. Permafrost is overlain generally by a surface layer that thaws during the warmer seasons of the year. This is termed the active layer. In areas of permafrost, waste disposal, well drilling, mining activities, and construction activities generally are seriously impeded. Permafrost has preserved the carcasses of extinct Ice Age mammals, including mammoths, dating from at least 10 000 years ago.

permanent hardness See HARDNESS.

permanent open freshwater WETLANDS encompassing areas of permanent water, either natural or human-made, often greater than 1 m in depth.

permeability The capacity of a porous material for transmitting a fluid; it is expressed as a velocity. Permeability is largely dependent on the size and shape of the pores and, in sedimentary rocks, the packing arrangement of the grains.

peroxyacetyl nitrate (PAN) A component of PHOTOCHEMICAL SMOG, PAN injures sensitive plants and causes eye irritation, and other health effects. (See LOS ANGELES SMOG.)

persistence The ability of a chemical to retain its molecular integrity and hence its physical, chemical, and functional characteristics in the ENVIRONMENT through which such a chemical may be transported and distributed for a considerable period of time. CHLORINATED HYDROCARBONS (organochlorines) have a high persistence over time and tend to migrate up the food chains, with more serious effects.

pest Any organism or creature considered to be a threat to human beings or their interests. However, in the identification of pests there is an element of subjectivity. Australia's national symbol, the kangaroo, is considered by many Australians in country areas to be a pest, needing an annual culling. The cane toad, once a blessing, is now a pest. Common pests include rats and mice, snails, flies, and many insects. The presence of one bat, possum, rat, mouse, cockroach, mosquito, insect, silverfish, flea, moth, bedbug, earwig, or slater, is enough to qualify it as a household pest, to be rapidly eradicated. (See PESTICIDE.)

pesticides A collective name for a variety of insecticides, fungicides, herbicides, fumigants, and rodenticides. All pesticides interfere with normal metabolic processes in the PEST. Insecticides may be classified as: (1) CHLORINATED HYDROCARBONS (organochlorines); (2) ORGANOPHOSPHORUS types; (3) CARBAMATES; and (4) others. Pesticides have contributed greatly to human health and welfare, increased PRODUCTIVITY, and food production throughout the world, but excessive use, misuse, and environmental contamination have allowed these often persistent chemicals to spread well beyond their targets and to pass up the FOOD CHAIN. (See DDT; PERSISTENCE.)

pest-risk analysis (PRA) The systematic assessment and management of risks due to domestic or exotic pests and diseases. Typically, it has three stages: (1) risk identification; (2) risk assessment; and (3) risk management. The analysis of pest biology and control methods using PRA provides estimates of the level of risk under different management options, encouraging informed decisions.

petrochemicals Chemicals manufactured from the products of oil refineries, based largely on ethylene, propylene, and butylene produced in the cracking of petrol (gasoline) fractions. It is the unsaturated olefin compounds that are used to produce chemicals. The saturated paraffin compounds (methane, ethane, propane, and butane) are used as refinery fuel or sold as bottled gas.

petrol (gasoline) A complex mixture of light hydrocarbon blending stocks, the characteristics of which are varied to suit all engines, seasons, and climates. The 1980s witnessed many countries introducing lead-free petrol as an environmental measure. Lead had traditionally been added to reduce the tendency to knock and improve the octane number. Improved engine design and performance eliminated the need for this.

petroleum According to generally accepted theory, a substance derived from the remains of plant and marine life that existed on the Earth millions, and in some cases, hundreds of millions of years ago. A complex and variable substance, petroleum ranges from solid bitumen, through liquid oils, to highly volatile natural gases such as METHANE. All are essentially mixtures of compounds made up from the elements hydrogen and CARBON, known as natural hydrocarbons.

pH value The measure of the degree of acidity or alkalinity of a liquid or effluent, where pH 1 is highly acidic and pH 14 is highly alkaline; pH 7 is neutral. Unpolluted rivers generally vary in pH value from about 5.0 (acid moorland peaty streams) to about 8.5 (chalk streams). Acid water draining from an abandoned coal mine may have a pH value as low as 2.0. In order that the normal biological and chemical processes that occur in a river or during sewage treatment should not be adversely affected, the river water of sewage should not vary greatly from neutral. The pH is usually measured with a pH-meter.

phon A dimensionless unit, used to express the loudness level of a given sound or noise. (See ACOUSTICS.)

phosphates Essential nutrients for plants and normal constituents of the food of humans and animals. The principal human sources are sewage, agricultural runoff, and detergent 'builders'. In excessive amounts they are considered a major factor in the process of EUTROPHICATION resulting in excessive growth of aquatic plants, depletion of OXYGEN, loss of fish, and general degradation of water quality.

phosphatic fertilizers Fertilizers rich in PHOSPHATES intended to overcome deficiencies of phosphorus in certain soils. As such soils may also be deficient in SULPHUR, superphosphates may be applied to soils, these containing both sulphur and phosphorus.

phosphorus Occurring in various forms, the commonest is white phosphorus, a waxy solid emitting a greenish glow in the air, burning spontaneously to phosphorus pentoxide, and very toxic. It is used widely in fertilizers. (See PHOSPHATIC FERTILIZERS.)

phosphorus cycle The circulation of PHOSPHORUS atoms brought about mainly by living things. Phosphorus is an essential element in ribonucleic acid (RNA) and deoxyribonucleic acid (DNA), which are the bearers of genetic information and play other important parts in the processes that take place in living cells. Phosphorus becomes available to plants from a variety of natural and artificial sources. The process of decay releases phosphates to the SOIL, completing the phosphorus cycle.

photochemical Pertaining to the permanent chemical effects of the interaction of radiant energy (especially light) and matter. (See LOS ANGELES SMOG.)

photochemical smog A type of smog first observed and named in Los Angeles, California, shortly after the Second World War; photochemical smog has since been identified in many cities, most notably Tokyo. It is characterized by a whitish haze. It causes eye irritation and damage to certain kinds of vegetation and to rubber. Smog occurs in warm sunny weather, where there are no strong winds to disperse its precursors, namely nitrogen oxides and hydrocarbons. Smog forms in the ATMOSPHERE through the action of sunlight on these primary pollutants, which are emitted mainly by motor vehicles. The reactions are highly complex. (See LOS ANGELES SMOG.)

photolysis The degradation of a chemical in water by light.

photosynthesis The creation of organic compounds in green plants in the presence of light. Green plants take up CARBON DIOXIDE from the ATMOSPHERE and release OXYGEN while creating CARBOHYDRATES. Water also plays an essential part in this process, while CHLOROPHYLL, the green pigment, absorbs the light energy which is converted into chemical energy. Complex organic compounds such as starch are synthesized in this process.

photovoltaic conversion A technique for generating electricity directly from solar energy utilizing, usually, silicon cells. Silicon cells have a virtually unlimited life. The process is silent, static, and without waste discharge. High costs have so far restricted their use to satellite communications systems, spacecraft and experimental vehicles.

phreatophyte A plant with roots long enough to reach the WATER TABLE and which can draw its water from the GROUND WATER supply.

phylogeny The evolutionary history of a group of organisms, as distinct from the development of the individual organism.

phylum A principal grouping of organisms in the animal kingdom, equivalent to a division in the plant kingdom. (See TAXONOMIC.)

physical science The systematic study of non-living matter, as distinct from the study of living organisms, the province of biological science. Physical science consists of four broad areas: astronomy, physics, chemistry, and the earth sciences.

physiology The study of the functioning of living organisms, human, animal, or plant, or of the functioning of their constituent tissues or cells.

phytobenthos Higher aquatic plants that are fixed to the sea bed or the bottom of lakes or ponds. These plants are divided into: hydrophytes (plants submerged in water), amphiphytes (amphibious plants), and marsh plants. Through the process of

PHOTOSYNTHESIS, the surrounding water is freed from CARBON DIOXIDE and supplied with free OXYGEN.

phytophagus Descriptive of animals that feed on plants, that is, herbivorous animals.

phytoplankton The primary basis of animal life in the sea consisting of passively floating microscopic plants, chiefly diatoms (microscopic unicellular brownish ALGAE). The phytoplankton carry out PHOTOSYNTHESIS which requires solar energy that penetrates only the surface layer of the sea. Many ocean fish feed on phytoplankton, which is also the basic food of ZOOPLANKTON.

phytotoxicant Poisonous to plants.

Pigou, A. C See MARGINAL SOCIAL AND PRIVATE NET PRODUCTS.

Pinatubo, Mount A volcanic mountain, 90 km northwest of Manila, Philippines, that erupted in 1991 after 600 years of inactivity; the months April to June were characterized by several enormous volcanic blasts followed by mudflows and floods. Over 800 people died with two dozen towns destroyed; the homes of 1.2 million people were lost and 81 000 ha of farmland buried. Over half-a-million refugees fled to Manila. The plumes of ash rose to 14 900 km in the atmosphere, spreading ash around the world. It was the largest volcanic disaster of the century.

Pinchot, Gifford (1865–1946) A pioneer of US forestry management and conservation, he developed the US forest service and founded the Yale School of Forestry. (See FOREST SERVICE, US.)

pink noise NOISE exhibiting constant power per constant frequency interval, for example, per octave interval.

pit voids Voids or large depressions that remain after completion of surface coal mining. The shape of a void is often rectangular, perhaps 3 or 4 km long and up to 300 m deep. Slope angles may be quite steep. Pit voids not infilled with overburden may be utilized as water storage for public recreation, or as water supplies for farms. Water-filled voids may also be used as wildlife habitats to increase the diversity of flora and fauna. Landscaping encourages recreational activities.

pitch The frequency of a tone or SOUND, depending on the relative rapidity of the vibrations by which it is produced. In low-frequency sounds, the vibrations are relatively far apart; in high-frequency sounds they are close together. While the greater the frequency, the higher the pitch, the relationship is not a linear one.

placental mammals Those mammals in which the young develop inside the mother's uterus and are connected to the mother's blood supply through a placenta. The majority of the world's mammals are placental, including humans. (See MONOTREMES.)

plain Any relatively level area of the Earth's surface exhibiting only gentle slopes with occasional relief. Plains vary enormously in area, the largest covering hundreds of thousands of square kilometres, such as the Great Plains of North America. They occupy slightly more than one-third of the terrestrial surface of the Earth. With certain exceptions such as the Sahara and most of the interior of Australia, plains have often become the sites of major centres of population, industry, and commerce.

plankton A collective name for marine and freshwater organisms that exist in a drifting floating state, either because they are non-motile, or too small or weak to swim against the current. Such organisms include ALGAE, BACTERIA, protozoans, crustaceans, molluscs, and coelenterates. (See NEKTON; PHYTOPLANKTON; ZOOPLANKTON.)

planning The plotting of a course of action involving the allocation of resources; a characteristic of most though not all human activity whether individual, family, collective, corporate, or at various levels of government. Planning, often sophisticated, is often the hallmark of successful corporations and public agencies and, indeed, of individual activity. The role of planning in the public arena remains highly controversial and political, with opinion ranging from a belief that the free market should determine most things to a belief that government should have a large and important role in managing the economy. Most countries fall somewhere within this spectrum, while all countries have accepted the need for some measure of government intervention in the field of ENVIRONMENTAL PLANNING.

planning blight Descriptive of a situation when land becomes unsaleable, or is devalued, because of a belief that it will be required for, or affected by, some public project. Blight may result from a town planning scheme that indicates future needs and requirements, or as a result of the announcement of a large public undertaking. In some cases compensation may be payable to those whose land is acquired for a project, but without acquisition compensation is not usually paid. In a worst-case scenario, a decision to construct an airport at a particular site may result in a lengthy debate, placing the owners of properties possibly affected under a cloud of uncertainty for a considerable period of time. In the end, the decision may be reversed by the same or a different government, without compensation, even after years of delay. (See INJURIOUS AFFECTION.)

planning standards Regulatory or prescriptive standards that relate to development

plant mix

projects, imposed by local planning authorities in conjunction with zoning restrictions. Standards are applied in many countries in relation to some or all of the following: residential density, lot ratio, daylight and sunlight, car parking, highways, visual privacy, traffic noise, amenity space, restrictions on building heights, access, public transport, recreational facilities, safety and security, and standards of construction.

plant mix The proportions of different kinds of plant in a productive system. In respect of electricity generating plant, the different types of plant are usually classified as coal, oil, natural-gas, nuclear, hydro-electric, pumped-storage, gas-turbine, wind, tide, or solar. The first four are simply labelled by the fuels they use and are all steam-electric plant.

plant tissues These are made up of complex compounds such as carbohydrates (for example, cellulose, lignin, starch, and sugar) and proteins, fats, oils, resins, and waxes. The carbohydrates, which form a high proportion of most plant tissues, consist of CARBON, HYDROGEN, and OXYGEN; the proteins are compounds of carbon and nitrogen, with little or no oxygen. Plants synthesize their component tissues by PHOTOSYNTHESIS and other chemical reactions using mineral salts taken in with water from the SOIL.

plant virus Any of a number of agents that cause plant disease; many plant viruses are rodlike and contain RNA (ribonucleic acid) or DNA (deoxyribonucleic acid). Symptoms of virus infection include colour changes, dwarfing, and tissue distortion.

plantation forests Forests that have been created artificially through the systematic planting and nurturing of seedlings, usually of one species only, and are being managed intensively for timber production. Plantation forests are very productive and an efficient way of harvesting high-quality timber. Plantations may be for softwood or hardwood species. In plantation forests all the trees are of the same age. (See NATURAL FOREST.)

plate count The number of bacterial colonies that grow on a plate of nutrient agar to which a sample has been applied and incubated at a specified temperature for a specified period, under laboratory conditions.

plate tectonics The theory that the Earth's surface is made up of massive plates supporting the continents which have moved slowly but constantly during geological history, thus changing the positions of continents and oceans. (See LITHOSPHERE.)

plume (1) The discharge of waste gases from a chimney, stack, or vent, that is initially visible, becoming slowly invisible to the eye as it disperses. (2) The discharge of heated liquid effluent from, say, a power station into a river, harbour, or ocean, being slowly cooled to ambient temperature.

plume rise The extent to which a gaseous PLUME, emitted from a stack into the general ATMOSPHERE, will rise while it slowly disperses. Some rise will be ensured as a result of the velocity of discharge alone (the efflux velocity) while a further rise is ensured through buoyancy, assuming that the emitted gas is warmer than the general atmosphere. The total effective height of emission is therefore: stack height, plus velocity rise, plus buoyancy rise. The buoyancy rise is directly affected by wind speed and turbulence, the presence of a TEMPERATURE INVERSION, and by DOWNDRAFT and DOWNWASH.

plutonium A radioactive chemical element. The most important plutonium isotope is plutonium-239 because it is fissionable; it has a relatively long half-life of 24 360 years. It is also used in the manufacture of weapons. Plutonium is a radiological poison because of its high rate of alpha emission and the absorption of alpha particles into bone marrow. (See ALPHA PARTICLE.)

plutonium economy Descriptive of an economy in which the basis of electricity generation is the fast breeder reactor using PLUTONIUM as the fuel, the plutonium being recovered from the spent fuel elements of nuclear reactors using natural or enriched uranium. The economy then hinges on the use of plutonium, and becomes increasingly dependent upon supplies of it; the risks of the diversion of plutonium into the manufacture of nuclear weapons are thus significantly increased.

podzolic soil A SOIL that usually forms in a broad-leaved forest and is characterized by moderate leaching which results in an accumulation of clay and perhaps iron. Podzolic soils may have a soil layer cemented together by iron. A podzolic soil may be distinguished from a PODZOL SOIL because the main effect of the humic acid is clay eluviation, whereas in podzol soils it is iron eluviation or the leaching of all weathering products or both, which leaves a bleached ashy horizon.

podzol soil A SOIL that typically forms in very cold climates in coniferous forests, heaths or tundra. The HUMUS formed is very rich in slightly polymerized substances such as fulvic acid, which percolates through the soil and forms complexes with the clay minerals and iron and leaches out all the weathering products. (See PODZOLIC SOIL.)

point sources In respect of POLLUTION, end-of-pipe or chimney-stack sources; all other sources are non-point or diffuse. Point sources such as industrial chimneys, motor-vehicle exhausts, waste-water discharge pipes, drains and sewers have received much attention from pollution-control agencies in the past. Increasing attention is now being given to DIFFUSE SOURCES, such as multiple small sources spread over a considerable area, for example, urban runoff, road traffic noise, chemically induced OZONE depletion, soil erosion, inland water salinity, agricultural runoff, river sedimentation, and waterway algal blooms.

polishing   See TERTIARY TREATMENT.

polishing processes   Additional treatment designed to produce a better effluent, for example, tertiary treatment for sewage effluent.

Pollutant Release Inventory, Canada   A national inventory to provide comprehensive data on releases of specified substances to air, water, and land. The National Pollutant Release Inventory has operated since 1993 under the Canadian Environmental Protection Act. Major manufacturing facilities are required to report to Environment Canada once a year. It is envisaged that the scheme will evolve over time. (See TOXIC RELEASE INVENTORY, US.)

polluter pays principle (PPP)   (1) A principle which equates the price charged for the use of environmental resources with the cost of damage inflicted on society by using them. (2) A principle that asserts that the full cost of controlling pollution to an adequate degree shall be carried out by the polluter without public subsidy or tax concessions. The potential cost of pollution to society is translated into pollution-control costs which are 'internalized' and reflected in the costs of production.

pollution   Any direct or indirect alteration of the physical, chemical, biological, thermal, or radioactive properties of any part of the ENVIRONMENT by discharging, emitting, or depositing wastes or substances so as to affect any BENEFICIAL USE adversely; to be hazardous or potentially hazardous to public health, safety, or welfare; or to be detrimental to FAUNA or FLORA or ENDANGERED SPECIES. (See AIR POLLUTION; LAND POLLUTION; NOISE; WATER POLLUTION.)

pollution charges   Charges applied to discharge rates from individual point sources, or to non-point sources. The main purpose of emission or effluent charges is to encourage a reduction in rates of discharge; Dischargers face the option of paying the charge or reducing the quantities emitted. Some of the first examples of effluent charges were in respect of water quality in France, Germany, and the Netherlands; economic instruments are now widely applied in many countries. (See also EFFLUENT CHARGES.)

pollution-control charges   The costs incurred by an enterprise to prevent or restrict the emission of potentially harmful pollutants to air, land, and water; in addition to pollution-control technologies increasing attention is being given to cleaner production technologies, new processes, and recycling techniques. Pollution-control costs in various industries have reportedly ranged from 10 to 20% of the initial capital cost of plant.

pollution-control strategy   Measures adopted to combat POLLUTION both at its source and after release. For the community, such a strategy involves: appropriate legislation covering the whole spectrum of pollution; the setting of standards and targets to be achieved by individual sectors of the economy, including motor vehicles, in respect of AIR POLLUTION, NOISE, LAND POLLUTION, WATER POLLUTION, and RADIOACTIVITY; the creation of a competent environment protection agency with adequate powers; consultation with all relevant branches of industry and commerce and the public; consultation between all levels of government; annual reports on progress towards the objectives, and the integration of those objectives and achievements with the objectives of SUSTAINABLE DEVELOPMENT.

pollution credits   A provision in US clean air legislation allowing utilities to accumulate credits for keeping air-pollution emissions within limits and to sell excess credits. In particular, this gives utilities an incentive to reduce emissions of sulphur dioxide, a by-product of fossil-fuel burning that contributes to the formation of acid rain. In 1991, the Chicago Board of Trade voted to create a marketplace where utilities and speculators could buy and sell the right to pollute.

polychlorinated biphenyls (PCBs)   Any of a class of highly stable organic compounds resulting from the reaction of CHLORINE with biphenyl. PCBs have been useful as lubricants, heat-transfer fluids, and fire-resistant fluids in transformers and capacitors; they have also found application in paints, paper coatings, and packaging materials. PCBs have proved detrimental to the ENVIRONMENT as they remain in soil and water for many years, enabling them to accumulate and enter the FOOD CHAIN. They have proved particularly toxic to fish and have caused liver dysfunction in humans. Steps have been taken in many countries to restrict the use of PCBs; satisfactory substitutes are available in some areas.

polycyclic aromatic hydrocarbons (PAH)   A variety of chemical compounds such as benzopyrene, dibenzopyrene, and dibenzoacridine, exposure to which can cause cancer in humans. They are produced by incomplete COMBUSTION processes, including the operation of diesel- and petrol-powered engines, combustion of fossil fuels, forest fires, cigarette smoke, and the incineration of refuse.

polyethene   See POLYTHENE.
polyethylene   See POLYTHENE.

polythene (polyethene; polyethylene; high-pressure polythene) A compound of CARBON and HYDROGEN, the products of complete COMBUSTION being CARBON DIOXIDE and water only.

Pompeii One of several cities buried by the volcano Vesuvius, Italy, on 24 August AD 79 and the following days, remaining buried and forgotten for almost 2000 years. Today most of the city has been excavated revealing the details of life in that prosperous city at that time, and the fate of the inhabitants. It is a model of HERITAGE CONSERVATION.

pool fire Ignition of flammable vapour from a spill of liquid that has collected on the ground or in a container; the combustion of material evaporating from a layer of liquid at the base of the fire.

population density The number of organisms, species, or humans found in a prescribed area. (See CENSUS OF POPULATION; UN CONFERENCE ON POPULATION AND DEVELOPMENT; UN WORLD POPULATION CONFERENCE; WORLD POPULATION.)

population dynamics A study of the changes in the size, age, and sex composition of a population due to major BIOTIC and ABIOTIC factors.

positional good or service A good or service, the value of which in the market is socially influenced. Many things in society have hierarchical significance and personal satisfaction is often derived from being top of the tree in one respect or another, or in enjoying acclaim or prestige. To many people, position in the hierarchy matters very much, and goods and services associated with well-regarded positions become more highly valued as basic necessities are met. A general rise in real income can cause dissatisfaction among those whose progress has been less rapid than that of others. Thus economic growth may yield less satisfaction than might have been expected, for the distributional aspects may become more prominent.

post-project analysis A systematic examination or ENVIRONMENTAL AUDIT of a completed project, to ascertain whether all the environmental conditions imposed on the development consent have been complied with, attention being paid also to the accuracy of any predictions made in the original ENVIRONMENTAL IMPACT STATEMENT. (See ENVIRONMENTAL IMPACT ASSESSMENT.)

poverty See BRANDT COMMISSION; ECONOMIC AND SOCIAL COMMISSION FOR ASIA AND THE PACIFIC; UN CONFERENCE ON NUTRITION; UN FOOD AND AGRICULTURE ORGANIZATION; WORLD COMMISSION ON ENVIRONMENT AND DEVELOPMENT; WORLD FOOD PROGRAM.

power lines See ELECTROMAGNETIC RADIATION.

Poza Rica incident An AIR POLLUTION incident involving respiratory illness and death. At Poza Rica, Mexico, on 24 November 1950, some 22 persons died and 320 were hospitalized as a result of the malfunction of an oil refinery recovery unit; large quantities of HYDROGEN SULPHIDE were vented to the ATMOSPHERE. Meteorological data indicated that a pronounced low-altitude TEMPERATURE INVERSION prevailed at the time. Unlike other acute air-pollution incidents, a single source and an air pollutant were responsible.

PPP POLLUTER PAYS PRINCIPLE.

PRA See PEST-RISK ANALYSIS.

prairies Level or rolling grassland, especially that found in central North America. The vegetation is composed primarily of perennial grasses, with many species of flowering plants.

precautionary principle A principle adopted by the UN CONFERENCE ON ENVIRONMENT AND DEVELOPMENT (the Earth Summit) 1992, that in order to protect the environment, a precautionary approach should be widely applied. The Rio Declaration on Environment and Development, principle 15, interpreted the precautionary approach as meaning that where there are threats of serious or irreversible damage to the environment, lack of full scientific certainty should not be used as a reason for postponing cost-effective measures to prevent environmental degradation. Critics of this approach are concerned about large commitments of resources to deal with vaguely defined problems, but it should be noted that the reference to cost-effective measures implies a high degree of certainty about the nature of the problem.

precipitation All water particles that fall from clouds and reach the ground. These particles include drizzle, rain, snow, ice crystals, and hail.

precursor A substance which may participate in or influence a reaction in air or water to produce a new substance; for example, in the ATMOSPHERE, NITROGEN OXIDES and HYDROCARBONS with sunlight are the precursors of OZONE and PHOTOCHEMICAL SMOG. (See AIR POLLUTION.)

present value or worth The discounted sum of a series of future payments or receipts, discounted at a selected rate. The present-worth method offers a general and systematic approach to the problem of making comparisons between present proposals for investments, plant, programmes, or policies when future streams of costs and revenues are relevant to the comparison. The present worth of a future sum may be calculated by multiplying the sum by a present-worth factor. For an 18% discount rate, the present-worth factor is given by:

$$\frac{1}{1.18^m}$$

where m = the number of years to when the future sum is received or paid.

pressure burst The rupture of a vessel or system of pipes and tubes under pressure, resulting in the formation of a blast wave and projection of missiles, both of which may cause damage. (See BOILING LIQUID EXPANDING VAPOUR EXPLOSION; NOISE.)

pressure flow See MASS FLOW.

pressure group See LOBBYING.

price mechanism The system used in market economies for the distribution of resources through the agency of price. If too much of commodity A and too little of commodity B is being produced, the demand price for A falls and that for B rises. The signal is thereby given to the productive system to make the necessary adjustments. The price mechanism theoretically leads to a point of maximum efficiency. The price mechanism, even in conditions of imperfect competition, tends to perform better than any conceivable alternative in respect of most goods and services. In certain areas, however, it is deficient. It responds only to effective demand, not to need, and hence does not produce low-income housing, large-scale educational facilities, or infrastructure requiring huge expenditure. Further, in commodity markets a fall in price in one season may encourage producers to plant more for the next season and not less; thus a depressed price can increase supply.

primary treatment In respect of SEWAGE TREATMENT, a series of mechanical treatment processes that remove most of the floating and suspended solids, but which have a limited effect only on colloidal and dissolved matter. Mechanical separation is achieved by screening, the coarser materials only being removed.

primate In ZOOLOGY, a member of the mammalian order, the Primates consisting of two suborders: (1) the prosimians (lemurs, lorises, and tarsiers); and (2) the anthropoids (monkeys, apes, and humans). According to fossil records, primates originated within the late Cretaceous period (up to about 100 million years ago) as forest-dwelling creatures. A suggestion that modern humans are descendants of these early primates was first made by Charles Darwin (1809-82) in his *Origin of Species* (1859).

principles of sustainability Key principles embodied in a range of national and international conventions, agreements, and instruments, as follows:
(1) INTERGENERATIONAL EQUITY.
(2) INTRAGENERATIONAL EQUITY.
(3) PRECAUTIONARY PRINCIPLE.
(4) Conservation of BIODIVERSITY.
(5) INTERNALIZATION OF ENVIRONMENTAL COSTS.

(See AGENDA 21; CONVENTION ON CLIMATE CHANGE; CONVENTION ON PROTECTING SPECIES AND HABITATS; UN CONFERENCE ON ENVIRONMENT AND DEVELOPMENT; WORLD COMMISSION ON ENVIRONMENT AND DEVELOPMENT.)

private sector That part of the MARKET ECONOMY owned by private individuals, corporations, public and private companies, and multinationals, as distinct from the PUBLIC SECTOR, owned by government.

privatization The process of recent years, evident around the world, of selling government assets to the PRIVATE SECTOR. While governments acquire lump sums from sales, they also lose regular and sustained income from profitable enterprises; the unprofitable enterprises providing public services remain unsold.

producer goods In ECONOMICS, goods manufactured and used in further manufacturing processes or for resale; intermediate goods.

producers Organisms that produce their food from inorganic molecules by PHOTOSYNTHESIS or chemosynthetic processes that use external sources of energy. Most producers are green plants, or photosynthetic microorganisms such as PHYTOPLAKTON.

product A good or service available to the market, final goods and services being available to ultimate consumers.

product charges Charges imposed on inputs to economic activities as a means of indirectly controlling environmental impacts. In some European countries, charges are levied on fuels according to their sulphur or carbon content, as an incentive to reduce emission of SULPHUR DIOXIDE or CARBON DIOXIDE. Differential taxes may be applied to recycled paper to encourage reuse of paper, conserve timber supplies, and reduce waste disposal and litter.

production ecology The study of the more complex 'life communities' or BIOMES, considered as TROPHIC associations of food cycles. It is concerned essentially with the dynamic structure of the system whereby regulation is effected, rather than with the actual operation of the regulation processes on individuals within populations. The name 'production ecology' results from an increasing preoccupation with the supply, or production, of food and ultimately with the flow, or exploitation, of energy within trophic cycles.

productivity (1) The efficiency with which the FACTORS OF PRODUCTION are integrated; increasing productivity means the ability to produce goods at a progressively lower cost per unit of output while maintaining or improving quality. Improvements in productivity may come about through a more effective use of labour, better equipment and layout, or through better use of natural resources. It is difficult to attribute improvements in productivity to any one factor, but the relationship of physical output to hours of labour expended is often adopted as a rough guide. However, this ratio suggests an improvement in labour productivity whereas the improvement may have been attributable to capital investment alone. Improvements may come from a variety of additional sources such as economies of scale achievable in meeting a larger market demand, in improvements in the quality of raw materials, or in new management techniques such as just-in-time production.

(2) The rate at which biological matter (BIOMASS) is produced by an ECOSYSTEM, or part of an ecosystem. It is related to the rate of energy conversion within the system. Productivity is affected directly by ABIOTIC factors such as climate and SOIL. Terrestrial ecosystems in warm, moist regions, tend to be more productive than those in cooler, drier areas. Terrestrial ecosystems also tend to have higher productivity than systems in the open ocean, though coastal marine ecosystems are among the most productive ecosystems on Earth.

product liability A responsibility or onus imposed by consumer legislation on a manufacturer to warn consumers appropriately about the possible detrimental or harmful effects of a product and to foresee how it may be misused.

profundal zone The area immediately beneath the LIMNETIC ZONE of a lake; it extends downwards to the lake bottom. The penetration of sunlight to the profundal zone is very limited, hence green plant life is absent. Bacteria and fungi flourish in bottom ooze.

programme impact statements The adoption of ENVIRONMENTAL IMPACT ASSESSMENT procedures in respect to complete programmes, or in respect of the broader task of policy formulations. There have been numerous instances under the NATIONAL ENVIRONMENTAL POLICY ACT (US) in which impact statements have covered an overall programme of works, with additional statements being required for each component project. The notion has emerged of a tiered EIA process, with assessments undertaken in relation to the framing of broad policy objectives, public agency programmes, planning schemes, and with final decisions on specific projects.

propagation (1) In BOTANY, plant reproduction either sexual or asexual; (2) the breeding of offspring. In general, the 'propagation of the species'.

property value method See HEDONIC PRICE TECHNIQUE.

proponent The proposer of an activity, policy, plan, programme, or project in the private or public sectors. A proposal usually requires official approval or consent by some level of government. In the process of obtaining this, the public have increasing opportunities to voice opinions of support or objection.

protective covenant (restrictive covenant; restrictive agreement) A legal requirement attached by a vendor to a deed or statement of ownership, imposing duties and restrictions additional to formal local council requirements. Covenants may strengthen requirements in terms of size, type, quality, and architectural design. The development of an estate is thus held to a standard.

protein A complex biological molecule containing CARBON, HYDROGEN, OXYGEN, NITROGEN and usually sulphur. Proteins are present in all organisms, being essential for life.

proton A positively charged particle which is the constituent of the nucleus of all nuclides. The number of protons in the nucelus determines the chemical properties of an element, and hence is characteristic of each of the chemical elements.

protosoil A SOIL at a preliminary stage of development.

Protozoa A group, or PHYLUM of animals that consist of one cell only with at least one well-defined nucleus. It comprises the classes Mastigophora, Ciliata, Sarcodina, and Sporozoa. The division is not always clear-cut, as both fungi and ALGAE also include unicellular forms.

Pseudomonas aeruginosa A bacterium common in faeces, soil, water, and sewage. Its presence in drinking water indicates a serious deterioration in bacteriological quality.

psychometric chart A chart that defines the relationships between dry-bulb temperature, wet-bulb temperature, moisture content, relative humidity, and heat in the ATMOSPHERE.

psychosocial stressors Stimuli suspected of causing diseases that originate in social relationships or arrangements, and affect the organism or body through the medium of the higher nervous processes. (See ENVIRONMENTAL HEALTH IMPACT ASSESSMENT; SOCIAL IMPACT ASSESSMENT.)

psychosomatic Related to mind (the psyche) and body (the soma); descriptive of diseases arising from, or aggravated by, a mind–body relationship. (See PSYCHOSOCIAL STRESSORS.)

psychosphere A concept of human psychology, thought, and culture viewed as an environmental phenomenon in the same manner as other concepts such as ATMOSPHERE, BIOSPHERE, HYDROSPHERE, LITHOSPHERE, and SOCIOSPHERE.

public facility Any use of land, whether publicly or privately owned, for transportation, utilities, or communication, or for the benefit of the general public, including streets, schools, libraries, fire and police stations, county buildings, municipal buildings, power houses, recreational centres, parks, and cemeteries.

public goods and services Goods and services distributed by the government free to its citizens, being indivisible benefits shared by everyone equally. No private entrepreneur could provide and sell such goods and services efficiently on an individual basis. Examples include external defence, the police, the courts, regulation of industry, overseas representatives, tax collection, and emergency services.

public health The art and science of preventing disease, prolonging life, and promoting HEALTH and wellbeing through public administration and community efforts. These efforts are directed to the control of communicable diseases, promotion of personal hygiene, provision of SANITATION, environmental management, refuse disposal services, safe water supplies, adequate ventilation, vector control, food hygiene, the removal of slums and rehousing of the inhabitants, quarantine, clean air, and community services for tackling promptly potential outbreaks and epidemics of disease, and administering vaccination services. (See AIR POLLUTION; CHLORINATION; CONTAMINATED SITES; DRINKING-WATER SUPPLY AND SANITATION DECADE; ENVIRONMENTAL HEALTH IMPACT ASSESSMENT; ENVIRONMENTALLY HAZARDOUS CHEMICALS; NOISE; NUISANCE; WATER POLLUTION; WORLD HEALTH ORGANIZATION.)

public inquiry or hearing An avenue for the public, voluntary bodies and government agencies to express opinions before an independent and impartial commissioner of inquiry, to enable issues concerning a controversial proposed development to be fully discussed. The usual outcome is the submission of a report by the commissioner with recommendations to a decision-making body or minister, the report becoming immediately a public document. The success of the public inquiry hinges upon the choice, integrity, and independence of the commissioner, and upon a political and social context which encourages full participation by all citizens, without fear of reprisal or discrimination. The public inquiry often stands at the apex of EIA processes. (See Box 28.)

public local inquiry, UK A long-established form of public ENVIRONMENTAL PLANNING inquiry conducted in Britain under housing, energy, and environmental legislation. Many pieces of legislation allow for appeals to the central government or more specifically to the ministry or minister involved against the decisions of local councils and planning bodies in respect to a wide range of matters. It is customary for a public local inquiry to be conducted by one or more of Her Majesty's Inspectors (HMIs), specially appointed for the conduct of inquiries. Objections are carefully heard on a fairly formal basis, the HMI reporting to the minister in respect of findings and recommendations. In some instances the HMI will be assisted by technical assessors with expertise in the area under consideration. The subjects of objection including proposed power stations, transmission lines, housing developments, redevelopment schemes, slum clearance schemes, smoke-control areas, zoning and rezoning proposals, airports and airport extensions, industrial installations of many kinds, pipelines, storage facilities, and docks. HMIs enjoy a high reputation as a body of independence, integrity, and objectivity.

public participation See ENVIRONMENTAL IMPACT ASSESSMENT; ENVIRONMENTAL IMPACT STATEMENT; NATIONAL ENVIRONMENTAL POLICY ACT, US; PUBLIC INQUIRY OR HEARING; PUBLIC LOCAL INQUIRY, UK; SCOPING.

public sector Those national assets owned and operated by government, the remainder being part of the PRIVATE SECTOR. The public sector may encompass some banks and financial bodies, INFRASTRUCTURE, transport systems, electricity supply systems, docks and harbours, major redevelopment schemes, prisons, defence establishments, some shipping, and all nationalized industries.

puddling (1) The mixing of water with clay or chalk to increase the plasticity of the mixture. (2) The use of puddled clay or chalk to line the base of landfill sites so as to prevent LEACHATE from solid waste penetrating strata and contaminating GROUND WATER.

Pudong Development Area, Shanghai A newly developing area of Shanghai some 350 km^2 in extent; development of this area began in 1990. It has three components: (1) Lu Jia Zui (Financial and Trade Centre); (2) Golden Bridge (New Industrial Town); and (3) Wai Gao Qia (The Free Port). By 1995, a large number of companies had established themselves in Pudong.

pulp The raw material used in paper-making; it contains vegetable, mineral, or human-made fibres. Rags and other fibres have been used as paper pulp but nearly all papers are made of vegetable (cellulose) fibres. The most abundant source of cellulose is forest timber. Wood pulps may be mechanical or chemical. Chemical pulps are classified as: (1) unbleached sulphite; (2) bleached sulphite; (3) bleached and unbleached sulphate (kraft); and (4) soda. One pollution problem has been that high levels of chlorinated organic compounds in the effluents from pulp mills, caused by CHLORINE bleaching, do have serious effects on the ENVIRONMENT downstream of the mill. In an improved bleaching system, the chlorine is completely replaced by chlorine dioxide, hydrogen peroxide, and oxygen applied in special sequences.

pulp logs

> **Box 28 Public inquiry procedures, New South Wales, Australia**
>
> (1) Minister for Planning directs that a Commission of Inquiry be held into a proposed development and appoints a Commissioner of Inquiry for this purpose.
> (2) Notice of the Commission of Inquiry appears in newspapers indicating where and when the inquiry will commence; known interested parties are advised directly. Procedures to be followed are included in the notice.
> (3) Persons seeking to make submissions to the inquiry are required to register by lodging a primary submission with the Registrar of the Office of Commissioners of Inquiry.
> (4) Before the inquiry and the expiry date for primary submissions any person may examine the development application, the EIS, the initial EIA, and other related documents at specified venues.
> (5) Questions in written form should be available at the commencement of the inquiry. Further questions during the inquiry are dealt with at the Commission's direction.
> (6) Primary and subsequent submissions are to be made available to all parties.
> (7) Proceedings are generally as follows:
>   - Opening statement by Commissioner
>   - Preliminary matters such as procedures and personal difficulties
>   - Primary submissions (in stated order)
>   - Questions and replies to questions, in writing
>   - Submissions in reply (in reverse order).
> (8) Inquiries are conducted in accordance with the rules of natural justice; each person is treated on an equal footing whether legally represented or not; evidence is not generally on oath; cross-examination is rarely allowed.
> (9) Adjournments, usually of short duration, may be granted following an application from a party and argument by other parties.
> (10) All communication with the Commission is public and queries are through the Registrar. No private communication with a Commissioner may occur.
> (11) Following the conclusion of the inquiry, the Commission prepares its report to the Minister, setting out and discussing the issues, summarizing the views of the parties, concluding with findings and a specific recommendation. Options may be discussed but only one is recommended as a course of action for government. The report must take account of criteria laid down in the legislation, state environmental policies, the stipulations of planning instruments, and the public interest as perceived by the Commission.
> (12) The Commission presents its report to the Minister, usually personally and with a verbal briefing. However, the report is final and becomes a public document the moment it passes over the Minister's desk.
> 13) The Minister is not in any way bound by the recommendation of the report and may consult others; however, departure from such recommendations is rare, including the often quite stringent conditions laid down in respect of projects thought fit to proceed.

pulp logs Logs that are processed into woodchips or pulp for wood-based panels, paper, and paper products.

pulverized fuel Fuel, usually coal, finely ground; modern power stations often use bituminous coal in this way. Pulverized fuel ash (PFA) is a finely divided greyish ash resulting from the COMBUSTION of pulverized coal. Formerly regarded solely as a waste product, PFA is used today in building materials of various kinds.

pumped storage A means of 'storing' electricity in the form of water by pumping it to a high level, when there is electricity available to drive pumps, then letting the water flow down through water turbines to generate electricity at times of peak demand. The addition of a pumped storage station to a system saves the installation of an equivalent capacity of conventional generating plant otherwise required to meet peak demand.

pumping station An installation comprising a wet-well, pump, and pressure pipe, often located underground and operated for the purpose of raising SEWAGE to a higher elevation. A pumping station, located at a low-point in a CATCHMENT may be used to despatch sewage to a treatment plant elsewhere.

putrefaction Decomposition of organic matter in an ANAEROBIC state releasing foul-smelling gases and incompletely oxidized products. It is important to distinguish between putrefaction and the process of ANAEROBIC DIGESTION; an anaerobic digestion process permanently removes the offensive odour of many organic wastes so that they can be used on agricultural land without causing NUISANCE.

putrescible wastes Wastes that consist mainly of plant or animal residues and which undergo degradation by bacterial action; for example, abattoir refuse, animal residues, cannery wastes, fats, fish residues, vegetable wastes, and waxes.

pyramid of numbers The numerical relationship between life forms in different niches. (See NICHE.)

pyrolysis (destructive distillation) The reforming of material in an inert ATMOSPHERE to produce combustible gases, volatile fluids, tar, and charcoal. To achieve this, solid domestic refuse is placed in a retort and heated without additional air to temperatures between 500 and 1000°C, its weight being reduced by 90%. The solid residue remaining is then disposed of at LANDFILL sites. Industrial wastes can be reduced by up to 65% in volume.

Q

QRA QUANTIFIED RISK ASSESSMENT.

quality of life A concept embracing a miscellany of desirable things not always recognized, or adequately recognized, in the marketplace. It embraces such highly relevant matters as real income, housing and working conditions, health and educational services, and recreational opportunities, which may be regarded as the general standard of living. Other highly relevant matters include community relationships, race relationships, civil liberties, compassion, justice, freedom and fair play, safety and security, law and order, and environmental conditions.

quantified risk assessment (QRA) The quantification of the total risk to life, property, or production, associated with any hazardous situation. In respect of an individual plant, there may be a preliminary analysis at the conceptual stage, followed by a more detailed analysis during the design stage. Risk assessment can be applied to general policy decisions as well as to single projects. It has proved to be the most useful technique in the understanding and control of industrial hazards.

quantitative chemical analysis A branch of chemistry that deals with the determination of the amount or percentage of the constituents of a sample. This must be distinguished from qualitative chemical analysis that deals with the identification of elements or constituents present in a sample. (See CHROMATOGRAPHY; SPECTROPHOTOMETRY.)

quarantine A period of time during which a person, animal, plant, or object suspected of carrying a contagious disease is detained in isolation at a port of entry to prevent disease from entering the country.

Queenstown Founded in 1897 after the discovery of gold, silver, and copper, a town in western Tasmania, Australia. Before new ore-treatment methods were introduced in 1952, the surrounding hills were stripped of their timber for fuel, while fumes from the smelter killed whatever vegetation remained. Coupled with erosion, this has left a lunar-like landscape around the town. The regeneration of plant life has been very slow. However, in the meantime the 'moonscape' has proved an immense tourist attraction.

Quaternary period An interval of geological time, the youngest of the 11 periods in the history of the Earth. Its boundaries have been defined as 1 600 000 years ago to the present day. It is subdivided into the Pleistocene epoch and the Holocene epoch. The Quaternary period is characterized by major cyclical changes of climate on a global scale, with repeated invasions of mid-latitude North America and northwestern Eurasia by ice sheets; it is often referred to as the Great Ice Age. As a result of these climatic changes, there were substantial and repeated shifts of vegetation belts and their associated FAUNA, particularly in the Northern Hemisphere. Another major feature of the period has been the evolution and dispersion of human beings. The drastic changes of CLIMATE and ENVIRONMENT during the period led to rapid rates of evolution and extinction, particularly among the larger mammals. (See ICE AGE.)

R

radiation (1) The transmission of heat from one point to another without affecting the temperature of the medium through which it passes. (2) IONIZING RADIATION. (3) Radiation from radioactive substances. (See RADIOACTIVITY.)

radiation absorber A substance that absorbs atomic particles and radiation. The control of nuclear reactors is commonly effected by the use of control rods which contain materials, particularly boron and cadmium, being good absorbers of thermal neutrons.

radiation biology The study of how living things are affected by radioactive (ionizing) emissions, and by electromagnetic (nonionizing) radiation. (See ELECTROMAGNETIC RADIATION; RADIOACTIVITY.)

radiation pyrometer A device for measuring temperatures, utilizing the emission of radiant energy from a hot body. There are several kinds of radiation pyrometer available.

radiation sickness Ill-health resulting from overlong exposure to RADIOACTIVITY, including X-rays, gamma-rays, neutrons, and other nuclear radiation. Such radiation may cause nausea, vomiting, diarrhoea, and other symptoms. Leukaemia and genetic changes may be induced.

radioactive decay The process of continuous disintegration undergone by the nuclei of radioactive elements, such as radium, various isotopes of uranium, and the transuranic elements. The radiation consists of alpha rays, beta rays, or gamma rays, or a combination of these. Disintegration takes place with a characteristic HALF-LIFE.

radioactive waste Any waste that emits radiation in excess of normal background level, including the toxic by-products of the nuclear energy industry. Reactive waste is of three types: (1) 'high-level' spent fuel being fuel removed from a nuclear reactor; (2) intermediate waste, that may be long- or short-lived; and (3) 'low-level' waste from a reactor which has only short-lived RADIOACTIVITY. Disposal by burial on land or at sea, raises problems of environmental pollution, safety, and security. (See DEEP REPOSITORIES; HIGH-LEVEL WASTE; LOW-LEVEL WASTE; YUKKA MOUNTAIN.)

radioactivity The spontaneous emission of ionizing particles following the disintegration of certain atomic nuclei, or as a consequence of a nuclear reaction. RADIATION SICKNESS results from overexposure.

radiological guidelines Standards that may be set for radiological quality; that is, limits for gross (total) alpha- and beta-radioactivity concentrations. In respect of drinking water, radiological contamination could result from naturally occurring radioactive isotopes in the water source, or from mining and processing sources, or from the medical and industrial use of radioactive materials. Radiological standards are thus universally set for drinking water.

radon A radioactive gas that is released from the SOIL into the ATMOSPHERE. It is also released during the COMBUSTION of coal, and for this reason there is often more radon in the air over industrial cities than in country air.

rainforest A dense luxuriant closed mesomorphic community; a global vegetation type containing many tree species associated with high rainfall and humidity. There are many types of rainforest, though it is usual to consider three major divisions: the tropical, subtropical, and temperate rainforests. Other kinds of rainforest include monsoon forests, and mangrove forests occurring along estuaries and deltas. Equatorial rainforests occur in the lowlands of the Amazon, the Congo, Indonesia, and in several islands in the Pacific. Subtropical rainforests are found in Vietnam, the Philippines, Central America, the Caribbean islands, Madagascar, and parts of Brazil. Most of the world's rainforest is in poor regions such as Asia, Africa, and Latin America; three countries, Brazil, Indonesia and Zaire, have half the world's rainforest. The greatest loss of rainforest is happening in Brazil and Indonesia. The preservation of the different kinds of rainforest has become an important social and political issue, nationally and internationally. (See CONVENTION ON PROTECTING SPECIES AND HABITATS; PRINCIPLES OF SUSTAINABILITY.)

rain shadow The lee side of a mountainous (orographic) barrier, that receives considerably less PRECIPITATION than the windward side. (See OROGRAPHIC PRECIPITATION.)

Rainbow Bridge National Monument Established in 1910, a national monument in southern Utah, USA, comprising a natural rainbow-shaped bridge of pink sandstone spanning a canyon of 88 m above a creek that winds towards the Colorado River. It is the largest natural bridge in the world.

Rainbow Warrior The flagship of GREENPEACE which, in 1985, sank in Auckland Harbour, New Zealand after explosives planted by agents of the French Government were detonated, killing a photographer. The *'Rainbow Warrior'* was to have led a flotilla of boats to MURUROA ATOLL in French Polynesia to protest against France's programme of nuclear testing.

rainout A process in which particles in the higher TROPOSPHERE may act as nuclei on which water droplets form, later to fall as PRECIPITATION.

Ramsar convention See CONVENTION ON WETLANDS OF INTERNATIONAL IMPORTANCE, ESPECIALLY AS WATERFOWL HABITAT.

random noise A fluctuating quantity of sound or electronic interference whose amplitude distribution with time is Gaussian; that is, forming a normal curve of distribution, most readings being around the average.

range (1) The area over which a SPECIES occurs; that is, where the environmental conditions and habitats suit the species concerned. The range of a species may be changed by human activities such as urban sprawl or the building of dams. (2) The maximum or effective distance that a sound, radio signal, radioactive particle, or pollutant may travel.

rangeland Any extensive area of land that is occupied by native herbaceous or shrubby vegetation that is grazed by wild or domestic HERBIVORES. Such vegetation may comprise tall-grass prairie, steppe, desert shrubland, shrub woodland, savannah, tundra, and chaparral. Rangeland occupies about one-half of the land area of the Earth. Rangeland is more generally confined to areas of marginal or submarginal agricultural land or to areas that are entirely unsuitable for permanent cultivation.

Ranger Uranium Environmental Inquiry A public inquiry set up by the Australian Government in 1975 to investigate the environmental aspects of proposed uranium mining in the Northern Territory. The final report was presented in 1977, recommending that the mining and export of uranium proceed subject to a whole range of conditions, including the careful selection of countries to which uranium exports should be allowed.

rare An infrequently occurring uncommon SPECIES, not necessarily an ENDANGERED SPECIES.

rare-earth element Any of a large family of chemical elements consisting of scandium, yttrium, cerium, and some 14 other rare elements. Not all the rare earths are quite as rare as first thought and they are plentiful in meteorites and on the Moon.

rational use of resources See ALLOCATIVE EFFICIENCY, SOCIAL; INDUSTRY, OPTIMAL LOCATION OF; OPTIMIZATION; PARETO OPTIMALITY.

RCI RESPONSIBLE CARE INITIATIVE.

'real effects' See TECHNOLOGICAL EFFECTS.

rebound The elevation of a part of the surface of the Earth during or following the removal of overlying material or the filling of an underground cavity.

receptor (target) The organism, population, or resource exposed to specific risks from POLLUTION or other adverse influences.

recharge area An area where water is absorbed and added to an AQUIFER.

recommended limit An occupational exposure limit for hazardous substances, representing good practice and a realistic criterion for the control of plant design, engineering measures, and the selection and use of personal protective equipment.

recycling The return of discarded or waste materials to the production system for utilization in the manufacture of goods, with a view to the conservation as far as practicable of non-renewable and scarce resources, promoting SUSTAINABLE DEVELOPMENT. Recycling goes beyond the reuse of a product (such as glass milk bottles) and involves the return of salvaged materials (such as paper, metals, plastics, or broken glass) to an early stage (pulping or melting) of the manufacturing process. Some recycling has always been profitable in certain industries such as the return of steel scrap to the steel industry, glass cullett to the glass industry, and aluminium drink cans to the aluminium industry. However, developments in recent years have raised the level of recycling in commercial and domestic premises through financial incentives to local councils and collection bodies. The principle of recycling has been extended in actual manufacturing processes, such as the recycling of water and other inputs within the plant itself. Recycling has a key role in SUSTAINABLE DEVELOPMENT. (See Figure 22.)

red algae See RED TIDE.

red sea See RED TIDE.

red-tide (red algae; red sea) The descriptive effect of the red–brown colour of dense populations of marine protozoa which can occur under exceptionally favourable conditions of water temperature, salinity, and/or nutrient salts. The accumulation of metabolic wastes in these conditions may cause massive fish mortality.

reducer In BIOLOGY, a heterotrophic organism that utilizes the chemical energy of organic matter while breaking it down or reducing it to more simple processes.

Redwood National Park Established in 1968, in the northwestern corner of California, USA, a national park preserving virgin groves of ancient redwood trees, including the world's tallest tree at 112 m. It is included in the WORLD HERITAGE LIST.

REF REVIEW OF ENVIRONMENTAL FACTORS.

reforestation The replanting of trees on cleared or destroyed forest areas. (See PLANTATION FORESTS.)

refuse The complete range of unwanted or undesirable material generated in the course of

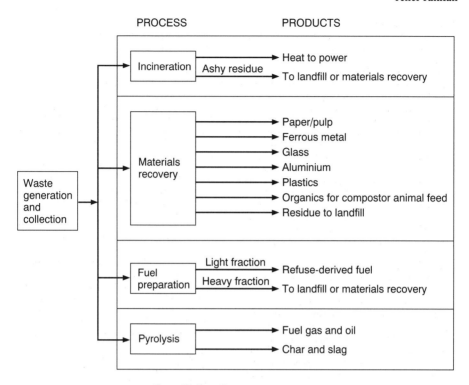

RECYCLING: Figure 22. Recycling or resource recovery process.

producing, processing, and consuming useful products.

refuse disposal See INCINERATION; LAND-FILL.

Regional Seas Program A programme developed by UNEP to reduce pollution in regional seas that are particularly vulnerable. Agreements were sought between the governments in each region. (See CONVENTION FOR THE PROTECTION OF THE MEDITERRANEAN SEA AGAINST POLLUTION; UN ENVIRONMENT PROGRAM.)

regrowth forests Native forests that have been partially cleared or logged or have suffered high-intensity forest fires in the past and are showing regrowth. (See NATURAL FOREST; PLANTATION FORESTS.)

regulatory impact statement (RIS) As distinct from an ENVIRONMENTAL IMPACT STATEMENT, an assessment of the likely costs and benefits of regulations created under a wide range of legislation. In this way an RIS for a CARBON TAX, or any other environmental or economic instrument, could be presented to the public for comment prior to a final decision.

rehabilitation The treatment of degraded or disturbed land to achieve a level of production and stability at least equal to that which existed before the degradation or disturbance, or to an alternative level of capability. Such rehabilitation may involve the reshaping of the land surface, the spreading of available top SOIL, the construction of soil-conservation works, revegetation, and the establishment of LAND-USE practices that will ensure continued productivity or recreational use.

relative humidity The ratio between the actual weight of water vapour in a given volume of air and the amount which would be present if the air was saturated at the same temperature, expressed as a percentage.

reliability In relation to industrial plant, the probability that an item of equipment is able to perform a required function under stated conditions for a stated period of time or for a stated demand; for example, water-pump reliability.

reliability studies In respect of individual items of industrial equipment, a statistical analysis of failure rates for a critical component, with a view to optimizing programmes for maintenance and redundancy to minimize costs.

relief rainfall Rainfall that is related to the relief of the land such as hills and valleys. When moisture-laden air is forced upwards over an elevated area of land, such as a range of hills, it is cooled; unable to retain as much moisture, this then falls as relief rain.

remnant vegetation Those parts of a natural system, or a natural species, remaining after a major change to the ECOSYSTEM or ENVIRONMENT.

remote sensing The use of satellite sensors to provide data on many of the Earth's features such as topography, soil type, vegetation, surface water, coastal resources, the oceans, atmospheric conditions, and pollution. Remote sensors record the electromagnetic energy emitted or reflected from Earth's surface. Since the 1970s, rapid advances in remote-sensing technology have improved the value of satellite data to resource managers. Further, the surveying of large areas by sensing is claimed to be more cost-effective than by using ground-based techniques. Several countries now have Earth observation and communication satellites in orbit, including the USA, Europe, Russia, and Japan.

renewable resource Those resources with short cycling times; that is, the length of time required to replace a given quantity of a resource that has been used with an equivalent quantity in a similar form. NON-RENEWABLE RESOURCES are those with very long cycling times. Examples of renewable resources include the following. (1) Pasture grasses. If care is taken to protect fertility, soil structure, with enough seeds and roots, a grass crop can be removed by grazing or mowing each year for an indefinite period. (2) Similarly, agricultural crops, animal forage, forest crops, wild and domestic animals, all can reproduce and regenerate as long as environmental conditions remain favourable. All can be cropped or harvested without diminishing the supply, provided that the cropping does not exceed the reproduction or growth rate. (3) Sheep in a mountain pasture are a renewable resource as long as the land produces enough vegetation that will nourish and support the sheep; the sheep cease to be a renewable resource if the land is overgrazed, the vegetation impaired, and the soil eroded. (4) Trees are a renewable resource if there is enough time for them to grow to maturity, and to produce seeds from which a new crop can be grown. Certain conifers can yield useful timber in less than 30 years. However, old-growth redwood trees can be considered a non-renewable resource requiring from 500 to several thousand years; (5) Air, water, and solar energy. (See BIOMASS; GASOHOL; GEOTHERMAL ENERGY; HOT DRY ROCK SYSTEMS; HYDROELECTRIC POWER; HYDROGEN ECONOMY; MAGMA BODIES; OCEAN THERMAL ENERGY CONVERSION; PHOTOVOLTAIC CONVERSION; RESOURCE; SOLAR ENERGY; SUSTAINABLE DEVELOPMENT; TIDAL ENERGY; WAVE ENERGY; WIND ENERGY.)

reovirus A virus widespread in the ENVIRONMENT; it can cause serious gastroenteritis in children and the elderly. It may enter water through faecal material such as sewage contamination.

reservation The withholding from use and exploitation of areas and resources that, it is judged, should be held back either against future needs or because they are best kept as they are. Areas so treated are called reserves; for example, forest reserves or game reserves. (See MARINE PARK; NATIONAL PARK.)

reservoir An artificial lake where water collects so that it may be drawn off for use. Usually the aim is to accumulate water at times of high natural flow from streams, and then release the collected waters gradually at a rate appropriate to their use. Reservoirs are ordinarily formed by the construction of dams across rivers and valleys. The largest of these may displace large numbers of people. Dam construction has been a source of much dispute in recent years. (See ASWAN HIGH DAM; GORDON-BELOW-FRANKLIN DAM; PAK PUN HYDROPOWER PROJECT; SARDAR SAROVA DAM PROJECTS; THREE-GORGES WATER CONSERVATION AND HYDROELECTRIC PROJECT, CHINA; WORLD BANK.)

residual A quantity of material or energy which is left over or 'wasted' in industrial processes and other human activities. Examples include waste heat and gaseous pollutants from electricity generation, slag from metal-ore refining, and garbage. Outputs and inputs have prices in existing markets; residuals do not. A residual becomes an output or input when a technological advance creates economic opportunities for the 'waste'. RECYCLING converts waste into inputs; while the use of power station flyash converts a waste into an output.

residual risk The remaining risk after every proposed improvement to an industrial facility has been made.

resource A general term for anything that can be used to provide the means to satisfy human needs and wants. A resource may be purchased, acquired, or inherited; it may be a common property resource, or a personal or corporate possession, or a collective possession; it may lie in the PRIVATE SECTOR or the PUBLIC SECTOR; and it may be well-used or misused for good or bad purposes, for war or peace. A resource is simply a service for noble or base purposes. (See CAPITAL; COMMON PROPERTY RESOURCES; FACTORS OF PRODUCTION; LABOUR; LAND; MINERAL; NATURAL RESOURCE; NON-RENEWABLE RESOURCE; RENEWABLE RESOURCE; SOIL; SUSTAINABLE DEVELOPMENT.)

resources, allocation of See ALLOCATIVE EFFICIENCY, SOCIAL; OPTIMIZATION.

resource-use analysis (RUA)   An assessment itemizing all the resources that will be consumed during the construction, operation, and decommissioning of a project, or the lifetime of an activity, including all market and non-market resources, and the implications for sustainability.

responsible care initiative (RCI)   A voluntary initiative of the chemical industry beginning in the 1980s requiring member firms to follow codes of conduct regarding such matters as toxic materials, waste reduction, chemical-accident minimization, worker safety, and community consultation. This initiative began in Canada, spreading to the US, Europe, and Asia. The British RCI was launched in 1989 by the Chemical Industries Association, together with an insurance fund known as the CIA environmental impairment liability insurance facility to provide members with funds to cover the legal liabilities of environmental claims.

restoration   (1) A concept of natural re-establishment. For example, the restoration of a forest by regrowth after it has been cut or burned; the revegetation of a barren mined area; or the augmentation of depleted populations of animals or plants. (2) The restoration of historic buildings and residences, otherwise threatened with demolition. Much valuable work has been done in many countries to preserve and restore the finest of the past for the enjoyment of present and future generations. Not only individual buildings, gardens, and contents have been so restored and preserved, but whole towns and districts have been conserved. (See WILLIAMSBURG RESTORATION.)

restrictive covenant   See PROTECTIVE COVENANT.

restrictive agreement   See PROTECTIVE COVENANT.

review of environmental factors (REF)   An approach often adopted when the environmental impact of a proposal is readily shown to be minor. Consequently, a full ENVIRONMENTAL IMPACT STATEMENT is not justified. The review should describe: the nature of the proposal; the zoning and general nature of the ENVIRONMENT affected; the safeguards for the control of pollution and protection of the environment; and aspects of the proposal that may enhance the environment; and aspects of the proposal that may adversely affect the environment with reasons why they cannot be avoided. (See ENVIRONMENTAL IMPACT ASSESSMENT; SUSTAINABLE DEVELOPMENT.)

ribbon development   The practice of building along both sides of a highway; the developments may be only one building wide, with open country behind. The visual impression from the highway is of each town being connected directly with its neighbour. At best, the edges of town and country are blurred; at worst they are eliminated. Also frequently, large areas of backland cannot be developed due to lack of road access.

Richter scale   A scale for measuring the magnitude of earthquakes developed by the American physicist and seismologist Charles F. Richter (1900–85). Richter also mapped out quake-prone areas in the USA, but avoided attempts at earthquake prediction. He collaborated with Beno Gutenberg (1889–1960).

right of common access   In principle, a right of entry or access by the public over all types of land, on foot. The right has been recognized in much Scandinavian legislation, and is prevalent in Scotland, though more restricted in England. Also known as a right-of-way, it may be public or private, be of various kinds, and limited in various ways. (See EASEMENTS.)

rights of nature   A view that creatures have an intrinsic value and that the ENVIRONMENT should at least be respected for its own sake, and not simply as something useful directly or indirectly to human beings. It is a non-anthropocentric ethic. The welfare of creatures who share the Earth with us ranks therefore as an ecological, political, ethical, and philosophical issue, currently debated. It has particular relevance for RARE and ENDANGERED SPECIES. (See ANIMAL RIGHTS.)

rill   A small stream or rivulet. In soil erosion, rills appear in the topsoil as the water moves across the SOIL in a series of channels, pushing soil particles ahead of it.

Ringelmann smoke chart   A smoke chart devised by Professor Ringelmann of France in the early 20th century; a specification for the chart and its use was issued by the British Standards Institution. As many air-quality regulations strictly control the emission of smoke from chimneys both in terms of time and density, the chart has had widespread use by environmental health officers. (See Figure 23.)

Rio Declaration on Environment and Development   See UN CONFERENCE ON ENVIRONMENT AND DEVELOPMENT.

riparian   Living or situated on the bank of a watercourse such as a river or stream; hence the terms 'riparian owner' and 'riparian interests'. Riparian rights may give owners contiguous to water equal rights to that water; the landowner does not own the water but rather the right to use it on his/her riparian land, at least in a reasonable manner and in such limited quantity that does not cause suffering and deprivation to downstream riparian owners.

riprap   A wall of broken rock thrown together irregularly so as to protect embankments from erosion; the term is also used of the rock itself.

RIS

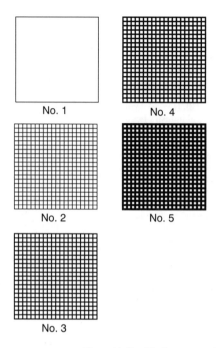

RINGLEMANN SMOKE CHART: Figure 23. The Ringlemann smoke chart.

RIS REGULATORY IMPACT STATEMENT.

risk and hazard assessment An essential component of the ENVIRONMENTAL IMPACT ASSESSMENT relating to any major project. It embraces those potentially adverse effects of a project or programme involving fire, heat, blast, explosion, or flood, arising from plant or associated transport systems. An assessment should reveal the hazards to life and limb, and to property expressed in the form of a risk probability; that is, the expected frequency of undesirable effects arising from a given event.

riverine wetlands WETLANDS formed from the seasonal flooding of rivers. Land inundated by floodwaters may form important wetland areas with characteristic flora and breeding grounds for waterbirds.

River Narmada Irrigation and Hydroelectric Project, India See SARDAR SAROVA DAM PROJECTS.

River Thames clean-up A programme inaugurated in the 1960s to effect an improvement in the quality of the River Thames, London, in its lower reaches which had been seriously polluted since early times. There can be no doubt that improvements at sewage works and in the treatment of other effluents has led to a marked improvement in the condition of the Thames. By 1969, the sand goby, flounder, and smelt were being found above London Bridge. In 1985, a seal reached central London in search of fish.

road-pricing schemes Schemes to reduce demand for road space at times of acute congestion. On the basis of successful pricing schemes in Singapore and Oslo, many transport planners believe that charging for road use is the best way of tackling urban congestion. Such schemes may take a variety of forms; for example: (1) a charge or toll levied on all private cars entering a delineated central city zone at the point of entry; or (2) an automatic system based on cars equipped with electronic devices of some sort, charges being based on distance travelled within congested areas. The costs of any of these systems depends on their complexity. More cheaply, traffic congestion may be eased by high parking charges and limited parking facilities, achieving a balance of supply and demand through the medium of price and alternatively, a marked increase in pedestrian-only precincts. Another approach is to provide ample and inexpensive parking at the fringes of congested areas, with frequent public transport by road and rail into the city itself. In Britain, doubling the real price of petrol (gasoline) over the next decade has been recommended by the ROYAL COMMISSION ON ENVIRONMENTAL POLLUTION, though this may not reduce urban congestion at critical times.

road traffic noise NOISE generated by road vehicles of all kinds and in all modes, detrimentally affecting the public at home, at work, or outside. Noise generated by traffic is now one of the most pressing of urban environmental problems, particularly affecting those living or working near major roads. Measures to control noise must begin with engineering design imposing performance standards on vehicles of different types; on the selection of appropriate road surfaces to minimize noise generation; and appropriate noise bunds, barriers, and buffer zones. (See ACOUSTIC SITE PLANNING; NOISE-REDUCTION TECHNIQUES.)

Rocky Mountain National Park Established in 1915, a national park in north-central Colorado, USA, containing numerous high peaks as well as broad valleys, gorges, alpine lakes, and plunging streams. The flora includes over 700 known species; the fauna includes bighorn sheep, deer, mountain lions, and a variety of birds.

rodenticide Any substance that is used to kill rats, mice, and other rodent pests. Warfarin, 1080 (sodium fluoroacetate), ANTU, and red squill are commonly used, as well as fumigants such as SULPHUR DIOXIDE, CARBON MONOXIDE, hydrogen cyanide, and methyl bromide. Other poisons are also used mixed with bait.

Rondane National Park Established in 1962, a Norwegian national park embracing 10 major peaks in a high alpine region. The flora embraces mainly sparse tundra; the fauna includes reindeer, roe deer, elk, Arctic fox, red fox, wolverine, lynx, lemming, martens, and many birds, including eagles, owls, and hawks.

root In BOTANY, that part of a plant usually underground; its primary functions are anchorage of the plant, absorption of water and dissolved minerals, and storage of reserve foods.

Roskill Commission An environmental planning inquiry conducted in Britain with the task of identifying the best site for a Third London Airport. The commission reported in 1971 in favour of Cublington, Buckinghamshire, with one commissioner favouring Foulness on environmental grounds. The bulk of air transport continues to be divided between Heathrow and Gatwick.

Ross Ice Shelf The world's largest body of floating ice, lying at the head of the Ross Sea, itself a major indentation in the continent of Antarctica.

rotavirus A virus widespread in the ENVIRONMENT which can cause serious gastroenteritis in children, the elderly, and hospital patients. The virus may enter water through faecal material or sewage contamination.

Royal Commission on Environmental Pollution A commission set up in 1970 as a

---

Box 29 Russia: evolution of environmental policy
1960 USSR Council of Ministers adopt proposals to conserve water resources and protect those resources against pollution
1970 Supreme Soviet adopts water management legislation
1972 Supreme Soviet adopts measures to improve nature conservation and promote the rational utilization of natural resources; issues an appeal to the nations of the world to combat pollution and protect the environment; approves decrees to protect the Volga and Ural Basins, the Black and Azov Seas, and Lake Baika; US–USSR joint committee for environment protection established
1975 Supreme Soviet issues decrees for the protection of mineral wealth and the natural environment
1976 Guidelines issued for the development of the national economy dealing in detail with measures to protect the environment, control pollution, and manage natural resources
1977 Amendment of the Constitution incorporating a new Article 18 to promote the rational use of the land and its mineral and water resources, protect flora and fauna, prevent pollution, and ensure sustainable development
1993 Explosion at the Siberia Chemical Centre plant near Tomsk releases radioactive material to the atmosphere; the Ministry of Ecology and Natural Resources reports excessive air pollution levels affecting over 50 million people, and many examples of severe incidents
1994 Following reports on the deterioration of Lake Baikal, the closure of the Baikal pulp and paper mill

standing body to advise the British Government 'on matters, both national and international, concerning the pollution of the environment; on the adequacy of research in this field; and the future possibilities of danger to the environment'. Thus, the Royal Commission is intended to serve as a general watchdog on pollution and environment protection. The Royal Commission has reported and published on a whole range of environmental issues over the years, with a report in 1994 on transport and the environment recommending increases in fuel taxes, stricter enforcement of pollution control regulations, and tax incentives to persuade operators of older petrol and diesel vehicles to switch to newer models equipped with catalytic converters.

RUA RESOURCE-USE ANALYSIS.

runoff Fresh water from PRECIPITATION that flows across the surface of the ground into nearby streams and rivers, or into stormwater drains, Such runoff may become contaminated as it passes over urban impervious surfaces, or as it flows over agricultural land.

runoff coefficient The ratio of the rate of RUNOFF of rain water from the ground to the rate of rainfall or precipitation. The ratio indicates the proportion of the rainfall that is actually contributing to the runoff, and as such, the coefficient is always less than 1.0

Russia, evolution of environmental policy See Box 29.

Ruwenzori National Park Established in 1952, a national park in southwestern Uganda embracing rolling hills and many volcanic craters. The flora consists mainly of thickets of various types of small trees, including acacias and evergreens. The fauna includes the chimpanzee, leopard, lion, elephant, hippopotamus, water buffalo, antelope, and many species of kingfisher.

S

sacred sites Those sites, ceremonial grounds, buildings, or structures regarded as sacred by religious, racial, or cultural groups. Many nations protect sacred sites by legislation on behalf of indigenous peoples or minorities.

safety audit A critical examination of all or part of an operating industrial system, with relevance to SAFETY PRECAUTIONS. Safety audits should be conducted on an annual basis. (See ENVIRONMENTAL AUDIT.)

safety precautions All those measures, techniques, activities, and procedures that seek either to minimize or eliminate risks that may impair HEALTH or cause bodily injury or death. (See ENVIRONMENTAL IMPACT ASSESSMENT; ENVIRONMENTAL IMPACT STATEMENT; OCCUPATIONAL HEALTH AND SAFETY.)

Sahel The semiarid region of western Africa extending from Senegal to the Sudan; a traditional belt that borders the edges of the arid Sahara desert. The Sahel has natural pasture with short grass and tall, herbaceous perennials; it also provides forage for animals including camels, grazing cattle, and sheep. The area has suffered overstocking and overfarming. At least eight months of the year are normally dry, while intermittent severe droughts worsen the situation. The loss of human life by starvation and disease has been calamitous, some 100 000 dying in the 1970–73 drought. Severe drought and famine again afflicted the Sahel in 1983–85. Malnutrition remains endemic.

Saint Helens, Mount A volcanic peak in the Cascade Range, southwestern Washington, USA that erupted on 18 May 1980 in one of the greatest volcanic explosions ever recorded in North America; debris was carried more than 20 km from the summit with ash falling as far east as central Montana. Further eruptions have occurred since then.

saline wetlands WETLANDS in which SALINITY exceeds 1000 ppm, total dissolved solids (1000 parts per million) throughout the year. Permanent saline wetlands include coastal wetland areas and areas of intertidal zones.

salinity The total content of dissolved mineral constituents, of all kinds, in water. Rain and other forms of precipitation are normally free from minerals. Only when these waters come into contact with the Earth does the process of mineralization begin. Natural sources of salinity include springs passing through rock formations which contain soluble salts. Other sources of salinity include domestic and industrial wastes, oil-well brines, and mine water. Salinization has become a major environmental problem in the irrigated areas of the world. In some instances irrigated land has been abandoned with a decline in food production. SOIL acquires salt mainly by salt going into solution from certain types of bedrock. Salinity is often expressed in parts per thousand; sea water is approximately 35 parts per thousand (35 000 ppm). (See DEAD SEA.)

salinization The accumulation of salts in SOIL to an extent that causes degradation of soils and vegetation.

Salmonella A genus of bacteria which may enter water through faecal contamination; it can cause outbreaks of gastroenteritis.

sampling See ISOKINETIC; MONITORING.

sand mining The extraction of sand by mining for building purposes and for the extraction of heavy minerals such as rutile and zircon. (See MINERAL SANDS.)

sanitary landfill See LANDFILL.

sanitation An important health-related branch of development embracing drainage and sewerage, sewage and sullage treatment and effluent disposal, safe and adequate domestic water supplies, avoidance of public nuisance and uncontrolled tipping, and drainage facilities for floodwater and surface runoff. Few countries renowned for high-tech achievements have been able to resolve the basic requirements of sanitation, relying on primitive methods or none.

Santa Barbara blow-out An incident that occurred in 1969, during the drilling of the sea bed from an oil rig located about 13 km from the shore of the Santa Barbara Channel near the coast of California, USA. A blow-out of crude oil continued for 10 days, causing much environmental damage along the coastline, polluting the beaches, and destroying wildlife.

saprogenic Causing or resulting from organic decay.

saprophyte An organism that lives on dead or decaying organic matter; a plant or other organism using dead organic material as food, and commonly causing its decay. In contrast, a PARASITE is an organism that feeds on living tissue.

Sardar Sarova Dam Projects A major Indian water project involving the resettlement and rehabilitation of large numbers of displaced persons, perhaps 100 000 in all. The project was to be financed in part by the World Bank. However, an independent review conducted for the World Bank concluded that the projects were flawed, that resettlement and rehabilitation of all those to be displaced was not possible, and that the environmental impacts of the projects had not been properly considered or adequately addressed. In October 1992, the

World Bank gave India six months to rectify the flaws found; at the end of that time the Indian Government announced it would not seek further World Bank funding for the project.

Sarek National Park Established in 1909, in northwestern Sweden, encompassing most of the Sarek mountain range; it adjoins two other national parks, Stora Sjöfallet and Padjelanta. The region is almost inaccessible, characterized by high peaks, deep valleys, wide plateaus, and glaciers. Sarek is rich in vegetation and is the habitat of various wildlife, including bears, wolves, wolverines, and lynx.

savannah Any tropical or subtropical GRASSLAND characterized by scattered trees or shrubs with a pronounced dry season. Savannah vegetation is found primarily in Africa and South America, with smaller areas in Australia and Madagascar. Large herds of grazing animals, once common to all savannah areas, are now found only in Africa.

sawlogs Logs for processing into sawn timber, veneer, poles, and sleepers.

scalding The removal of topsoil, exposing a naturally saline subsoil. (See SALINITY.)

scavenging Removal of a solid, liquid, or gaseous contaminant from a liquid or gas by the adherence of the contaminant to a third substance.

schistosomiasis (bilharziasis) A parasitic disease of humans caused by infection with a genus of blood-flukes comprising *Schistosoma mansoni*, *S. japonicum* and *S. haematobium*. Regarded as a major health problem in certain parts of the world, notably parts of Africa (such as South Africa and Egypt), Central and South America, and southern and eastern Asia, schistosomiasis is commonly transmitted by bathing in polluted canal water. An increase of irrigation favours its spread. The intermediate host of the worm is the freshwater snail. The cercaria (larval form) of the fluke enters the host (humans) through the mouth or directly through the skin. The infection gives rise to severe disease of the urinary tract, liver, and lungs, which can be fatal. The maturation of the worms in the human system leads to the excretion of eggs. On passing into water the cycle is repeated. Treatment is dominated by the broadly effective drug praziquantel. There is still no vaccine for schistosomiasis. (See ASWAN HIGH DAM.)

scoping A procedure for attempting to ensure that a REVIEW OF ENVIRONMENTAL FACTORS or an ENVIRONMENTAL IMPACT STATEMENT focuses on the key environmental issues associated with a development proposal, omitting irrelevant material. Beyond discussions between the PROPONENT and the planning bodies, the public should be brought into the process.

scrubber (wet collector) Gas-cleaning equipment in which a scrubbing liquid, usually water, is used to achieve collection of the particulate matter in the gas stream. The scrubbing liquid is usually dispersed in a spray or spread in a film over the internal surfaces of the scrubber. Downstream of the collection zone, some form of spray eliminator is often required. Wet collection is advantageous when high removal efficiencies are required to remove coarse dusts and meet effluent standards.

scrubland Areas of low, evergreen, leathery-leaved, and often aromatic shrubs, often described as Mediterranean vegetation. Scrubland is confined primarily to coastal regions that have hot dry summers and cool wet winters.

Sea, Law of the See CONVENTION ON THE LAW OF THE SEA.

sea level The position of the air–sea-level interface, to which all terrestrial elevations and submarine depths are referred. As the sea level constantly changes at every locality, sea level is better defined as 'mean sea level', the height of the sea surface averaged over all stages of the tide over a very long period of time. Global mean sea level rose at an average rate of about 1.2 mm per year over much of the 20th century.

SEC SECURITIES AND EXCHANGE COMMISSION.

Secchi disc A matt black disc used to measure the clarity of water.

Second Law of Thermodynamics See ENTROPY.

secondary treatment A series of biochemical, chemical, and mechanical processes used in SEWAGE TREATMENT to remove, oxidize, or stabilize non-settleable, colloidal, and dissolved organic materials found in sewage, following primary treatment.

Second World Conservation Strategy See WORLD CONSERVATION STRATEGY.

secure landfill A LANDFILL constructed specifically for the disposal of chemical and hazardous wastes. A secure landfill requires special precautions against the release of leachates; impervious linings are essential.

Securities and Exchange Commission (SEC) Established by the US Congress in 1934, a supervisory body to help protect investors. It has broad authority over the US stock exchanges, and seeks to ensure that investors have enough information in relation to new security issues to allow rational decision-making. It requires continuous disclosure by publicly quoted companies of relevant facts. In 1994, stringent environmental reporting standards took effect; companies must now disclose in their financial statements the specific costs of complying with environmental laws, including ongoing expenses, remediation expenses, capital costs, and contingent liabilities. In environmental law, such liability arises mainly under Superfund and to a lesser degree under the Resource Conservation and Recovery Act. (See CONTAMINATED SITES; CONTINGENT LIABILITY.)

sediment (1) Finely divided solid matter suspended in or falling to the bottom of a liquid or gas. (2) Material such as rocks and sands deposited by glaciers, wind, or water.

sediment control The minimization and control of SEDIMENT generated in industrial and disposal operations by limiting its occurrence at or near its source, by stabilizing and sealing disturbed areas, and by controlling RUNOFF. These measures consequently reduce the size and maintenance costs of a final treatment facility. (See MINE DRAINAGE.)

sediment trap A structure, usually relatively small, designed specifically to collect sedimentary material in a drainage channel. Sediment traps (like grease traps) should be regularly cleaned to maintain their efficient use.

sedimentary cycle The circulation of nutrients in an ECOSYSTEM involving geological weathering and erosion, with the eventual recovery of the elements by the uplift of marine sediments to form land masses.

sedimentary rock Rock formed at or near the surface of the Earth by the accumulation of SEDIMENT, or by the precipitation of material from solution. Sedimentary rocks are the most commonly exposed rocks, though the Earth's crust is dominated by igneous and metamorphic rocks. (See IGNEOUS ROCK; METAMORPHIC ROCK.)

sedimentation The removal of settleable solids within primary- or secondary-treatment plants using settling tanks. The low velocity of (say) sewage flowing through the system allows particles to gravitate to the bottom, permitting ready removal as SEWAGE SLUDGE.

sedimentation tank (settlement tank) A tank in which water or sewage containing SEDIMENT is retained for a sufficient time at a suffiently low velocity to remove the sediment by gravity.

sedimentology In GEOLOGY, a scientific discipline that is concerned with the physical and chemical properties of sedimentary rocks, and the processes involved in their formation.

seepage (1) The gradual flow of GROUND WATER to the surface over a wide area. (2) A slow leakage of a SEPTIC TANK into the surrounding soil. (See LEACHING.)

seepage pit A buried perforated tank or gravel-filled cavity allowing effluent to seep into the surrounding soil.

seismology A scientific discipline that is concerned with the study of earthquakes and the propagation of seismic waves within the Earth; it is a branch of GEOPHYSICS. In recent years increasing attention has been directed to earthquake prediction, and ascertaining seismic risks at different geographical sites in an effort to reduce the hazards of earthquakes. Attention has been paid to earthquakes that might be induced by human activities such as impounding waters behind high dams, injecting fluids into deep wells, and detonating underground nuclear explosions. (See KOBE EARTHQUAKE; MURUROA ATOLL; RICHTER SCALE.)

selective forestry See SELECTIVE HARVESTING.

selective harvesting (selective forestry) The removal of a part only of the growing stock at short intervals, allowing time for new seedlings to regenerate the stock. (See FORESTRY.)

selective logging The cutting of selected individual trees, or small stands of trees, for timber.

self-cleansing The general tendency of a body of air to recover naturally from contamination by waste gases. Pollution in its many forms is continually being introduced into the ATMOSPHERE, while much is removed by gravitation, PRECIPITATION, and chemical interactions.

self-purification The general tendency of a body of water to recover naturally from contamination by organic wastes. This process of self-purification depends very largely upon biochemical reactions in which BACTERIA, in the presence of sufficient dissolved OXYGEN, use the organic matter as food, breaking down complex compounds into simpler and relatively harmless products. Other factors such as dilution, sedimentation, and sunlight also play an important part in the self-purification process.

Selous Game Reserve A game reserve in southeastern Tanzania, East Africa, the largest in the world; it bestrides a complex of rivers. The vegetation is woodland with some of the finest virgin bush left in Africa. The reserve holds concentrations of big-tusked elephants, large-maned lions, buffalo, leopards, rhinoceros, zebra, antelopes, and numerous bird species. It has been included in the WORLD HERITAGE LIST.

semiarid A term applied to areas that have somewhat more rainfall than arid areas, though insufficient for most crops.

separate system A drainage system in which foul sewage and surface water are conveyed in separate pipes.

septic A biochemical condition depending on anaerobic bacterial activity, characterized by putrefaction.

septic tank A tank, usually underground, into which sewage flows, the deposited matter being wholly, or partially, broken down through anaerobic action. The final effluent may be allowed to soak into the ground through a system of agricultural drains, if the

Sequoia National Park

SOIL is suitable. Alternatively, the tank must be emptied at regular intervals by a special road-tanker.

Sequoia National Park Established in 1890, a national park in the Sierra Nevada of California to protect groves of large trees (*Sequoiadendron giganteum*), which are among the world's largest and oldest forms of life. The General Sherman Tree is the largest tree in the park, being some 3000 to 4000 years old. Fauna in the park includes black bears, deer, foxes, and squirrels.

service activities Useful functions performed by persons or organizations within the framework of an economic system. Examples include: the transport and distribution of food, energy, and water; the provision of managerial, legal, administrative, organizational, and educational services; and services relating to security, health, communication, entertainment, and finance.

settlement tank See SEDIMENTATION TANK.

severance The effect, for example, when a new motorway or highway acts as a barrier to the movement of people, animals, and goods. Thus, communities on either side are separated from each other.

Seveso incident, Italy On 10 July 1976, a major incident at a herbicide plant at the village of Seveso, near Milan, Italy. A build-up of pressure in the process plant led to the release into the atmosphere of a vapour cloud containing a highly toxic dioxin known as 2,3,7,8-tetrachlorodibenzo-p-dioxin (TCDD). Within three weeks, 700 people had to be evacuated; no fewer than 500 of these showed evidence of poisoning. Many domestic animals were destroyed. All contaminated crops were destroyed by fire over a zone extending 8 km south of the factory. There were no human deaths. The European community issued a directive in 1981 that obliged industries to notify the authorities, their own workers, and local residents of dangerous chemicals held, while also laying down rules for the storage of specified chemicals. The directive is known as the Seveso Directive.

sewage The contents of sewers carrying the water-borne soil wastes of a community. Sometimes the term 'foul sewage' is used to distinguish between sewage, as defined here, and the contents of sewers carrying surface or storm water only.

sewage farming An alternative method of SEWAGE TREATMENT, in effect, the use of sewage for irrigation. This method of treatment may be seen in Melbourne, Australia, where the sewage from a substantial part of the city is treated at a sewage farm at Werribee, about 30 km from the city centre. The methods used include land filtration, grass filtration, and lagooning, while the runoff passes into Port Phillip Bay, the nutrient load being reduced by 50%.

sewage fungus Unsightly slimes resulting sometimes from the pollution of rivers by discharges of organic matter.

sewage gas METHANE produced naturally during the process of digestion in sludge tanks at sewage works. In large works, it is utilized for heating the tanks to accelerate sludge digestion processes, power generation, space heating and lighting, hot-water supplies, and laboratory work.

sewage overflow A relief point in the SEWERAGE system to avoid the backflow of sewage into homes; an emergency relief.

sewage sludge Accumulated solids produced during the treatment of SEWAGE; the solid matter that settles to the bottom of a sedimentation tank during SEWAGE TREATMENT. This material may then be processed into a material (biosolid) that can be beneficially used. Biosolids may be added to COMPOST and used for landscaping, added to commercial fertilizers, added to soil to hold water, that is, used as a soil conditioner.

sewage treatment The modification of SEWAGE to make it more acceptable to the ENVIRONMENT. Sewage treatment may be divided into four main stages: (1) primary treatment, the removal of suspended matter by mechanical means (screening, grinding, flocculation, or sedimentation); (2) secondary treatment, the removal of finely suspended solids and colloidal matter, and the stabilization and oxidation of these substances; (3) tertiary treatment, the attainment of higher effluent standards for many purposes; and (4) SEWAGE SLUDGE disposal, that is, the disposal of the suspended matter removed. A decision as to the stages of treatment to be adopted depends on what is to become of the final effluent.

sewerage A physical arrangement of drains, pipes, sewers, and plant for the collection, removal, treatment, and disposal of liquid and solid soil wastes.

sewerage backlog The absence or shortfalls in the sewering of urban areas. Many cities and urban areas throughout the world, including technologically advanced societies, are devoid of SEWERAGE systems or such systems are confined to the richer classes. Night-soil collection remains a normal practice, or alternatively, direct discharge by open channels into rivers and streams. Many western cities have virtually complete sewerage systems that developed slowly throughout the 19th and 20th centuries. Once a sewer has been laid, connection to that sewer by adjacent property owners often becomes compulsory.

shadow price A price or value imputed to unpriced social or environmental benefits or losses, or to resources that are not satisfactorily priced in commercial markets. The concept is much used in COST-BENEFIT ANALYSIS. In evaluating any project, the economist may adjust a number of market prices and attribute prices to unpriced gains and losses likely to arise.

shale oil   An oil obtained from the destructive distillation of oil shale. Oil shale resembles low-grade black coal; there are large deposits in various parts of the world. Generally, oil from this source is economically unattractive, but it represents a large available natural resource.

shallow freshwater marshes   WETLANDS that are usually dry by mid-summer, filling again with rain water in winter. The soils, however, are waterlogged throughout the year.

sheet erosion   Erosion that occurs when a sheet of water moves across SOIL on a broad front, pushing soil particles ahead of it.

Shell smoke number   An instrument for measuring stack solids developed by the Shell Oil company for use with oil-fired installations. Shade 9 is the equivalent to chimney emissions of less than shade 1 on the RINGELMANN CHART, while shade 6 represents the lowest concentration of stack solids just visible to the human eye.

shelter belt   A belt of trees purposely planted to provide protection against prevailing winds; SOIL EROSION is retarded by this technique and crop yields generally improved.

Shenandoah National Park   Established in 1935, a national park in the Blue Ridge sections of the Appalachians, in northern Virginia, USA. Noted for its scenery, the park offers some of the widest views in the eastern States. The park is heavily forested with both hardwoods and conifers.

shift and share analysis   A technique using a comparison of the growth rate of some element within a system with the growth rate of some relevant larger aggregate; for example, the growth rate of a region with the growth rate of the national economy.

shifting cultivation   See SLASH-AND-BURN AGRICULTURE.

Shigella   A genus of highly infectious BACTERIA that causes bacillary dysentery. Its presence in water indicates recent faecal contamination.

shockwave   A pressure pulse formed by an explosion in which a sharp discontinuity in pressure is created as the wave travels through a fluid medium at greater than sonic velocity.

short-term exposure limit   A time-weighted average concentration of a pollutant or contaminant, usually averaged over 10 minutes, aimed at avoiding acute effects.

SIA   SOCIAL IMPACT ASSESSMENT.

Sierra Club   Founded in 1892, a US voluntary organization for the conservation of natural resources; its headquarters are in San Francisco, California. Its first president was the naturalist John Muir (1892-1914), involving the club in political action to further conservation of natural resources. The Sierra Club has branches in all 50 American states. It continues to lobby for environmental legislation and to educate the public on environmental matters.

Silent Spring   See CARSON, RACHEL.

silt   An unconsolidated SEDIMENT or inorganic granular material that has been transported mechanically and deposited. The particles are between sands and gravels in size and are capable of settling.

siltation   The process whereby fine particles of sand, mud, and other material picked up from a source by moving water are deposited to form a SEDIMENT; deposition occurs as the velocity of the water declines. Sediments may arise from activities such as dredging and farming, natural erosion of land surfaces, and disturbances of sand and gravel beaches by natural currents.

silviculture   A branch of FORESTRY focusing on the cultivation of trees.

single-stack concept   The concept of directing all gaseous wastes to a single stack or chimney, of considerable height, instead of a multiplicity of stacks of varying height. The single stack may be of multiflue design, serving an entire industrial complex. Single-stack arrangements are common with fossil-fuelled power stations, serving perhaps four 500 MW or 660 MW electricity generators. The single stack ensures maximum efflux velocity and plume rise, and effective dispersal of gases. Stacks of 200 m or more are common in the power industry.

sink   (1) In AIR POLLUTION, a reception area for the absorption of material from the ATMOSPHERE; for example, the absorption of CARBON DIOXIDE by the oceans, or the absorption of carbon dioxide by photosynthetic plants. (2) In WATER POLLUTION, the assimilative capacity of bodies of water for THERMAL POLLUTION and other pollutants. (See SELF-CLEANSING; SELF-PURIFICATION.)

sinkholes   Depressions in a limestone area formed by the solution of surface limestone, or by the collapse of underlying caves.

site   Any LAND on which DEVELOPMENT is carried out, or is proposed to be carried out, or could be carried out,

site coverage   That portion of a SITE actually covered by a building, a fixed structure, or an outdoor storage area, excluding unroofed parking areas.

site planning   The arrangement of buildings and other structures on LAND to achieve function and efficiency and, as far as practicable, harmony; SITE plans are necessary at an early stage of design, whenever substantial groups of buildings or structures are to be erected.

site rehabilitation objectives Objectives defined for industrial sites or exploration sites that have been cleared of vegetation, or otherwise disturbed, to ensure satisfactory REHABILITATION; these objectives normally require approval and endorsement as conditions attached to a development consent. Rehabilitation normally requires the clearing of the site of industrial debris, the filling in of excavations wherever possible, and recovering and revegetation. It may be possible to achieve a near restoration of the original ENVIRONMENT. However, alternatives may be considered such as recreational facilities, redevelopment, or parkland.

skyways and skywalks Different names for elevated, enclosed walkways connecting buildings in a city core or CENTRAL BUSINESS DISTRICT. Crossings above congested streets offer pedestrians a safer, weather-protected alternative to city pavements or sidewalks.

slash-and-burn agriculture (shifting cultivation; swidden agriculture) A method of cultivation often used by tropical-forest root-crop farmers around the world, and by dry-rice cultivators of the forested hill country of southeast Asia. Areas of the forest are burned and cleared for the planting of crops, the ash providing some fertilization. After several years, the fertility declines and the area is then left fallow and reverts to secondary forest or bush. Cultivation than shifts to another locality. Most agriculturists consider slash-and-burn agriculture to be wasteful and inefficient, while large amounts of land are required to operate it. It is an unsustainable activity.

slow sand filter A shallow basin partly filled with sand and provided with an underdrainage system which is used for the purification of surface waters intended for domestic water supplies.

sludge See SEWAGE SLUDGE.

sludge cake The residue left on a filter after the filtration of SEWAGE SLUDGE.

sludge digestion A final biochemical reduction stage in SEWAGE TREATMENT. Raw SEWAGE SLUDGE is a semi-liquid whose solid content varies with the source. The primary purpose of sludge digestion is to change the offensive fresh sludge into an innocuous residue, the organic matter in the fresh sludge being broken down and stabilized by BACTERIA and other microorganisms. The products of digestion are withdrawn as gas, supernatent liquor, and digested sludge.

slurry A mixture of solid and liquid; a free-flowing suspension.

smallpox (variola) Until 1977 one of the world's most dreaded plagues, when it was declared eradicated. It is an acute infectious disease, characterized by fever and skin eruptions. Smallpox was estimated to have caused 2 million deaths in 1967. Routine smallpox vaccination has been discontinued in most countries.

smog A term first used by Dr. H. A. Des Voeux, founder-president of the British National Smoke Abatement Society in 1905 to describe the mixture of smoke and fog characteristic of AIR POLLUTION at that time. The term now embraces any objectionable mixture of air pollutants, including PHOTOCHEMICAL SMOG or LOS ANGELES SMOG, which contains neither smoke nor fog.

smoke A common carbonaceous contaminant of town air; the visible products of incomplete COMBUSTION. (See BEAVER REPORT.)

smoke control area Introduced under the British Clean Air Act 1956, a concept of a geographical area proclaimed smoke free. The emission of smoke from chimneys within the designated area is an offence.

smoke testing (1) A technique for testing whether properties have stormwater drains illegally connected to a sewer. (2) A technique for testing the soundness of drains and sewers. Normally smoke rockets are used to generate smoke.

Smoky Mountain A huge rubbish tip outside Manilla in the Philippines, where hundreds of people regularly scavenge or tat for useful items, a dangerously unhealthy way of earning a living.

snig track The track along which logs are pulled from the point of felling to the log dump.

Snowdonia National Park Established in 1951, a national park located in Wales, Britain. It is a wild mountain region extending southeastward from the Snowdon peaks to Cardigan Bay. It includes a great range of FLORA, including maple and beech woodlands, while FAUNA include otter, grey seal, polecat, pine marten, voles, squirrel, brown hare, and numerous birds, among them the kingfisher, great crested grebe, whitethroat, golden plover, raven, and red grouse. Snowdon peak is the highest in England and Wales.

social benefits The value of gains accruing to a community from the establishment of industries and facilities, though not necessarily constructed with that aim or purpose. Benefits may include better opportunities for employment, improved shops and amenities, better roads and transport, improved recreational facilities, enhanced environmental assets such as parks and open space, reduced noise, improved air and water quality, readier access and reduced congestion, or the removal of unsightly existing premises.

social Darwinism A theory that individuals, groups, and races are subject to the same laws of natural selection as Charles Darwin

(1809-82) had perceived in plants and animals; a struggle for existence ruled by the 'survival of the fittest', a phrase coined by the British philosopher and scientist Herbert Spencer (1820-1903). Societies, like individuals, were viewed as surviving in this way. Hence, the poor should not be assisted.

social evolution   As distinct from biological evolution, the evolution of human society with languages, laws, customs, histories, and settlements; such evolution is comparatively short, records extending back only a few thousand years, there being no trace of humans prior to the QUATERNARY PERIOD. The history of humanity is essentially barbaric, marked by instances of rare achievment; human beings are the only mammals to systematically attempt to destroy each other. Most governments and societies are autocratic, with democracy making limited inroads. The next phase of social evolution may be one of collective cooperative action for the common good, requiring far-reaching ATTITUDINAL CHANGES.

social exclusion   The marginalization of people because of sex, race, social standing, disability, or age. These issues were closely examined at the UN WORLD SUMMIT FOR SOCIAL DEVELOPMENT in 1995. (See INTRAGENERATIONAL EQUITY.)

social forestry   The growing of trees for fuel as well as raw materials. The WORLD BANK has financed a variety of social forestry projects.

social impact assessment (SIA)   A subset of ENVIRONMENTAL IMPACT ASSESSMENT, an appraisal of the effect on people of major policies, plans, programmes, activities, and developments. Social impacts or effects are those changes in social relations between members of a community, society, or institution, resulting from external change. The changes may be physical or psychological involving: social cohesion; general life style; cultural life; attitudes and values; social tranquillity; relocation of residents; and severance or separation. For example, in the construction of large hydroelectric dams, large populations are relocated into unfamiliar environments. The consequences have been social discontent, unhappiness, increased illness, and a loss of productivity and income.

social indicator   An attempt to measure by means of an index the degree of human welfare or QUALITY OF LIFE in a given area, national or regional. (See HUMAN DEVELOPMENT INDEX.)

socialization of losses   See CAPITALIZATION OF GAINS, AND SOCIALIZATION OF LOSSES.

social planning   An aspect of ENVIRONMENTAL PLANNING concerned with notions of equity, redistribution, social justice, and social planning. Social planning is primarily concerned with the welfare of the aged, disabled and socially disadvantaged. (See SOCIAL IMPACT ASSESSMENT.)

social rank   The order of individuals within a community or population in a dominance hierarchy.

social science   Any discipline or branch of science that deals with the economic, social, and cultural aspects of human behaviour. The social sciences embrace ECONOMICS, SOCIOLOGY, political science, social psychology, cultural anthropology, and comparative law.

social time preference rate   The relative value that the community as a whole assigns to present as opposed to future consumption.

social welfare programmes   Any of a variety of government programmes designed to provide a 'safety net', protecting citizens from the more common economic risks and insecurities of life. The most common types of programmes provide benefits to the retired and elderly, dependents, mothers, the unemployed, the sick and work-injured, and families. (See WELFARE STATE.)

societal risk criteria   Criteria relating to the likelihood of a number of people suffering a specified level of harm in a given population from the occurrence of specified hazards.

sociobiology   The study of the biological basis of social behaviour, including the application of population genetics to evolution of behaviour; it does not assume that all behaviour is genetically determined.

sociology   The science of human behaviour that investigates the nature, causes, and effects of social relations among individuals, and between individuals and groups. These studies embrace customs, structures, and institutions, and the interactions between individuals and groups. (See SOCIAL SCIENCE.)

sociosphere   The area of study of the social scientist; it is analogous to the HYDROSPHERE as the area of study of the oceanographer, the BIOSPHERE as the area of study of the biologist, among many others, and the LITHOSPHERE as the area of study of the geologist. The sociosphere embraces people, their roles and patterns of behaviour, their organizations and groups, and their interactions. (See SOCIAL SCIENCE; SOCIOLOGY.)

sodium fluoracetate (1080)   A RODENTICIDE marketed under the designation '1080'. It is manufactured from ethyl chloroacetate and potassium fluoride, which react to give ethyl fluoracetate. This is then hydrolysed with sodium hydroxide. Highly toxic, 1080 is very stable; strict regulations govern its use.

Sofia Protocol   See CONVENTION ON LONG-RANGE TRANSBOUNDARY AIR POLLUTION.

softwood   Timber obtained from coniferous trees, that is, cone-bearing trees, primarily of the pine and fir families. Softwood accounts

soil

for over 80% of the world's production of timber. (See HARDWOOD.)

soil The thin veneer of comparatively unconsolidated material covering large areas of the Earth's surface; soil is a dynamic medium in which occurs many physical, chemical, and biological processes. Soils comprise five main components: mineral particles; organic materials; water; air; and living organisms. Soils differ from one another because each has different proportions of these components, or because the components are arranged in different ways. Soils usually have a well-defined uppermost layer or topsoil, which merges gradually into other layers or horizons; the distinct layers are known as the soil profile. (See ALFISOL; ANDO; ARIDOSOL; BRUNISOLIC; CHERNOZEMIC; CINNAMONIC; HUMUS; INCEPTISOL; KAOLISOL; LATERITE; LOAM; MOLLISOL; MULCH; ORGANIC MATTER; OXISOL; PED; PEDOLOGY; pedogenesis; PODSOL; PODSOLIC; PROTOSOIL; SOIL DEGRADATION; SOIL EROSION; TILTH; TOPSOIL; ULTISOL; VERTISOL soils.)

soil acidification The process by which SOIL becomes more acidic, so that soil pH decreases. This has a detrimental effect on soil growth and crop yields; most plants can only tolerate a limited range of soil pH. Soil pH can be raised again by adding alkaline substances such as lime and dolomite.

soil biota The living organisms, mainly microorganisms and microinvertebrates which live within the SOIL, and which are largely responsible for the decomposition processes vital to soil fertility.

soil chemistry A discipline embracing all chemical and mineralogical compounds and reactions occurring in soils and soil-forming processes.

soil cohesion The mutual attraction between SOIL particles as a result of molecular forces and moisture.

soil conservation Practical measures taken to retard or inhibit SOIL EROSION or SOIL DEGRADATION, such as improved agricultural practices, flood mitigation measures, and soil rehabilitation measures

soil creep The gradual movement of SOIL down a slope, being dislodged by PRECIPITATION.

soil degradation SOIL impaired as a consequence of human activity. A study financed by UNEP, reporting in 1992, found that more than 1.2 billion ha (3 billion acres) of land (about 10.5% of the world's vegetative surface) had been seriously damaged by human activity since 1945. The study found that much of the damage had been masked by a general rise in global agricultural productivity resulting from expanded irrigation, better plant varieties, and greater use of production inputs, such as fertilizers and pesticides. More than one-third of the damaged land was in Asia, almost one-third in Africa, and a quarter in central America. Some land had been damaged beyond restoration. The greatest sources of soil degradation were overgrazing, unsuitable agricultural practices, and deforestation.

soil erosion The loss of SOIL as a result of natural and human activities. Natural erosion is the starting point of pedogenesis (creation of soils); on the other hand, accelerated erosion due to bad soil management, deforestation, brush and forest fires, overgrazing, and poor agricultural practices destroys the soil, with far-reaching consequences. Among the most important measures for combating soil erosion are re-afforestation, reduction of overgrazing, return of manure to the land, terracing of mountain slopes, and contour ploughing.

soil mechanics A branch of engineering that studies the nature and properties of SOIL. Soil is investigated before construction work is undertaken to ensure that it has the mechanical properties to support the foundations planned.

soil organism See SOIL BIOTA.

soil plasticity The property of a SOIL that allows it to be deformed or moulded in a moist condition without cracking or falling apart.

soil profile See SOIL.

soil stripping The removal of topsoil and subsoil from land.

soil-structure decline The loss of SOIL structure and texture that occurs when soils become compacted.

solar energy The energy of the sun, that reaches the surface of the Earth in the form of visible light, short-wave radiation, and near ultraviolet light. After penetrating the ATMOSPHERE, the energy heats the surface of the Earth while part of it is re-radiated into the atmosphere in the form of long-wave radiation and absorbed by CARBON DIOXIDE and water vapour in the atmosphere. The utilization of solar energy for the generation of electricity using photovoltaic cells has been developed in recent years, providing power for houses and satellites. In another application, many solar-energy heating and hot-water units have been installed in domestic homes in countries with appropriate climates. Solar steam generators have been built using moveable mirrors to concentrate large amounts of solar radiation to convert water to steam. Biological systems use sunlit ALGAE to convert carbon dioxide and water into OXYGEN and protein-rich carbohydrates. Solar energy is also used to produce salt from sea water by evaporation. Solar energy is non-polluting and constantly renewable. (See GREENHOUSE EFFECT.)

solar oil A description of liquid hydrocarbon fuels produced from crops on a renewable basis.

solar pond A natural or artificial 'pond' such as the DEAD SEA, in which bottom temperatures may reach 100°C.

solid-waste All material of a solid or semisolid character that the possessor no longer considers of sufficient value to retain.

solid-waste management The purposeful, systematic control of the generation, storage, collection, transport, separation, processing, recycling, recovery, and disposal of solid wastes in a sanitary, aesthetically acceptable, and economical manner. (See CAR SCRAPPING SCHEME; CONVENTION ON THE CONTROL OF TRANSBOUNDARY MOVEMENT OF HAZARDOUS WASTES AND THEIR DISPOSAL; CONVENTION ON THE PREVENTION OF MARINE POLLUTION BY THE DUMPING OF WASTES AND OTHER MATTER; DEEP REPOSITORIES; HAZARDOUS WASTES; INCINERATION; INTRACTABLE WASTES; KEEP AMERICA BEAUTIFUL; KEEP AUSTRALIA BEAUTIFUL COUNCIL; KEEP SINGAPORE CLEAN CAMPAIGN; LANDFILL; LEACHATE; LITTER; ORGANIC WASTE; SECURE LANDFILL; TRADE WASTES; WASTE REDUCTION AND RESOURCE CONSERVATION STRATEGIES. See Figure 24.)

solvent extraction A process in which a dissolved substance is removed from a liquid by bringing it into contact with another, immiscible (non-mixable) liquid.

Recoverable

Reuseable

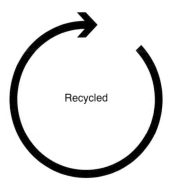

Recycled

SOLID-WASTE MANAGEMENT: Figure 24. International symbols used in solid-waste management.

sonic boom (supersonic bang) Caused by supersonic aircraft 'crashing the sound barrier', that is, flying faster than the velocity of SOUND (about 1200 km per hour).

sorptive Capable of absorbing or adsorbing suspended solids, sediments, or organisms in water.

sound A physiological sensation experienced by the ear, originating in a vibration (a pressure variation in the air) which travels in every direction spreading out as an expanding sphere. The LOUDNESS of a sound depends primarily on the amplitude of the vibration of the air.

sound barrier A sharp increase in aerodynamic drag that occurs as an aircraft approaches the speed of sound. (See SONIC BOOM.)

sound level The level of frequency-weighted and exponentially time-weighted sound pressure as determined by a SOUND-LEVEL METER. The unit symbols are dB, dB(A), dB(B), dB(C), and dB(D). The decibel A-scale is the most popular of the scales, correspondingly closely to the human ear, while the others tend to have specialized purposes such as measuring aircraft noise. (See DECIBEL; DECIBEL (A-SCALE).)

sound-level meter An instrument consisting of a microphone, amplifier, and sound-indicating device, having a declared performance.

sound shadow The acoustical equivalent to a light shadow which may be created by an acoustical screen. However, the benefit may be partially offset by diffraction which diverts the direction of travel of a SOUND WAVE.

sound wave An audible vibration in the air; at 20°C, a sound wave travels at 344 metres per second.

source control The elimination, before or during the consumption of a product, of potential contaminants from the raw materials or the process, thus preventing the emission of contaminants at source.

sources and sinks The origins and destinations of substances. For example, the burning of fossil fuels is a CARBON DIOXIDE source, while a growing forest is a carbon dioxide sink.

South Pacific Nuclear-Free Zone Treaty The Treaty of Rarotonga, signed by 13 South Pacific nations; the Treaty is designed to prevent the dumping of nuclear waste and nuclear testing within the zone, and prohibits the holding and testing of nuclear weapons. The treaty came into effect in 1986.

South Pacific regional environmental programme Launched at Rarotonga in the Cook islands in 1982, a cooperative programme aimed at developing a comprehensive environmental management programme in the region. The participating bodies include UNEP and ESCAP.

South Spitzbergen National Park Established in 1973 in Norway, a national park and bird sanctuary on the island of Spitsbergen. The park has four separate bird sanctuaries.

Southern Ocean Sanctuary A sanctuary for whales nominated by members of the INTERNATIONAL WHALING COMMISSION (IWC), embracing waters from Bass Strait to Antarctica and encircling the globe in a vast swathe; it is designed to keep commercial whaling out of the largest remaining habitat of the world's whales. The sanctuary was approved by the IWC in 1994. Japan and Russia have objected to the sanctuary.

South-west National Park Initially created in 1968 in southwestern Tasmania, Australia, out of the amalgamation of Lake Pedder National Park and the Huon Serpentine Impoundment which had inundated Lake Pedder as a consequence of hydroelectric construction. In 1976 the park was almost doubled with Precipitous Bluff added to it. In 1981, it was enlarged again, now merging with the adjacent Franklin-Lower Gordon Wild Rivers National Park. These parks are now embodied in the Tasmanian World Heritage Area, included in the WORLD HERITAGE LIST.

space biology A more recent discipline in the science of BIOLOGY arising through the activities of scientists and engineers concerned with the exploration of space.

space law That body of conventions, agreements, treaties, and regulations that governs international conduct in areas of space beyond the lower strata of the Earth's ATMOSPHERE. The concept was initiated by the USA before the United Nations in 1957; in 1959, a permanent Outer Space Committee was formed for the purpose of maintaining the UN Charter and other international law in space. The major powers recognized that it was not possible to extend sovereignty upwards into space, the upper atmosphere becoming a COMMON PROPERTY RESOURCE. In 1963, the Nuclear Test Ban Treaty was signed, followed by a prohibition of nuclear testing in space. In 1967, an Outer Space Treaty was ratified by the UN; this document was a landmark in the development of international space law.

Spaceship Earth A concept evolving in the 1960s, that the planet Earth may be regarded as a spaceship dependent essentially upon its own resources for survival.

special-use corridors Areas of LAND set aside in environmental plans for future use by public authorities and private enterprise to accommodate major roads, pipelines, transmission lines, and for other public uses and regional open space. Corridors also serve as physical and visual boundaries between existing and future urban areas.

speciation The process by which one SPECIES splits into two or more species; one of the fundamental processes of evolution.

species Organisms of the same kind that are capable of breeding and producing fertile offspring; the smallest unit of classification commonly used. A subspecies is a segment or subdivision of a species, being a group whose members resemble each other in certain characteristics, but do not embrace the whole range of characteristics of the species. Generally, a species is unable to breed with organisms of other species; that is, species are reproductively isolated units.

spectrophotometry A technique of chemical analysis based on the absorption or attenuation by matter of electromagetic radiation of a specified wavelength or frequency. (See CHROMATOGRAPHY.)

speleology The scientific study of caves, their origins, evolution, structure, FLORA, FAUNA, exploration and surveying.

spillway The passage for surplus water over or around a dam when the reservoir itself is full. Spillways are a particularly important safety feature for earth dams, protecting the dam and its foundation from erosion. (See DAMS.)

spore In BIOLOGY, a reproductive cell capable of developing into a new individual without union with another reproductive cell. Spores differ from gametes, which are reproductive cells that must fuse in pairs in order to create a new individual. Spores are therefore asexual, being produced by bacteria, fungi, and green plants.

stabilization lagoon See LAGOON.

stakeholder Any individual, group, or body whose interests in a narrow or broad sense are linked with the future of an enterprise; the concept includes shareholders, creditors, directors, managers, employees, customers, suppliers, members of the local community, and possibly government at all levels. Some will have a defined contractual relationship; others a financial interest with certain claims in the event of liquidation; many, perhaps, with direct personal concerns by way of employment, career prospects, job satisfaction, and the ability to support a family. Other stakeholders include, for example, suppliers who have invested in new plant as an act of faith in the future of the enterprise; customers who seek reasonable prices, quality, reliability, and courtesy; and local and state government which may have encouraged the enterprise, installed infrastructure at public expense, waived planning restrictions, provided subsidies and concessions, and at times other means of financial support. Some of these interests take the form of explicit contracts probably enforceable; others the form of implicit contracts or understandings not enforceable; and the remainder, a conglomerate of interests in the success of the enterprise. Adverse trading conditions, high-interest rates, errors of business judgement, changes in taxation, variations in levels of protection if any, and mergers and takeovers endanger the interests of all stakeholders. In the event of a merger or takeover, all non-enforceable contracts may count for nothing; the extinguishment of implicit contracts and understandings is a cost to be taken into account when weighing the merits and demerits of mergers and takeovers, both specific and in general.

state of environment reports Reports prepared by government departments or agencies on the state of the environment within their respective geographical jurisdictions. Such reports may appear annually or at regular less frequent intervals. They require input from the community, voluntary conservation organizations, other government departments, universities and other educational institutions, the private and public sectors, and individual authorities in the various segments of the subject. Government ministers, it may be noted, are not always happy with such far-reaching and independent reports, fearing political embarrassment.

statutory planning A traditional term for the role of government concerned with guiding and controlling the use and development of land; the purpose of statutory planning is to provide a framework within which public and private interests may be as far as practicable reconciled, the interests of the community safeguarded, and a wide range of physical, social, economic, and environmental objectives pursued. (See ENVIRONMENTAL PLANNING; ZONING.)

Stava tailings dam disaster In July 1985, the collapse of a tailings dam resulting in a devastating mud slide, killing more than 250 people in the mountain villages of Stava and Tesero in northern Italy.

Steppe That broad band of grassland that begins in Hungary, extending eastwards across the Ukraine and southern Russia into Central Asia and ultimately, Manchuria. Migrations across the Eurasian Steppe have greatly affected the settled populations of Europe, the Middle East, India, and China.

Stockholm Conference See UN CONFERENCE ON THE HUMAN ENVIRONMENT.

stomata The microscopic pores or holes in the surface (epidermis) of plants, particularly leaves, through which plants take in CARBON DIOXIDE from the air and give out OXYGEN. Water vapour is lost through stomata in the process of TRANSPIRATION. The singular of stomata is stoma.

storm tank A tank, usually located at a sewage treatment works provided for the storage and partial treatment of excess storm sewage prior to discharge to a watercourse or other body of water.

storm overflow A device on a combined or partially separate SEWERAGE system, for relieving the system of excessive flows during storms, the excess being discharged to a convenient watercourse.

storm sewage Foul sewage mixed with relatively large quantities of storm water.

storm water runoff That portion of rainfall that flows directly from the land surface into streams and lakes, as distinct from rainfall that evaporates or seeps into the SOIL to recharge GROUND-WATER supplies. Developments increase the amount of impervious cover, this cover reducing the amount absorbed into the ground and increasing the amount of runoff. Storm water flushes more salts, oils, and other pollutants from urban areas into the receiving waters.

strategic EIA The application of ENVIRONMENTAL IMPACT ASSESSMENT not only to individual projects, but also to policies, plans, programmes, activities, and regional land-use objectives. There is a growing conviction that matters cannot be completely resolved at project level when many matters have been decided already at a higher level. Matters difficult or impossible to settle at the project level relate to the cumulative effects of other projects within the same or related programmes; to transportation decisions governing the modal split between road and rail movement; to energy policies relating to power generation; to greenhouse strategies; and to natural resource conservation and management.

stratosphere That layer of the ATMOSPHERE immediately above the TROPOSPHERE, beyond 10 km the Earth, wherein the temperature is constant. The stratosphere is warmer than the upper troposphere, containing the OZONE LAYER.

streamflow regulation A method of water quality control in which good-quality stored water is added to a stream at times of deteriorating stream-water quality.

street improvement schemes Schemes, often voluntary, involving the redecoration of property, traffic flow changes, creation of pedestrian precincts, the removal of unsightly clutter, the replacement of confusing or unsatisfactory signs, and the improvement of lighting and street furniture.

streetscape A complete collection of buildings along a street frontage; a streetscape can be as visually distinctive as individual buildings.

stress The non-specific response of a body or organisms to any demand made upon it, primarily preparing the organism for physical activity; for example, fight or flight. A certain level of stress is probably healthy and stimulating; excessive stress is suspected of leading to disease or impairment in modern life.

strip cropping A land management practice of growing crops and fallowing land in alternate strips on the contours of the land, intended to reduce erosion by speading runoff and reducing its velocity.

structural change In ECONOMICS, changes in the structure of the economic system by way of rationalization, reconstruction, responses to changes in the legislative framework, and repositioning following changes in overseas and domestic markets. Some structural changes will inevitably flow from the application of the principles of SUSTAINABLE DEVELOPMENT.

structural conservation That aspect of CONSERVATION concerned with works and practices aimed at improved management of the ENVIRONMENT; for example, soil and water conservation measures, coastal defence works, land reclamation measures, and flood control.

structural geology A subdiscipline of GEOLOGY, concerned with rock deformation on both a macro- and micro-scale.

stubble burning A management practice in which the stubble from one crop is burnt after harvest and before the commencement of fallowing for a subsequent crop.

subdivision A division of LAND into parts for essentially residential purposes, with provision for roads, schools, shopping centres, and recreational facilities. Most countries have adopted regulations requiring the approval of plans for proposed subdivisions with the setting of minimum standards.

subsidence The lowering or falling of an area of ground. Subsidence may occur when rock below is removed as a result of mining, when the ground is compressed following the removal of water, or because of other Earth movements.

subsidence inversion A temperature INVERSION in the ATMOSPHERE usually well above the Earth's surface, formed as a result of the slow descent of air which becomes warmer through adiabatic compression.

subsistence farming A level of farming when the produce is enough to feed only the farmer and family.

subsoil water Water occurring naturally in the subsoil. (See GROUND WATER.)

subspecies A segment or subdivision of a SPECIES, being a group whose members resemble each other in certain characteristics, but whose range of variation falls short of the total range of variation of the species as a whole.

substitution effect (1) The change in the quantity demanded of a good or service resulting solely from a change in its price relative to the prices of other goods and services. The rate of substitution depends on the closeness of the substitutes. (2) The substitution of one substance or material for another as a result of legislation; for example, the displacement of ASBESTOS and LEAD by other materials. (3) The environmental consequences of substituting one form of pollution control for another; for example, when wet scrubbers convert an air-pollution problem into a liquid-waste problem, or when domestic refuse-disposal units also convert a solid-waste disposal problem into a liquid-waste disposal problem. (4) The potentially large-scale substitutions likely to flow from the application of the principles of SUSTAINABLE DEVELOPMENT.

succession In ECOLOGY, the slow orderly progression of changes in communities of fauna and flora in a particular area over time. Often, a community becomes increasingly more complex until it reaches a climax community.

sullage water Waste water from kitchens and bathrooms, as distinct from SEWAGE.

sulphur A non-metallic element occurring in nature both in a free state and combined as sulphides and sulphates. It is present in coal and oil, being derived from the substances from which they were formed. During COMBUSTION, most of the sulphur is released into the ATMOSPHERE as SULPHUR DIOXIDE and SULPHUR TRIOXIDE.

sulphur cycle The circulation of sulphur atoms brought about mainly by living things; sulphur circulates globally between air, land, and sea. A large part of the sulphur in the global atmosphere (as distinct from the atmosphere of towns) is emitted originally as HYDROGEN SULPHIDE from natural sources, being converted later into SULPHUR DIOXIDE. Sulphur is removed from the ATMOSPHERE by PRECIPITATION and gaseous absorption in the oceans.

sulphur dioxide A colourless pungent gas released when sulphur burns in air. Emissions of sulphur dioxide in urban areas come from industrial, commercial, and domestic sources. During COMBUSTION all the sulphur in oil, and most of the sulphur in coal, is emitted from stacks into the general ATMOSPHERE as sulphur dioxide with a small proportion as SULPHUR TRIOXIDE. The discharge into the atmosphere of large quantities of oxides of sulphur and nitrogen has led to the problem known as ACID RAIN. (See CONVENTION ON LONG-RANGE TRANSBOUNDARY AIR POLLUTION.)

sulphur trioxide A constituent of flue gases of sulphur-bearing fuels, frequently to the extent of 3 to 5% of the SULPHUR DIOXIDE present. The reduction of excess air tends to inhibit the formation of this corrosion-promoting gas.

superfund See COMPREHENSIVE ENVIRONMENTAL RESPONSE, COMPENSATION AND LIABILITY ACT 1980; CONTAMINATED SITES.

superhighway A growing international system of interconnected computers (such as Internet) enabling academics and students to search libraries and institutions worldwide from home; facilitating environmental studies, ENVIRONMENTAL IMPACT ASSESSMENT, and SUSTAINABLE DEVELOPMENT.

supersonic bang See SONIC BOOM.

supply and demand, laws of The basic laws in respect of the marketing of goods and services in a MARKET ECONOMY. Five laws may be identified: (1) If, at the price ruling in the MARKET, demand exceeds supply, the price tends to rise; (2) conversely, when supply exceeds demand, the price tends to fall; (3) an increase in price tends to lead to a contraction of demand, ceteris paribus; and (4) conversely, a fall in price tends to lead to an expansion of demand and a contraction of supply; and (5) prices tend to move towards a level at which demand is equal to supply, that is, to an equilibrium price. The term 'demand' refers to effective demand, or ability-to-pay, and not to need. Short-term exceptions may occur in the operation of these laws. For example, during a period of growing shortages an increase in price may be accompanied by an increase in demand, due to a fear of scarcity; also a fall in share prices might not be accompanied by an immediate increase in demand, if potential buyers are confident that prices will fall still lower.

surface mining Mining operating from the surface of the Earth, as distinct from underground mining. The rock and SOIL (the overburden) over the coal or ore-bed is removed to expose the mineral, the coal or ore being readily removed by excavation. Surface-mining methods include stripping, contour stripping, and auger mining. Area stripping is generally practised on terrain that is gently rolling or relatively flat, utilizing drag lines, shovels, and bucket-wheel excavators. Surface or open-cut mining presents problems of pollution control, disposal of MINING WASTES and REHABILITATION.

surface water Water that is lying on the surface of the ground, as distinct from GROUND WATER, such as water in streams, lakes, rivers, and wetlands. It describes water from a perennial flowing stream; water stored behind a rock bar, weir, or dam placed across a stream; and water collected from a small drainage area such as a roof, being stored in a tank for domestic use.

surfactant A surface active agent or DETERGENT.

survivorship That proportion of individuals in a community or population surviving to a particular age.

suspended solids Solids in a liquid that can be removed through sedimentation or filtration.

sustainable development DEVELOPMENT that provides economic, social, and environmental benefits in the long term, having regard to the needs of living and future generations. Defined by the WORLD COMMISSION ON ENVIRONMENT AND DEVELOPMENT as 'development that meets the needs of the present without compromising the ability of future generations to meet their own needs'. Thus, the satisfaction of human needs and aspirations is the major objective of development. Sustainable development considers both the living and non-living resource base with regard for conservation and the advantages and disadvantages of alternative courses of action for future generations. It allows the use of depletable resources in an efficient manner, with an eye to the substitution of other resources in due course. Sustainable development calls for much more emphasis on conserving natural systems and the resource base on which all development depends; a greater regard for equity within society at present and between rich and poor nations, with particular regard to the world's poor; and a planning-horizon that goes well beyond the needs and aspirations of those alive today. It requires an integration of environmental, social, and economic considerations in decision-making. (See ECONOMIC DEVELOPMENT; ECONOMIC EFFICIENCY; ECONOMIC GROWTH; ECONOMIC IMPACT ASSESSMENT; ECONOMIC SYSTEM, FUNCTIONS OF; ECONOMIC WELFARE; ENVIRONMENT; ENVIRONMENTAL IMPACT ASSESSMENT; ENVIRONMENTAL MANAGEMENT; ENVIRONMENTAL PLANNING; FACTORS OF PRODUCTION; GROSS DOMESTIC PRODUCT; HUMAN DEVELOPMENT INDEX; INTERGENERATIONAL EQUITY; INTRAGENERATIONAL EQUITY; PER CAPITA INCOME; PRODUCTIVITY; QUALITY OF LIFE; UN CONFERENCE ON ENVIRONMENT AND DEVELOPMENT; WORLD BANK; WORLD CONSERVATION STRATEGY; WORLD POPULATION; Box 30.)

Sustainable Development Commission See COMMISSION FOR SUSTAINABLE DEVELOPMENT.

Sustainable Development, President's Council on An advisory body set up to advise the US President on SUSTAINABLE DEVELOPMENT, following the UN CONFERENCE ON ENVIRONMENT AND DEVELOPMENT of 1992.

sustainable global society A wholistic view that SUSTAINABLE DEVELOPMENT and SUSTAINABLE YIELD need to be considered within an international framework, rather than a nation-state framework, or regional, or local context.

sustainable yield The use of living resources at levels of harvesting and in ways that allow those resources to supply products and services indefinitely. Sustainable yield means living off the interest, rather than the CAPITAL, of a resource base. Being concerned

---

Box 30 Guiding principles for ecological sustainable development (ESD)

(1) Decision-making processes relating to development should take into account the short-, medium- and long-term effects of an economic, social, equity, and environmental nature in a local, regional, national, international, and global context.

(2) Development should be defined as all physical development and social activities with likely environmental effects resulting from domestic investment or foreign aid.

(3) The implications of development for the use and misuse of natural and human resources.

(4) The arrangements for the involvement of the public, both individuals and groups, government at all levels, and stakeholders generally in the process leading to a development decision.

(5) Avenues of appeal by all involved parties against development decisions, or conditions attached to those decisions.

(6) Recognition that a strong, highly competitive economy, increases the capacity for responsible environmental management and enhances the real income of the community.

(7) The sustainability of the resources supporting the enterprise, either through renewability or substitutability.

(8) The intragenerational and intergenerational implications of the enterprise.

(9) The implications of the enterprise for the greenhouse effect and the ozone layer.

(10) The transboundary implications of the enterprise.

> **Box 31 Sweden: evolution of environmental policy**
>
> | | |
> |---|---|
> | 1964 | Nature Conservancy Act, superseding earlier laws in 1909 and 1952 |
> | 1967 | National Environmental Protection Agency created |
> | 1968 | Environmental Advisory Committee appointed. Sulphur Content in Fuel Oil Act |
> | 1969 | Environment Protection Act; the use of DDT in Sweden stopped, with some exemptions until 1975 |
> | 1971 | Marine Dumping Prohibition Act; insecticides aldrin and dieldrin totally banned |
> | 1973 | Products Hazardous to Health and the Environment Act |
> | 1975 | Vehicle Scrapping Act |
> | 1976 | Sulphur Content in Fuel Oil Act amended |
> | 1978 | Environment Monitoring Programme introduced |
> | 1979 | Forest Conservation Act; Cleansing Act |
> | 1980 | The Riksdag voted to phase out nuclear power by the year 2010; Water Pollution from Vessels Act |
> | 1985 | Chemical Products Act |
> | 1986 | Building and Planning Act; Natural Resources Act; lead-free petrol introduced; Environmental Damage Act |
> | 1987 | Hunting Act |
> | 1988 | National phase-out plan adopted for CFCs |
> | 1989 | Catalytic conversion for car exhaust gases compulsory from 1989 models; Environment Protection Act amended |
> | 1990 | Environmental impact assessment requirements extended |
> | 1991–92 | EIA Regulations introduced |
> | 1994 | Environmental Code introduced by the Association of Swedish Chemical Industries to promote better industrial environments |

with living resources, sustainable yield is a concept somewhat narrower than that of SUSTAINABLE DEVELOPMENT. More specifically, sustainable yield aims to: maintain essential ecological processes and life-supporting systems; to preserve GENETIC DIVERSITY; and to maintain and enhance environmental qualities relevant to productivity. In common with sustainable development, it seeks not to disadvantage future generations. Sustainable yield presents particular problems worldwide in relation to WORLD POPULATION, FORESTRY, agriculture, FISHERIES, and SOIL; INTERGENERATIONAL EQUITY; and INTRAGENERATIONAL EQUITY.

Sweden, evolution of environmental policy See Box 31.

swidden agriculture   See SLASH-AND-BURN AGRICULTURE.

Sydney International Airport Third Runway   A controversial development in Sydney, Australia, with the construction and operation of a third runway. The effect was to much reduce aircraft noise levels for those living under the east–west flight path, while substantially raising noise levels for those living in suburbs north of the airport. Noise mitigation measures have been slowly introduced, while the construction of a new airport at Badgery's Creek, near Sydney, has been accelerated.

symbiosis   Two or more species living together to the benefit of each other. The terms symbiosis and mutualism are often used to mean the same thing.

synergism   An effect whereby two substances together have more of an impact than anticipated. Synergistic effects occur in the formation of photochemical smogs. The formation of highly toxic methyl MERCURY through the interaction of organic matter and mercury is another example.

synroc   A synthetic rock designed to integrate high-level radioactive wastes into its stable crystal structure. Britain, Japan, and Australia have been involved in its development.

systematics   A study of the diversity of living organisms, often used synonymously with taxonomy.

systems engineering   The use of knowledge from various branches of engineering and science in the development of technological innovations.

T

Tablas de Damiel National Park Established in 1973, a nature reserve located about 30 km northeast of the city of Ciudad Real, south-central Spain. It lies at the confluence of the Guadiana and Ciguela rivers, with fresh and brackish waters supporting both migratory and resident waterfowl year-round.

Tādoba National Park Established as a wildlife santuary in 1935 and later expanded, a national park in the Chandrapur district, Mahārāshtra state, western India, consisting of dense forests, interspersed with lakes and plains. The park has tiger, panther, leopard, jackal, bison, antelope, bear, and crocodile.

tailings Solid or fluid material of no economic or environmental value separated from useful material during processing.

tailings dam structures Embankments constructed to contain the fine mineral and coal particles (for example) from washeries. Two kinds of dam are used: the evaporation-decantation dam, pond, or lagoon; and the filtration dam, pond, or lagoon. Careful design with adequate safety factors is necessary to avoid the risk of pollution and structural failure. (See STAVA TAILINGS DAM DISASTER.)

Taiwan, evolution of environmental policy See Box 32.

Talloires Declaration In 1990, an environmental declaration by 22 presidents, rectors, and vice-chancellors of universities around the world meeting at the Tufts European Centre in Talloires, France. The declaration is a statement of commitment to environmental responsibility by university campuses. Since 1990, 125 university presidents from 32 countries as well as the Conference of European Rectors have formally endorsed the principles. The declaration recognizes that universities have a major role to play in education, research, policy formation, and information exchange in promoting a sustainable and equitable future for all humanity in harmony with nature, while establishing programmes for resource conservation, recycling, and waste reduction at the universities.

tanker ballast-water treatment See BALLAST WATER.

Tarawa Declaration A declaration resulting from the 20th South Pacific Forum held in 1989, aimed at banning drift net fishing in the South Pacific. (See CONVENTION ON THE PROHIBITION OF FISHING WITH LONG DRIFT NETS IN THE SOUTH PACIFIC; DRIFT NET FISHING.)

target See RECEPTOR.

Tasmanian World Heritage Area An area of southwest Tasmania, Australia, included in the WORLD HERITAGE LIST in 1989. The area contains large pristine tall-forest ecosystems, together with superlative examples of wilderness, rainforest, alpine ecosystems, and glacial landscapes. It includes the Lemonthyme and Southern Forests, the Walls of Jerusalem National Park and the Central Plateau Conservation Area, and a number of sites of major Aboriginal cultural significance.

tax incentives Incentives provided by government through tax concessions, tax relief, tax breaks, accelerated depreciation, and sales tax relief, for the purposes of stimulating investment in pollution control equipment and environment protection measures. (See POLLUTER PAYS PRINCIPLE.)

taxonomic Describing the sorting of living organisms into related groups or types, such as species or genera.

TBT TRIBUTYL TIN.

TCM TOTAL CATCHMENT AREA.

TDS TOTAL DISSOLVED SOLIDS.

technological change Change in a whole system of scientific and technical knowledge, not simply changes in technique; change involving fundamental scientific discoveries coupled with innovative application and far-reaching economic, social, and environmental consequences. Such change characterizes much development over the past 250 years. Today, technological change springs from electronic computing, biotechnology, energy technology, and materials technology.

technological effects ('real' effects) Effects that alter the total production possibilities, or the total welfare opportunities, for consumers within an economy. The effects are described as economies when they are favourable, and diseconomies when they are unfavourable. Classical examples of external real diseconomies are air and water pollution, the effects being borne often by persons and firms other than those who cause the pollution, reducing their HEALTH and welfare opportunities.

technology transfer The transfer of development and design work: (1) from a parent company to a subsidiary, perhaps in another nation where it will be paid for in repatriated profits or royalties; (2) from one country to another, as a form of aid, to help promote development and sustainable growth. Many nations, such as Taiwan and South Korea, have made great progress on the strength of technology transfer.

telemetry Literally, measurement from a distance; highly automated communications systems by which measurements are made and data collected at remote or inaccessible points

Box 32 Taiwan: evolution of environmental policy

1967 Taiwan Government assigns air and water pollution control to the Taiwan Institute of Environmental Sanitation and the Taiwan Water Pollution Control Agency, respectively

1974 Water Pollution Control Act

1975 Water pollution control regulations announced for the Haintien River

1976 Water pollution control measures announced for Taiwan, including Taipei Municipality; water-pollution control regions announced for the Keelung and Tamshui Rivers

1979 Executive Yuan orders that government departments are to undertake EIAs in respect of proposed major projects, to reconcile growth and development with social and environmental considerations

1980 Executive Yuan instructs the Taiwan Power Company (Taipower) to undertake EIAs in relation to all future nuclear power plants

1981 Atomic Energy Council establishes a radioactive waste unit

1982 Department of Health establishes a division of toxic substances control; Cultural Heritage Preservation Act

1983 Executive Yuan directs that EIAs are to be conducted not only in respect of government projects but also for major private sector projects; National Parks Act stipulates the undertaking of EIAs; Taiwan Environment Protection Bureau established

1984 Kenting National Park established

1985 Taiwan Environment Protection Bureau formulates a national EIA programme for 1985–90; Yushan, Yangmingshan, and Taroko National Parks established

1986 Council of Agriculture begins designating protected areas for rare and endangered species; Hill Land Preservation Act; public demonstration against a proposed titanium dioxide plant at Lukang

1987 Environmental Protection Administration established (superseding the Bureau) to control air, water, and noise pollution, and hazardous wastes

1989 Wildlife Conservation Act

1991 Amendments to the Air Pollution and Noise Control Acts; paper-recycling plan a success; new exhaust standards for motorcycles

1992 Shei-Pa National Park established: afforestation campaign launched; sixth naphtha cracking plant endorsed after EIA; residents protest against fourth nuclear power station

1993 Kinment Islands proposed as sixth national park

1994 Decision to build fourth nuclear power station; conservation strategy announced; Coastal Protection Act

1995 Amendments to the Wildlife Conservation Law, increasing penalties for violations, while furthering compliance with the Convention on Internation Trade in Endangered Species (CITES)

and transmitted to receiving equipment for monitoring, displaying, and recording.

teleoclimate The microclimate at the boundary between living organisms and the ENVIRONMENT.

temperature inversion See INVERSION.

temporary hardness See HARDNESS.

Tennessee Valley Authority (TVA) A US government agency created in 1933 to control floods, improve navigation, produce electrical power along the Tennessee River, and generally improve the living standards of farmers struggling with the effects of the Great Depression. The TVA power system includes over 50 dams, together with coal-fired and nuclear-power plants.

Ten Thousand Smokes, Valley of See KATMAI NATIONAL PARK.

terpenes A group of volatile aromatic substances commonly released from the shoots of flowering plants in particular families.

terracing An embankment of earth, constructed across a slope in such a way as to control water runoff and minimize erosion. To be effective, terraces must check water flow before it attains sufficient velocity to loosen and transport SOIL.

territoriality The identification of an individual organism, population, or community, with a particular spatial area or volume.

tertiary treatment Any sewage purification process that is capable of removing most of the pollutants from sewage, following primary and secondary treatment. It is sometimes known as 'polishing'.

Test Ban Treaty A treaty signed by the USA, the former USSR, and Britain in 1963, to restrict the testing of nuclear weapons to underground locations only. Since then many other nations have signed the treaty. (See MURUROA ATOLL.)

tetra ethyl lead An additive which has been added to petrol (gasoline) (lead) to reduce the tendency to 'knocking' in car engines. As the combustion of leaded petrol has been a major source of lead in the ATMOSPHERE of urban areas and inhibits the effectiveness of catalytic converters for exhaust gases, the target in many countries is to eliminate leaded fuel completely, unleaded petrol being now generally available.

TFR TOTAL FERTILITY RATE.

Thailand, evolution of environmental policy See Box 33.

thalweg The deepest part of a stream channel which tends to meander across the stream between banks. A thalweg survey is the measurement of the bed level of a length of a stream along a line in the deepest part of the channel.

Box 33 Thailand: evolution of environmental policy

1914	Local Government Act providing limited environmental protection in respect of public lands
1921	Wild Elephant Conservation Act
1941	Forestry Act
1960	Wild Animal Reservation and Protection Act
1961	National Parks Act
1962	Khao Yai National Park established; now an ASEAN Heritage Site
1964	National Reserved Forest Act
1967	Mineral Act with provisions for pollution control
1975	National Environmental Quality Act
1976–81	Fourth five-year development plan incorporating resource protection, rehabilitation strategies, water and soil conservation
1981	EIA introduced as a tool for environmental planning and management
1981–86	Fifth five-year national development plan incorporating an integrated approach to natural resource development and socio-economic progress
1984	EIA reports to be prepared by consultant firms registered with the Office of the National Environment Board
1986–91	Sixth five-year national development plan recognizing that the depletion of the resource base could become a constraint to development
1989	Thailand's Natural Resource and Environmental Conservation Year
1991	Wildlife Conservation Act
1992	Thailand's Report to the UN Conference on Environment and Development (the Rio conference); National Environmental Quality Act, replacing the Act of 1975; Wildlife Conservation Act
1992–96	Seventh five-year national development plan setting definite targets for improving environmental quality throughout the country and promoting sustainable development
1993	Catalytic converters required on all new cars; unleaded fuel introduced, national ambient air quality standards established
1994	Plans to increase national parks, nature reserves and recreational areas from 12.7% of the country to 25%, as provided for in the seventh national development plan; modification of the Pak Mun hydroelectric project

thermal efficiency   The useful heat or energy produced by a plant or steam generator, expressed as a percentage of the heat value of the fuel or energy consumed. A modern power station achieves about 35% thermal efficiency overall.

thermal expansion   The increase in volume that occurs when gases, liquids, or solids are heated; or conversely, an increase in length, expressed as a coefficient of expansion.

thermal liquid   A liquid used as a heat-transporting medium in the field of process heating and cooling.

thermal oxide reprocessing plant (THORP) A controversial plant at Sellafield, Cumbria, England, reprocessing nuclear-reactor fuel returning fissionable uranium and plutonium, while retaining most of the waste for disposal in Britain. In 1992, concerns were expressed over emissions of krypton-85. The Inspectorate of Pollution issued a permit, allowing an increase in the emissions of some radioactive substances but reducing overall discharges. In December 1993, the British Government gave THORP the go-ahead.

thermal pollution   The discharge of heated effluents into bodies of water or watercourses at temperatures and in quantities that may prove detrimental to the ENVIRONMENT. Effluents having a significant effect on the temperature of a river may affect the oxygen content of the water and accentuate other adverse conditions. At sufficiently high temperatures, fish may die either as a direct consequence, or through the destruction of a food supply or spawn.

thermal radiation   The propagation of energy in the infrared region of the electromagnetic spectrum.

thermal shield   A metallic shield placed around a NUCLEAR REACTOR between the core and the concrete biological shield.

thermocline   A temperature gradient occurring in a layer of water, the temperature of the water decreasing with increasing depth but at a greater rate of decrease than that occurring in the water layers above or below the layer under consideration; the middle stratum of water in a lake, below the EPILIMNION and above the HYPOLIMNION, typified by a temperature gradient of more than 1°C per metre of depth.

thermonuclear fusion   See NUCLEAR FUSION.

thermosphere   A region of increasing temperature above the MESOSPHERE. The base of the thermosphere (the mesopause) is at an altitude of about 80 km, whereas its upper layer (the thermopause) is at about 450 km.

thermotolerant coliforms   A subgroup of COLIFORM BACTERIA found predominantly in the intestine and faeces of humans and other warm-blooded animals. The presence of thermotolerant coliforms in a water sample means that the water is probably contaminated with faecal material. These organisms should not be detected in any sample of drinking water.

therophyte   An annual plant; that is, a plant that completes its life cycle within one season, surviving the subsequent unfavourable period as a dormant seed.

thio alcohols   See MERCAPTANS.

Third World   Developing countries defined by the WORLD BANK as the world's 100 poorest countries as measured by GROSS DOMESTIC PRODUCT (GDP) per capita; they are concentrated in Africa, Asia (including China and India), and Central America. Problems associated with Third-World countries have included high population growth, high infant mortality, and poor education and health facilities. Yet the Third World accounts for over 75% of arms imports. (See BRANDT COMMISSION; POVERTY; SAHEL; UN CONFERENCE ON POPULATION AND DEVELOPMENT; UN WORLD SUMMIT ON SOCIAL DEVELOPMENT.)

THORP   THERMAL OXIDE REPROCESSING PLANT.

threatened species   See ENDANGERED SPECIES.

Three-Gorges Water Conservation and Hydroelectric Project, China   A major development project in China planned to yield 40 000 MW of electricity, as well as help to control destructive floods and introduce new agricultural land. It involves the damming of the mountainous three gorges of Tsyuytan, U, and Silin, located in the middle course of the Changjiang (Yangtze) River, the longest river in both China and Asia. The total length of the reservoir created by the dammed section of the river will be 600 km; consequently, some 725 000 people need to be resettled. The major construction programme is expected to take 15 years. The issue illustrates the difficulties of balancing short- and long-term losses against prospective benefits. The Three Gorges are one of the 10 most famous scenic regions in China.

Three-mile Island nuclear incident   An incident in 1979 involving the release of radioactive material to the ATMOSPHERE from the Three-mile Island nuclear power station located some 16 km from Harrisburn, the capital of Pennsylvania, USA. A series of mechanical failures occurred in one of the reactors and a core meltdown followed; a massive hydrogen explosion was feared. There were no casualities from this incident but much damage to the reputation of the nuclear industry. (See CHERNOBYL NUCLEAR CATASTROPHE.)

threshold costs   The costs of overcoming a physical, quantitative, or structural limitation to growth, or a higher level of production or sales; the initial costs of entering a market, establishing a new town, or supplying electric power.

threshold limit values (TLVs) The concentrations of a substance to which reasonably healthy adult workers may be repeatedly exposed for eight hours a day over a 40-hour week without ill-effects.

threshold of hearing The pressure at which a sound source, in the absence of background noise, first becomes audible.

throughfall Rain water passing through a forest CANOPY.

tidal energy The use of the rise and fall of the tides to generate electrical energy. A significant commercial plant is a 240 MW power plant in the La Rance Estuary, France, which also functions as a dam and a bridge. Studies have indicated that changes in tidal height, periodicity, salinity, pollutant dispersion, and other factors could cause major alterations to the general ecology of an estuary or coastal habitat, with consequent effects on sea fisheries and wildlife in general.

tidal marsh Low flat marshlands traversed by interlaced channels and tidal sloughs while subject to tidal inundations; normally the only vegetation present consists of salt-tolerant bushes, grasses, and mangroves.

tides The periodic rise (flood) and fall (ebb) of the sea resulting from the water being drawn towards the sun and moon by their gravity.

tiering The coverage of general matters in a broader ENVIRONMENTAL IMPACT STATEMENT, such as a national programme or policy statement, with subsequent narrower statements of environmental analyses, such as regional or basinwide programme statements, or site-specific statements. The narrower EIS's would incorporate the general description by reference only and concentrate solely on the issues now arising.

tillage The unavoidable disturbance of SOIL in preparation for growing a crop. Tillage is the start of a sequence of events leading eventually to a crop being harvested. Tillage involves the ploughing, ripping, rolling, heavy discing or harrowing, and rotavation.

tilth An indication of the suitability of a SOIL for plant growth and cultivation. A soil having good tilth is composed of small crumbs of largely organic material that allows easy access of water, air, and plant roots. Generally, good tilth results from a soil initially having a well-developed structure. It may be improved with the application of organic matter such as COMPOST and animal manures.

timbre (1) The quality or characteristic of musical tones. (2) The characteristic tone or quality of a musical instrument or voice.

time-of-day pricing See PEAK-LOAD PRICING.

Times Beach An example of the consequences of environmentally hazardous chemicals, in this case DIOXIN. In February 1983, the US Government offered to buy out all 2400 residents of Times Beach, Missouri, after confirming that the town was too contaminated with dioxin to be safe for habitation; the dioxin level has been found to be 100 times what was considered safe. Times Beach was one of at least 100 communities in Missouri where dioxin mixed with waste was sprayed on dirt roads in the 1970s to keep down dust. Severe flooding had exposed the inhabitants to very high dioxin levels.

TLVs THRESHOLD LIMIT VALUES.

tone In ACOUSTICS, sound that can be recognized by its regularity of vibration. A simple tone has only one frequency; a complex tone consists of two or more simple tones, called partial tones. A combination of harmonic tones is pleasant to hear, being known as a musical tone.

Töpfer Law Named after Klaus Töpfer, Germany's environment minister, a law requiring waste collection to increase in stages until 80% of packaging waste was to be collected by 1995. Of the waste collected, 90% of the glass and metal was recycled, and 80% of the plastic and paper. Manufacturers and distributors have to take back their own packaging, or join Duales System Deutschland (DSD), an organization set up by German industry. Members of DSD have the right to put a green dot on their packaging; any dotted pack can be left with retailers in DSD bins.

topsoil That part of the SOIL profile, or top horizon, containing material that is usually more fertile and better structured than underlying layers. Topsoil is the most important part of the soil with respect to the growth of crops and pastures, and its loss or degradation represents the most serious aspect of soil erosion. The retention of topsoil is particularly important in the revegetation of exposed batters or earthworks.

Toronto Conference on the Changing Atmosphere Convened in June 1988, a conference of scientists and policy makers from 48 countries held in Toronto, Canada, which gave much emphasis to the seriousness of the GREENHOUSE EFFECT.

Torrey Canyon **incident** A major spill that occurred in March 1967 when the tanker, the *Torrey Canyon* went aground on the Seven Stones reef about 24 km northeast of the Scilly Isles, near Land's End, Britain. The ship was bound for Milford Haven with a cargo of 120 000 tonnes of Kuwait crude oil when wrecked. Much of the oil was carried south to the coast of France; the French Government acted promptly to safeguard the beaches and oyster fisheries of Brittany.

total catchment management (TCM) Catchment management central to maintaining water quality; it involves the control of any activities that contribute to point or diffuse sources of pollution, including human habitation, agriculture, industry, and mining. The clearing of vegetation should also be carefully controlled as this can result in soil erosion and increased water SALINITY. (See CATCHMENT.)

total dissolved solids (TDS) The residue of solids remaining after evaporating a sample of water or effluent, expressed in mg/l. Water with less than 500 mg/l TDS is of good quality; as TDS exceeding 1000 mg/l is likely to cause excess scaling, corrosion, and unacceptable taste.

total energy system A system that meets all the energy requirements of an industrial site by on-site generation of electricity, waste-heat recovery from the prime movers, and the provision of supplementary heat as required. The term is also applied more generally to combined heat and power systems, where on-site generation and heat recovery optimize the use of fuel within the site, usually to meet the bulk of energy requirements.

total fertility rate (TFR) An estimate of the number of live children the average woman will produce through all her child-bearing years (taken as ages between 15 to 45).

total quality management (TQM) An effective and efficient way of managing a water supply system; it involves establishing a regime whereby each step of system management and performance assessment is reliably carried out, thus ensuring good-quality water. (See Box 34.)

town planning See ENVIRONMENTAL PLANNING.

townscape The complex of built and unbuilt spaces that comprise the urban landscape or urban environment; the townscape is the visible expression of the collective activities and attainments of the inhabitants, a reflection of an enormous array of social forces.

toxicity A physiological or biological property that enables a chemical to do harm, or create injury, to a living organism by other than mechanical means; the ability of a chemical to cause poisoning when the chemical is administered to a living organism in an appropriate form and manner. Some chemicals have a high-toxicity potential, while some have a low-toxicity potential.

toxic metals See HEAVY METALS.

toxicology The study of poisons and their effects, particularly on living systems.

Toxic Release Inventory, US (TRI, US) A US federal inventory of toxic and hazardous releases to the environment, requiring annual reports from all major manufacturing industries; the inventory was introduced in 1986 under the US Emergency Planning

---

Box 34 Total quality management of a water supply system: essential components

(1) A regime in which all steps in system management and performance assessment will be reliably carried out.
(2) An agreed standard of service.
(3) Effective treatment processes, including disinfection such as chlorination.
(4) Regular inspection and maintenance of the system.
(5) Routine practices that quickly identify external sources of contamination.
(6) Monitoring programmes to assess water quality throughout the system and identify quickly any water-quality problem within the system.
(7) Validation procedures for sampling and testing programmes.
(8) The use of monitoring data to facilitate day-to-day management and assess performance over time.
(9) Procedures for immediate rectification of water contamination problems.
(10) Procedures to address longer-term supply problems.
(11) Clear lines of responsibility for remedial steps.
(12) Attention to the skill and training of personnel.
(13) Environmental auditing.
(14) Regular reports to consumers.
(15) Arrangements for discussions with the public on proposed changes and improvements to the system, particularly those changes involving development applications and environmental impact statements.

toxic wastes and Community Right-to-know Act. Releases include pollutant discharges to air, water, and by underground injection as well as transfers of wastes to waste contractors or to sewers. The data is collated on a geographical basis by the US EPA and made generally available in computerized form as the Toxic Release Inventory (TRI) for the United States. The purpose of the TRI is not only to inform the public but to aid the development of appropriate regulations, guidelines, and standards. The US EPA has released a series of manuals on recognized techniques for estimating the releases. The US TRI attracts about 85 000 reports on releases of listed toxic materials each year from about 24 000 facilities filing reports. The EPA has targeted some 17 chemicals out of this inventory for a special reduction programme. The TRI is undoubtedly one of the most successful and cost-effective tools among those available in the anti-pollution campaign in the USA. (See POLLUTANT RELEASE INVENTORY, CANADA.)

toxic wastes See CONVENTION ON THE CONTROL OF TRANSBOUNDARY MOVEMENT OF HAZARDOUS WASTES AND THEIR DISPOSAL.

TQM TOTAL QUALITY MANAGEMENT.

tradable emission rights See BUBBLE CONCEPT.

trade wastes Wastes of organic and inorganic origin discharged by industrial and commercial enterprises. Organic wastes are discharged on a considerable scale by the food industries: canneries, dairies, breweries, abattoirs, and fishmeat factories. Other contributors include paper mills, tanneries, petrochemical works, textile manufacturers, and laundries. Inorganic wastes include acids, alkalis, cyanides, sulphides, and the salts of arsenic, lead, copper, chromium, and zinc.

traditional inhabitant The 'original' inhabitants of an area usually before the arrival of Europeans or other foreign invaders; such inhabitants often display ancient cultures, languages, and art forms.

traffic impact study Often an important component of an ENVIRONMENTAL IMPACT STATEMENT, outlining and analysing the traffic implications of a proposal; the number of types and movements of the vehicles involved; the effects on the local and regional road systems; and the environmental implications.

traffic segregation The separation of different types of traffic, especially pedestrians from vehicles, by arranging roads and pedestrian ways that never cross, or by physical separation using overpasses, underpasses, and SKYWAYS AND SKYWALKS.

Tragedy of the Commons See COMMONS, TRAGEDY OF THE.

tramp material Undesirable items in a material being processed that are detrimental to the processing material or to the final product.

transfer facility A facility operated for the purpose of transferring refuse from collection trucks and other vehicles to larger-capacity trucks; the larger trucks transport the refuse to a disposal site. The purpose of this arrangement is to assist in minimizing overall collection and disposal costs per ton of refuse handled.

translocated herbicide A herbicide which, when absorbed into a plant via leaves or roots, moves within the plant, finally destroying it.

translocation The movement of soluble foods through the tissues of higher plants; for example, from leaves to actively growing parts and storage organs. Translocation is the botanical counterpart of circulatory systems in humans and animals.

transpiration The diffusion of water from a leaf; most of the loss occurs by the diffusion of water molecules through open STOMATA. The loss of moisture by a leaf must be balanced by an equivalent uptake of water from the roots, if the plant is not to dry out.

transportation planning Generally, planning aimed at: reducing the time and energy involved in the movement of people and goods; enhancing access to public facilities and places of employment; reducing traffic accidents; alleviating congestion; reducing the movement of heavy vehicles through city, town, and village centres; promoting pedestrian safety; and reducing environmental adverse effects, such as noise, air pollution, advertising, and litter. This is coupled with landscaping and buffer zones and the appropriate promotion of public transport.

tributyl tin (TBT) An additive used in antifouling marine paints; it has allegedly caused widespread damage to marine life including oysters, prawns, and crabs. It has been banned in several European countries, including Britain.

trickling filter (biological filter) An aerobic process used in SECONDARY TREATMENT plants for the processing of sewage. Trickling filters consist of beds of coarse material (say crushed stone) over which the sewage is sprinkled at a uniform rate. They are generally not less than 2 m thick and are circular in outline.

TRI, US See TOXIC RELEASE INVENTORY, US.

trophic Pertaining to nutrition.

trophic accumulation The passage of a substance through a FOOD CHAIN, accompanied by an increase of concentration in each organism in the food chain.

trophic level A particular step occupied by a population in the process of energy transfer within an ECOSYSTEM; the position of an organism in the FOOD CHAIN.

Tropical Timber Agreement See INTERNATIONAL TROPICAL TIMBER AGREEMENT.

tropopause The boundary between the TROPOSPHERE and the STRATOSPHERE.

troposphere The lowest layer of the ATMOSPHERE, bounded by the Earth beneath and the STRATOSPHERE above; the upper boundary is the TROPOPAUSE, about 10 to 13 km above the Earth's surface. The troposphere is marked by decreasing temperature with height; most of the clouds and weather systems are within the troposphere.

truck routeing The confining of heavy trucks and road tankers to a particular designated road system; separate haul roads are sometimes constructed for mines and quarries.

tundra Treeless, level, or rolling ground in polar regions (arctic tundra) or on high mountains (alpine tundra) characterized by bare ground and rock or by vegetation such as lichens, mosses, low shrubs, and small herbs.

tunnelling A form of erosion related especially to one kind of SOIL; the erosion goes on below the surface and may not show until the surface collapses into the tunnel. An entire hillside may become undermined.

turbidity Visible pollution due to suspended material in water causing a reduction in the transmission of light. (See SECCHI DISC.)

turbulence The random movements of the air which are superimposed upon the mean wind speed. Individual movement is called a turbulent eddy, which may have almost any size and may move in any direction at any speed.

TVA TENNESSEE VALLEY AUTHORITY.

typhoid An acute infectious disease or fever of humans caused by the bacterium *Salmonella typhi*; the bacterium usually enters the body through the mouth by the consumption of contaminated food or water. Most major epidemics of typhoid fever have been caused by pollution of the public water supply. Foodstuffs may be contaminated by a carrier working in the food industry or by flies. Shellfish, such as oysters, may become contaminated by polluted water. The prevention of typhoid fever depends primarily on adequate sewage treatment, the filtration and chlorination of water, the exclusion of carriers from employment in food industries and restaurants, and measures against fly breeding.

typhus An infectious disease of humans transmitted by insect carriers, such as lice and fleas. Epidemic typhus is transmitted by *Rickettsia prowazekii*, conveyed from person to person by the body louse *Pediculus humanus*. Epidemic typhus has been one of the great scourages of humanity. It has been associated with people crowded together in filth, poverty, and hunger. Despite techniques such as vaccination and delousing, typhus remains an ever-present threat in refugee camps and crowded impoverished cities.

U

Ujung-Kulon National Park Set aside as a nature reserve in 1921, becoming a national park in 1980, a remote area of low hills and plateaus, with lagoons and coastal dunes, on the island of Java, Indonesia. The park faces the Sunda Strait at the western tip of Java. The park contains the last remaining low-relief forest on Java. Animals found there include the Javan tiger, the one-horned rhinoceros, banteng, gibbon, leaf-monkey, crocodile, sea turtle, green peafowl, barking deer, and mouse deer.

ultisol A weathered, frequently reddish, acidic SOIL type of humid areas, in middle to low latitudes. A typical profile consists of an acidic clay horizon, a dark surface horizon rich in HUMUS, and a leached horizon. Used with organic fertilizers, the ultisols rank among the world's most productive soils.

ultrafiltration A liquid filtration process involving the use of membranes with pore sizes below 0.02 μ; it can be used to separate materials on the basis of their molecular weight.

ultraviolet-B radiation (UV-B radiation Radiation) invisible to the human eye that is higher in energy than visible light; too much exposure can cause human skin cancer.

Uluru (Ayers Rock-Mount Olga) National Park Established in 1977, in the Northern Territory, Australia, the park is famous for Ayers Rock and the Olgas, while the natural environment of the park contains representative ecosystems of the arid environment in central Australia. In 1988, the park was included in the WORLD HERITAGE LIST.

UMPLIS See INFORMATION AND DOCUMENTATION SYSTEM FOR ENVIRONMENTAL PLANNING.

unconfined vapour cloud explosion (UVCE) The explosion of a flammable mixture of vapour and air in the open air, as distinct from a confined explosion which occurs within a closed system, such as a vessel or building. The risks may be reduced by minimizing hazardous inventories, implementing good design codes and quality construction, proper testing, careful operation and maintenance, remotely operated isolation valves, and adequate separation distances. (See BOILING LIQUID EXPANDING VAPOUR EXPLOSION.)

underground water Water other than SURFACE WATER, being drawn from relatively shallow wells or boreholes, and from deep artesian bores. (See GROUND WATER.)

UN Capital Development Fund (UNCDF) A UN organization created in 1966; it aids the 30 least developed countries by means of grants and loans. Assistance applies to agriculture, agro-industry, drinking-water supply, health and nutrition, low-income housing, roads, and rural schools. The fund is governed by an executive board.

UNCDF UN CAPITAL DEVELOPMENT FUND.
UNCOD UN CONFERENCE ON DESERTIFICATION.
UNCLOS UN CONFERENCE ON THE LAW OF THE SEA.

UN Commission on Sustainable Development See COMMISION FOR SUSTAINABLE DEVELOPMENT.

UN Conference on Desertification (UNCOD) Held in Nairobi, Kenya, in 1977 the first worldwide effort to consider the problem of advancing deserts. The outcome of the conference was a World Wide Plan or Act to Combat Desertification, with 26 specific recommendations. A fundamental problem has been that governments with desertification problems have limited financial and human resources; insufficient financing has plagued the programme. (See China's GREEN GREAT WALL.)

UN Conference on Environment and Development An international conference with representatives from some 167 countries held in Rio de Janeiro, Brazil, in 1992. The purpose of this Third-World environment conference was to review progress since the earlier conferences in 1972 and 1982 in safeguarding the human environment and promoting human welfare. The result was: The Rio Declaration on Environment and Development; the adoption of AGENDA 21, representing a programme for the 21st century; the creation of a COMMISSION FOR SUSTAINABLE DEVELOPMENT; the adoption of a CONVENTION ON PROTECTING SPECIES AND HABITATS (the convention on biological diversity); and the adoption of a framework CONVENTION ON CLIMATE CHANGE. (See Box 35.)

UN Conference on Human Settlements (Habitat) An international conference held in Vancouver, Canada, in 1976 at which 131 governments were represented. The conference was concerned with the urgent problems of housing shortages, crises of urban and rural communities, the proper use of land, access to essential services such as clean and safe water, and public involvement in remedial action to improve the living conditions of people throughout the world. The conference produced a Vancouver Declaration on Human Settlements and a Vancouver Plan of Action, a set of 64 recommendations suggesting concrete ways in which people might be assured of the basic requirements of human habitation. The UN responded

Box 35 The Rio Declaration on Environment and Development

July 16, 1992

ADOPTION OF AGREEMENTS ON ENVIRONMENT AND DEVELOPMENT

Preamble

The United Nations Conference on Environment and Development:

- having met at Rio de Janeiro from 3 to 14 June 1992,
- reaffirming the Declaration of the United Nations, Conference on the Human Environment, adopted at Stockholm on 16 June 1972, and seeking to build upon it,
- with the goal of establishing a new and equitable global partnership through the creation of new levels of cooperation among States, key sectors of societies and people,
- working towards international agreements which respect the interests of all and protect the integrity of the global environmental and developmental system,
- recognizing the integral and interdependent nature of the Earth, our home,

Proclaims that:

Principle 1

Human beings are at the centre of concerns for sustainable development. They are entitled to a healthy and productive life in harmony with nature.

Principle 2

States have, in accordance with the Charter of the United Nations and the principles of international law, the sovereign right to exploit their own resources pursuant to their own environmental and developmental policies, and the responsibility to ensure that activities within their jurisdiction or control do not cause damage to the environment of other States or of areas beyond the limits of national jurisdiction.

Principle 3

The right to development must be fulfilled so as to equitably meet developmental and environmental needs of present and future generations.

Principle 4

In order to achieve sustainable development, environmental protection shall constitute an integral part of the development process and cannot be considered in isolation from it.

Principle 5

All States and all people shall cooperate in the essential task of eradicating poverty as an indispensable requirement for sustainable development, in order to decrease the disparities in standards of living and better meet the needs of the majority of the people of the world.

Principle 6

The special situation and needs of developing countries, particularly the least developed and those most environmentally vulnerable, shall be given special priority. International actions in the field of environment and development should also address the interests and needs of all countries.

continued overleaf

UN Conference on Human Settlements

> Box 35 *(continued)*
>
> ### Principle 7
>
> States shall cooperate in a spirit of global partnership to conserve, protect and restore the health and integrity of the Earth's ecosystem. In view of the different contributions to global environmental degradation, States have common but differentiated responsibilities. The developed countries acknowledge the responsibility that they bear in the international pursuit of sustainable development in view of the pressures their societies place on the global environment and of the technologies and financial resources they command.
>
> ### Principle 8
>
> To achieve sustainable development and a higher quality of life for all people, States should reduce and eliminate unsustainable patterns of production and consumption and promote appropriate demographic policies.
>
> ### Principle 9
>
> States should cooperate to strengthen endogenous capacity-building for sustainable development by improving scientific understanding through exchanges of scientific and technological knowledge, and by enhancing the development, adaptation, diffusion and transfer of technologies, including new and innovative technologies.
>
> ### Principle 10
>
> Environmental issues are best handled with the participation of all concerned citizens, at the relevant level. At the national level, each individual shall have appropriate access to information concerning the environment that is held by public authorities, including information on hazardous materials and activities in their communities, and the opportunity to participate in decision-making processes. States shall facilitate and encourage public awareness and participation by making information widely available. Effective access to judicial and administrative proceedings, including redress and remedy, shall be provided.
>
> ### Principle 11
>
> States shall enact effective environmental legislation. Environmental standards, management objectives and priorities should reflect the environmental and developmental context to which they apply. Standards applied by some countries may be inappropriate and of unwarranted economic and social cost to other countries, in particular developing countries.
>
> ### Principle 12
>
> States should cooperate to promote a supportive and open international economic system that would lead to economic growth and sustainable development in all countries, to better address the problems of environmental degradation. Trade policy measures for environmental purposes should not constitute a means of arbitrary or unjustifiable discrimination or a disguised restriction on international trade. Unilateral actions to deal with environmental challenges outside the jurisdiction of the importing country should be avoided. Environmental measures addressing transboundary or global environmental problems should, as far as possible, be based on an international consensus.
>
> ### Principle 13
>
> States shall develop national law regarding liability and compensation for the victims of pollution and other environmental damage. States shall also cooperate in an expeditious
>
> *continued overleaf*

Box 35 (continued)

and more determined manner to develop further international law regarding liability and compensation for adverse effects of environmental damage caused by activities within their jurisdiction or control to areas beyond their jurisdiction.

Principle 14

States should effectively cooperate to discourage or prevent the relocation and transfer to other States of any activities and substances that cause severe environmental degradation or are found to be harmful to human health.

Principle 15

In order to protect the environment, the precautionary approach shall be widely applied by States according to their capabilities. Where there are threats of serious or irreversible damage, lack of full scientific certainty shall not be used as a reason for postponing cost-effective measures to prevent environmental degradation.

Principle 16

National authorities should endeavour to promote the internalization of environmental costs and the use of economic instruments, taking into account the approach that the polluter should, in principle, bear the cost of pollution, with due regard to the public interest and without distorting international trade and investment.

Principle 17

Environmental impact assessment, as a national instrument, shall be undertaken for proposed activities that are likely to have a significant adverse impact on the environment and are subject to a decision of a competent national authority.

Principle 18

States shall immediately notify other States of any natural disasters or other emergencies that are likely to produce sudden harmful effects on the environment of those States. Every effort shall be made by the international community to help States so afflicted.

Principle 19

States shall provide prior and timely notification and relevant information to potentially affected States on activities that may have a significant adverse transboundary environmental effect and shall consult with those States at an early stage and in good faith.

Principle 20

Women have a vital role in environmental management and development. Their full participation is therefore essential to achieve sustainable development.

Principle 21

The creativity, ideals and courage of the youth of the world should be mobilized to forge a global partnership in order to achieve sustainable development and ensure a better future for all.

Principle 22

Indigenous people and their communities, and other local communities, have a vital role in environmental management and development because of their knowledge and traditional

continued overleaf

UN Conference on Nutrition

Box 35 (continued)

practices. States should recognize and duly support their identity, culture and interests and enable their effective participation in the achievement of sustainable development.

Principle 23

The environment and natural resources of people under oppression, domination and occupation shall be protected.

Principle 24

Warfare is inherently destructive of sustainable development. States shall therefore respect international law providing protection for the environment in times of armed conflict and cooperate in its further development, as necessary.

Principle 25

Peace, development and environmental protection are interdependent and indivisible.

Principle 26

States shall resolve all their environmental disputes peacefully and by appropriate means in accordance with the Charter of the United Nations.

Principle 27

States and people shall cooperate in good faith and in a spirit of partnership in the fulfilment of the principles embodied in this Declaration and in the further development of international law in the field of sustainable development.

---

by creating a Commission on Human Settlements, and a Centre for Human Settlements in Nairobi, Kenya. A second UN Conference on Human Settlements (Habitat II) was held in Istanbul from 3-14 June, 1996, to review progress and set priorities for the next decade.

UN Conference on Nutrition An international conference held in Rome in 1992 for the purpose of drawing up a global agenda for nutrition policy. Attended by delegates from over 150 countries, the conference attracted the attendance of consumer and other public interest groups, working to ensure that future UN policy would provide adequate safety, quality, and security for consumers throughout the world.

UN Conference on Population and Development An international conference with representatives from some 150 countries held in Cairo, Egypt, in 1994. The conference sought means for modifying the growth rate of the world's population which had reached 5.7 billion people and might well reach 7.9 billion to 12 billion by the year 2050. The conference promoted the provision of contraceptive and family planning services, sex education, and safe abortion, and sought a definite improvement in the status of women throughout the world. The conference followed two previous population conferences: one in Bucharest in 1974 and the second in Mexico City in 1984. (See UN WORLD POPULATION CONFERENCE.)

UN Conference on the Human Environment An international conference with representatives from some 113 governments and agencies held in Stockholm, Sweden in 1972. The purpose of this First-World environment conference was to encourage action and provide guidelines designed to protect and improve the human environment, by means of international cooperation. The conference achieved: a declaration on the human environment; agreement upon an extensive programme of international action; the establishment of a permanent environment secretariat in Nairobi, Kenya; and the creation of an environment fund. A second environment conference was held in Nairobi in 1982, to review progress. A third conference was held in Rio de Janeiro, Brazil, in 1992. (See UN CONFERENCE ON ENVIRONMENT AND DEVELOPMENT.)

UN Conference on the Law of the Sea (UNCLOS) International discussions between 1974 and 1982 aimed at establishing a revised legal regime for the oceans and their resources, while maintaining the rights of ships to free passage. The discussions embraced the concepts of an exclusive 200-mile economic zone, wherever practicable, and an international authority to administer the rules regarding the management and conservation of mineral and food supplies as yet unexploited, and the prevention of pollution and over-exploitation. The conference ended with a CONVENTION ON THE LAW OF THE SEA (the Montego Bay convention).

UN Conference on Trade and Development (UNCTAD)   A UN standing conference set up in 1964 to assist the less-developed nations towards improved rates of economic growth. UNCTAD endeavours to do this by arranging aid and finance, and by promoting trade. UNCTAD collaborates with the UN ENVIRONMENT PROGRAM (UNEP) in providing governments with information on the impact of environmental policies and measures on international trade and development. A Trade and Development Board is responsible for UNCTAD's functions.

UNCTAD   UN CONFERENCE ON TRADE AND DEVELOPMENT.

UN Department on Policy Coordination and Sustainable Development   A UN body created following the UN CONFERENCE ON ENVIRONMENT AND DEVELOPMENT 1992. Among other responsibilities, it supports the work of the COMMISSION FOR SUSTAINABLE DEVELOPMENT.

UN Development Program (UNDP)   A UN agency formed in 1965 to help build more productive societies and economies in low-income nations by helping them to develop their natural resources and human capabilities.

UNDP   UN DEVELOPMENT PROGRAM.

UN Educational, Scientific, and Cultural Organization (UNESCO)   A UN agency which came into being in 1946 to promote international collaboration in education, science, and culture. A major activity has been the 'Man and Biosphere' programme which was launched in 1970; the aim has been to develop the scientific basis for the rational use and conservation of the resources of the BIOSPHERE, and for the improvement of the global relationship between humankind and the ENVIRONMENT. The Belgrade International Workshop on Environmental Education was convened in 1975 under the UNESCO-UNEP programme; the outcome was the BELGRADE CHARTER. The Intergovernmental Oceanographic Commission (IOC) with UNESCO was created in 1962 to promote a better understanding of the nature and resources of the oceans; the preservation of the marine environment has been an important element in this programme. In 1972, the CONVENTION FOR THE PROTECTION OF THE WORLD CULTURAL AND NATURAL HERITAGE (the Paris convention) was adopted by UNESCO members. The convention established a World Heritage Committee and a WORLD HERITAGE LIST.

UN Environment Program (UNEP)   Created by the UN CONFERENCE ON THE HUMAN ENVIRONMENT, 1972, UNEP has been charged with implementing its recommendations and those of subsequent environmental conferences. Based in Nairobi, Kenya, UNEP is subject to a governing council; its activities are supported by an environment fund, to which nations contribute. UNEP's environmental assessment programme, Earthwatch, has four closely linked components: evaluation and review, research, monitoring, and exchange of information. Earthwatch contains such important and operational elements as the GLOBAL ENVIRONMENTAL MONITORING SYSTEM (GEMS), the INTERNATIONAL REFERRAL SYSTEM (INFOTERRA), and the INTERNATIONAL REGISTER OF POTENTIALLY TOXIC CHEMICALS (IRTC). There has been much successful activity in the area of regional seas, resulting in several conventions; for example, the Barcelona CONVENTION FOR THE PROTECTION OF THE MEDITERRANEAN SEA AGAINST POLLUTION. UNEP has also promoted the Vienna CONVENTION FOR THE PROTECTION OF THE OZONE LAYER, and the Basel CONVENTION ON THE CONTROL OF TRANSBOUNDARY MOVEMENT OF HAZARDOUS WASTES AND THEIR DISPOSAL. UNEP also provided support for two conventions on wildlife protection: the CONVENTION ON INTERNATIONAL TRADE IN ENDANGERED SPECIES OF WILD FAUNA AND FLORA (CITES) and the CONVENTION ON THE CONSERVATION OF MIGRATORY SPECIES OF WILD ANIMALS (CMS). In 1978, UNEP established an EIA division; the Governing Council of UNEP adopted the goals and principles of EIA in 1987, for adoption throughout the world. In 1982, the Governing Council adopted the Montevideo Program for the Development and Periodic Review of Environmental Law which has been a mainspring behind the development of more recent conventions and agreements. (See EUROPEAN UNION; UN CONFERENCE ON ENVIRONMENT AND DEVELOPMENT.)

UNEP   UN ENVIRONMENT PROGRAM.

UNESCO   UN EDUCATIONAL, SCIENTIFIC, AND CULTURAL ORGANIZATION.

UN Food and Agriculture Organization (FAO)   Established in 1945, a UN agency concerned with investment in agriculture and emergency food supplies. It is based in Rome, Italy.

UN Industrial Development Organization (UNIDO)   Established in 1967, a UN agency that aims to assist in the industrialization of the developing countries by coordinating other UN organizations devoted to this end. Its activities help to formulate industrial development policies and programmes, mostly in terms of information, education, and research.

UNICEF   UN INTERNATIONAL CHILDREN'S EMERGENCY FUND.

UNIDO UN INDUSTRIAL DEVELOPMENT ORGANIZATION.

UN International Children's Emergency Fund (UNICEF) Created in 1946, the UN International Children's Emergency Fund is devoted to assisting national efforts to improve the health, nutrition, education, and general welfare of children, particularly in less-developed countries. Other activities include the development of health services and the training of health personnel, construction of educational facilities, and teacher training. These activities are financed by government and private voluntary contributions.

UN Research Institute for Social Development (UNRISD) The only UN organization engaged exclusively in research on social development. The Institute's work is concerned primarily with the problems of poverty, inequality, and discrimination, while examining ways in which people's lives can be improved. Its research programme seeks to contribute to a deeper understanding of the social dimensions of environmental change.

UNRISD UN RESEARCH INSTITUTE FOR SOCIAL DEVELOPMENT.

UN Water Conference (UNWC) A UN conference held in Mar del Plata, Argentina, in 1977 that attempted to focus the attention of policy makers on the water needs likely to arise up to the year 2000, the steps that could be taken to meet them, and the difficulties likely to be experienced by those who failed to make adequate provision. The conference noted that less than one-fifth of the world's population can get water simply by turning a tap; for the remaining four-fifths, the getting of water is part of the daily struggle for existence. The conference urged better water management at local, regional, and national levels. It was noted that Israel probably led the world in the efficient use of scarce water. In 1992, the UN Conference on Water and the Environment was held in Dublin, Ireland, to review progress. It was announced there that a EU ecological water-quality directive would call for the quality of all streams, rivers, and lakes to be restored to the condition they should have been in had they been uninfluenced by human activities.

UNWC UN WATER CONFERENCE.

UN World Conference on Women A fourth world conference on women hosted by the UN, held in Beijing, China, in September, 1995. The Conference dealt with a whole range of issues affecting females, such as the marginalization of women in society generally, the abuse of women and girls at home and work, genital mutilation, prostitution, particularly forced prostitution, religious discrimination against women, female infanticide, women and children as casualities of war, the torture and ill-treatment of women, the promotion of women's rights, the education of women as individuals and not merely as servants, and women as a source of bigotry, discrimination, and intolerance.

UN World Population Conference An international conference with representatives from some 136 governments held in Bucharest, Romania, in 1974. The purpose of this first world population conference was to consider population policies and programmes needed to promote human welfare and development. The principal achievement of the conference was a World Population Plan of Action and a number of related resolutions on such matters as the status of women, the environment, and food production. Delegates from 132 countries attended a second World Population Conference held in Mexico City in 1984. A third conference was held in Cairo, Egypt, in 1994. (See UN CONFERENCE ON POPULATION AND DEVELOPMENT.)

UN World Summit for Social Development A UN conference held in Copenhagen, Denmark, in March 1995, essentially a follow-up conference to the UN World Summit for Children held in 1990. The achievements between the conferences have been outlined in the *State of the World's Children* (UNICEF 1995): malnutrition has been reduced, polio has been eradicated in much of the world, deaths from measles have been more than halved, and immunization levels have been maintained or increased; there has also been progress in preventing vitamin A and iodine deficiencies and in providing safe water. As a result of these improvements, 2.5 million fewer children will die in 1996 than in 1990; and at least 750 000 fewer children will be disabled, blinded, crippled, or mentally retarded. But there was little progress in South Asia (India, Bangladesh, and Pakistan), in sub-Saharan Africa, and in Indo-China. where there was a lack of resources to make improvements. Diarrhoea still kills about three million children a year, with pneumonia the biggest single killer of children. The 1995 conference continued to promote immunization, provision of vitamins and iodized salt, and rehydration for those affected by dysentery.

upwelling The process by which water rises from a deeper to a shallower depth, usually because of the divergence of offshore currents. Upwelling provides nutrient enrichment, fostering fish populations.

urban and regional planning That part of resource management and ENVIRONMENTAL PLANNING that is concerned with spatial ordering in urban and regional environments.

urban areas Areas of land or localities where the principal land uses are residential, industrial, business, and commercial, with public

infrastructure, special uses, and related open spaces, within which a comprehensive range of public utility services and educational and community facilities are available to most of the population.

urban consolidation A policy and programme intended to increase the density of dwellings or population or both in URBAN AREAS where services and facilities are already established. A variety of land-use measures and housing initiatives may be used.

urban ecosystems Towns and cities viewed as ecosystems, having an input of matter and energy, recycling within the system, and an output of matter and energy into the surroundings.

urban renewal The renovation or redevelopment of the decaying central areas of cities by the demolition or up-grading of existing dwellings and buildings, reduction of street congestion, and a general improvement in environmental conditions. (See STREET IMPROVEMENT SCHEMES.)

urban sprawl The outward growth of an urban area, where new suburbs are continually being developed on the edges of existing towns and cities. This may result in the loss of good agricultural land and natural environment. (See URBAN CONSOLIDATION.)

urbanization A process leading to societal change, characterized by the movement of people from rural to urban areas; a process that is intensifying throughout the world.

US Agency for International Development See AGENCY FOR INTERNATIONAL DEVELOPMENT, US.

US Bureau of Land Management See BUREAU OF LAND MANAGEMENT, US.

US Council on Environmental Quality See COUNCIL ON ENVIRONMENTAL QUALITY, US.

US Environmental Monitoring and Assessment Program See ENVIRONMENTAL MONITORING AND ASSESSMENT PROGRAM, US.

US Environmental Protection Agency See ENVIRONMENTAL PROTECTION AGENCY, US.

user pays Charging the consumers of services the whole marginal cost of providing that service, both capital and running. Each consumer theoretically pays the difference in the costs to the system incurred with and without that consumer. The concept has considerable merit in relation to the main body of consumers in ensuring a rational and efficient use of resources in respect of any service, for an undue disparity between costs and prices will lead to a distortion in demand and, if prices are too low, an undue allocation of community resources through an inflated demand. On the other hand, the concept may at times conflict with certain social objectives, for example, that all homes should, if at all practicable, be connected to an electricity supply, to sewerage, to a telephone service and be close to public transport. The provision of universal services in all or most locations, with high marginal costs beyond the reach of consumers must conflict with the user pays principle. The resolution of these conflicting principles is a matter for public policy.

US evolution of environmental policy See Box 36.

Box 36 US: evolution of environmental policy

1872	Creation of Yellowstone National Park
1890	Creation of Yosemite National Park
1899	Refuse Act
1908	State of the Union Message by President Theodore Roosevelt on the need for the conservation of natural resources
1935	Historic Sites, Buildings and Antiquities Act
1940	Bald and Golden Eagle Protection Act
1946	Start of Pittsburgh clean-up
1947	Los Angeles anti-smog programme launched; Federal Insecticide, Fungicide and Rodenticide Act
1948	Federal Water Pollution Control Act
1956	Water Pollution Control Act re-enacted as a permanent measure. Fish and Wildlife Act
1962	*Silent Spring* published by Rachel Carson; White House conservation conference
1963	Federal Clean Air Act
1964	Wilderness Act; national wilderness preservation system

continued overleaf

US evolution of environmental policy

Box 36 *(continued)*

1965	Federal Water Quality Act; Solid Waste Disposal Act; Further Clean Air Act; Anadromous Fish Conservation Act
1966	Environmental impact assessment policy established; Clean Water Restoration Act; National Historic Preservation Act
1967	Federal Air Quality Act
1968	National Trails System Act; Wild and Scenic Rivers Act
1969	National Environmental Quality Act; Council on Environmental Policy appointed
1970	US Environmental Protection Agency (EPA) created. Environmental Quality Improvement Act; Water Quality Improvement Act; Clean Air Amendment Act; Mining and Minerals Policy Act; President's Message on the Environment
1972	Noise Control Act; Coastal Zone Management Act; Clean Water Act; Marine Mammal Protection Act
1973	Endangered Species Act
1974	Safe Drinking Water Act; Solar Energy Research, Development and Demonstration Act
1975	Energy Policy and Conservation Act
1976	Toxic Substances Control Act; Resource Conservation and Recovery Act; Magnusan Fishery Conservation and Management Act; National Forest Management Act; Federal Land Policy and Management Act
1977	Environmental impact procedures strengthened; Clean Water Act; Clean Air Amendment Act; Surface Mining Control and Reclamation Act; Soil and Water Resources Conservation Act; President's environmental message
1978	Environmental impact assessment regulations promulgated; Renewable Resources Extension Act; Public Rangelands Improvement Act; Surface Mining Control and Reclamation Act; Cooperative Forestry Assistance Act; National Energy Conservation Policy Act; Solar Photovoltaic Research, Development and Demonstration Act; Uranium Mill Tailings Radiation Control Act; National Ocean Pollution Planning Act
1979	Introduction by the US Environmental Protection Agency of the 'bubble concept' for the management of pollution
1980	Alaska National Interest Lands Conservation Act; Comprehensive Environmental Response, Compensation and Liability Act (Superfund); Wind Energy Systems Act; Low-level Radioactive Waste Policy Act; Act to Prevent Pollution from Ships
1982	Coastal Barrier Resources Act; Reclamation Reform Act; asbestos-in-schools rule; Nuclear Waste Policy Act
1983	Times Beach found too contaminated with dioxins for human habitation
1985	International Security and Development Act
1986	Emergency Wetlands Resources Act; Right-to-Know Act
1987	Water Quality Act; Driftnet Impact Monitoring, Assessment and Control Act
1988	Ocean Dumping Ban Act
1989	First nationwide survey of more than 320 toxic chemicals released to air by industry; tanker *Exxon Valdez* goes aground in Alaska; North American Wetlands Conservation Act; Marine Pollution and Research and Control Act
1990	Clean Air Act to reduce substantially air emissions; California Air Resources Board introduces strictest vehicle-emission controls ever; Coastal Wetlands Planning, Protection and Restoration Act; Coastal Barrier Improvement Act; Oil Pollution Act; Food Security Act; Pollution Prevention Act; Antarctic Protection Act; Global Change Research Act; Food, Agriculture, Conservation and Trade Act; National Environmental Education Act
1991	US signs UN ECE convention on environmental impact assessment in a transboundary context
1992	US Congress requires the analysis of the environmental effects of major US federal activities abroad; National Geologic Mapping Act
1993	California Desert Protection Act

US Fish and Wildlife Service   See FISH AND WILDLIFE SERVICE, US.

US Forest Service   See FOREST SERVICE, US.

US Great Lakes Water Quality Agreement   See GREAT LAKES WATER QUALITY AGREEMENT, US-CANADA.

US Green Lights Program   See GREEN LIGHTS PROGRAM, US.

US National Environmental Policy Act (NEPA)   See NATIONAL ENVIRONMENTAL POLICY ACT, US.

US National Park Service   See NATIONAL PARK SERVICE, US.

US Toxic Release Inventory   See TOXIC RELEASE INVENTORY, US.

utility   (1) In ECONOMICS, the condition or quality of being useful, satisfying human wants. (2) The amount of satisfaction that a consumer derives from using a product or service, consumers displaying preferences in the marketplace. (3) a public service such as the supply of energy or water, or the provision of SEWERAGE.

UV-B radiation   See ULTRAVIOLET RADIATION.

UVCE   UNCONFINED VAPOUR CLOUD EXPLOSION.

V

vadose water Water that is present between the WATER TABLE and the surface of the Earth. (See GROUND WATER; SURFACE WATER.)

***Valdez* oil spill** See EXXON VALDEZ DISASTER.

value (1) The exchange value or price of a commodity or service in the open market. (2) The value of assets and services not priced in the open market such as environmental assets of many kinds, requiring a SHADOW PRICE. (3) The intrinsic value of personal possessions, which may have little or no market value. (4) The value attached to relationships and to society. (5) The value attached to the ENVIRONMENT and animal life for its own sake. (See QUALITY OF LIFE.)

value judgement A decision involving basic issues of fairness, reasonableness, justice, or morality. (See INTRAGENERATIONAL EQUITY.)

Vancouver Plan of Action A product of the UN CONFERENCE ON HUMAN SETTLEMENTS, the Vancouver Plan of Action urged all countries to create, as a matter of urgency, a national policy on human settlements, embodying the desired distribution of population with related economic and social activities

variola See SMALLPOX.

vascular Characterized by or containing vessels that carry circulating plant or animal fluids necessary to life, such as lymph, sap, or blood.

veering wind A clockwise change of direction of a wind; for example, from N to NE; an opposite change of direction to backing.

venturi scrubber A device for scrubbing industrial waste gases to remove dust before they pass to the ATMOSPHERE; water is injected into a venturi nozzle, the wet dust being removed from the gas by cyclonic action. This technique is highly efficient, figures as high as 99.8% being quoted, though power consumption is also high.

vertebrates Animals which have a backbone, such as fish, amphibians, reptiles, birds, and mammals.

vertisol A SOIL type characterized by a high content of swelling clays, which during the dry season cause deep cracks to form; during the onset of the wet season, the cracks close again. The high clay content makes vertisols very sticky when wet, and very hard when dry.

vibration A rapid back-and-forth motion of the particles or substance of an elastic body or medium. Severe vibrations may result from blasting operations in mines affecting both the ATMOSPHERE and the ground. (See BLASTING; DAMPING.)

vibrio bacteria BACTERIA that cause CHOLERA; associated with water-borne epidemics.

Victoria Falls National Park A national park located within Zimbabwe adjacent to the Livingstone Game Park in Zambia, both being centred on the spectacular Victoria Falls about midway along the course of the Zambezi River, Africa. The parks abound with large and small game and offer recreational facilities. The Victoria Falls park was established in 1931.

Vienna Convention for the Protection of the Ozone Layer See CONVENTION FOR THE PROTECTION OF THE OZONE LAYER.

Villach Conference on Climatic Change Convened in October 1985, a conference of scientists from 29 countries held in Villach, Austria at which scientific recognition was given to the GREENHOUSE EFFECT.

viruses Organisms capable of causing disease in humans, plants, and animals; different from BACTERIA, viruses are unable to multiply or show any signs of life unless they are within another living organism. Once within living cells, viruses can multiply rapidly and affect the organisms in which they are growing. In humans, viruses are responsible for mumps, measles, poliomyelitis, the common cold, and a range of other ailments.

visual pollution Visual squalor in an environment, including items such as: overhead wirescape; litter and unauthorized rubbish; abandoned cars and large items of equipment; derelict factory premises and abandoned industrial equipment; unattractive hoardings and advertisements; overhead highways; large commercial vehicles parked in residential streets; unmaintained residential property; squatter settlements; and some kinds of alternative life styles.

visual quality The degree to which attractive visual elements or components are present in a landscape or view, and are combined in such a way as to create impressions of harmony, contrast, and diversity.

viviparous Descriptive of an animal that produces living offspring instead of eggs; that is, nearly all mammals. Also descriptive of plants that produce bulbils or small plants instead of, or in addition to, seeds.

Völklingen ironworks Included in the WORLD HERITAGE LIST in 1995, an ironworks in the German Saarland. The ironworks was founded in 1873, dominating the Saarland iron and steel industry. The demise of the industry brought the close of the plant in 1986. The ironworks has been preserved as an industrial museum, being the last of its kind in Europe.

voluntary conservation organizations See NON-GOVERNMENT ORGANIZATIONS.

Voyageurs National Park Established in 1975, a national park in northern Minnesota, USA. It lies along the Canadian Border, east of International Falls. The park comprises a network of streams and lakes; it sustains stands of fir, spruce, pine, aspen, and birch. The park was named after the mostly French-Canadian frontiersmen who were involved in fur trading in the area during the late 18th and early 19th centuries.

vulnerable species See ENDANGERED SPECIES.

W

'Warning to Humanity' A petition sponsored by the Union of Concerned Scientists and signed by more than 1500 eminent scientists (including nearly 100 Nobel prize winners) sent to government leaders around the world on 18 November 1992. The petition called for more efficient use of resources, an end to activities that damage the ENVIRONMENT, the elimination of poverty, and the granting to women of control over their own reproductive decisions. (See UN CONFERENCE ON ENVIRONMENT AND DEVELOPMENT; UN WORLD CONFERENCE ON WOMEN.)

Washington convention CONVENTION ON INTERNATIONAL TRADE IN ENDANGERED SPECIES OF WILD FAUNA AND FLORA.

washout The process of scavenging the ATMOSPHERE by PRECIPITATION.

waste Any matter, whether liquid, solid, gaseous, or radioactive, which is discharged, emitted, or deposited in the ENVIRONMENT in such volume, concentration, constituency, or manner as to cause a significant alteration of the environment. The concept of waste embraces all unwanted and economically unusable by-products or residuals at any given place and time, and any other matter that may be discharged accidentally or otherwise to the environment.

waste incineration See INCINERATION.

waste management A comprehensive, integrated, and rational systems approach towards the achievement and maintenance of acceptable environmental quality and the support of SUSTAINABLE DEVELOPMENT. It involves preparing policies; determining environmental standards; fixing emission rates; enforcing regulations; MONITORING air, water, and soil quality; NOISE emissions; and offering advice to government, industry, land developers, planners, and the public (See SOLID-WASTE MANAGEMENT.)

waste minimization An approach to WASTE MANAGEMENT that stresses the minimization of wastes at source, by the adoption of techniques and processes that reduce the waste stream; by the improvement of well-established processes to minimize the problem; by increasing the rate of recovery and RECYCLING within the plant itself so as to minimize the waste-disposal problem; or by adopting product materials and standards that reduce the waste stream. (See SOLID-WASTE MANAGEMENT.)

waste reduction and resource conservation strategies A variety of approaches and methods for reducing the generation of waste materials at source, and for promoting their recovery, reuse, and RECYCLING. (See SOLID-WASTE MANAGEMENT; WASTE MANAGEMENT.)

water A substance composed essentially of the chemical elements HYDROGEN and OXYGEN and found in liquid, solid, and gaseous states. Water is plentiful though not always in the right places; it is vital to life participating in virtually every process that occurs in plants and animals, and in the ENVIRONMENT. As a liquid, water is colourless, tasteless, and odourless, with a marked ability to dissolve many other substances. Water is the working fluid of steam systems, and a heating or cooling medium. Impurities and pollutants may impair the efficient use of water and threaten public health. (See WATER POLLUTION; WATER VAPOUR.)

water-borne diseases Diseases such as cholera, typhoid fever, dysentery, gastroenteritis, hepatitis, and schistosomiasis, which are commonly transmitted through contaminated water supplies. While the classical water-borne diseases have been virtually eradicated from many of the developed countries, these diseases are still endemic in other parts of the world. (See SANITATION.)

water cycle See HYDROLOGICAL CYCLE.

water pollution Substances, bacteria, or viruses present in such concentrations or numbers as to impair the quality of WATER rendering it less suitable, or unsuitable, for its intended use and presenting a hazard of the ENVIRONMENT to people. Pollution may be caused by: (1) bacteria, viruses, and other organisms that can cause disease such as cholera, dysentery, and typhoid; (2) inorganic salts making the water less suitable for drinking, irrigation, and for many industries; (3) plant nutrients such as potash, phosphates, and nitrates which increase weed growth, promoting algal blooms; (4) oily materials that may be inimical to fish life; (5) specific toxic agents ranging from HEAVY METALS to complex synthetic chemicals; (6) waste heat that may render the river less suitable for certain purposes; (7) silt that may enter a river in large quantities, changing the character of the river bed; and (8) radioactive substances.

water purification The treatment of WATER to make it safe and acceptable for human consumption and use. (See CHLORINATION; DRINKING-WATER QUALITY GUIDELINES.)

water quality guidelines See DRINKING-WATER QUALITY GUIDELINES.

water resources The waters of the oceans, rivers, and lakes, SURFACE WATER, GROUND WATER, VADOSE WATER, WATER VAPOUR, DAMS, glaciers, and snowfields. A water-resource strategy is a spectrum of measures: to conserve surface and ground water; to ensure the continued availability of water for growing domestic, commercial, and industrial uses; to ensure sufficient water for natural ecosystems; and in support of SUSTAINABLE DEVELOPMENT.

watershed The elevated boundary line separating the headstreams that are tributaries to different river systems or CATCHMENT basins; also called a water parting or divide.

water softener A device for removing calcium and magnesium from WATER, to minimize the formation of insoluble scale in pipes and tanks, and to make better use of soaps and cleaners. Water softeners usually consist of zeolite or an ion-exchange resin, which can be regenerated.

water table The top level of WATER in the ground, occupying spaces in rocks or SOIL. When the water table rises above ground a spring, lake, or WETLAND is formed.

water treatment Measures to condition water so that it is suitable for use in, say, steam generators. The main aims are to: (1) 'soften' the water, removing or neutralizing the scale-forming salts; (2) achieve the correct alkaline condition; (3) remove excess OXYGEN and CARBON DIOXIDE from the water.

water vapour The chief greenhouse gas, existing in the ATMOSPHERE at a concentration of about 1%. Along with CARBON DIOXIDE at its natural background level, water vapour provides a natural GREENHOUSE EFFECT, without which we would freeze. There is no evidence that the concentration of water vapour in the atmosphere has changed significantly over time. It is not usually listed among the causes of the anthropogenic greenhouse effect.

wave-cut platform A gently sloping rock ledge that extends from the high-tide level at a steep cliff base to below low-tide level. It develops over time as a result of wave abrasion; beaches normally protect the shore from abrasion preventing the formation of platforms. Also called an abrasion platform.

wave energy The utilization of the energy of waves, as distinct from tides, for the generation of electricity. The energy of waves is very diffuse, and the harnessing of this energy requires large and robust structures to achieve even low conversion rates. Most proposals for using wave energy comprise a system of floats allowing the movement of waves to compress or elevate a fluid which drives turbo-generators. Several devices have undergone sea trials and prototype devices have been built in Norway and Britain. Wave energy could replace fossil fuels in time and have a positive influence on reducing the harmful GREENHOUSE EFFECT, while supporting SUSTAINABLE DEVELOPMENT.

wave motion A propagation of regular disturbances from place to place. The most familiar examples are surface waves on WATER, while both sound and light travel in a wave-like manner. The study of waves is central in physical science and engineering.

wavelength The distance between corresponding points of two consecutive waves, that is two points in the same phase.

WCMC WORLD CONSERVATION MONITORING CENTRE.

WCS WORLD CONSERVATION STRATEGY.

WCU WORLD CONSERVATION UNION.

wealth In ECONOMICS, goods and other assets in existence at any time which command a market value (a price) if offered for sale, and which are actually owned; also FREE GOODS which do not command a price and are not in individual ownership. In more recent years, economists have given greater attention to social costs and social benefits that embrace many matters not falling within the conventional measuring rod of money.

weather The condition of the ATMOSPHERE at a certain time or over a certain short period as described by various meteorological phenomena such as atmospheric pressure, temperature, humidity, rainfall, cloudiness, and wind speed and direction. (See CLIMATE.)

weathering The wearing down of rock through exposure to the WEATHER such as heat, cold, wind, and rain, and exposure to certain organisms such as lichens; the chemical process of rock decomposition.

WEC WORLD ENERGY COUNCIL.

welfare state A nation that provides minimal standards in respect of housing, health, education, and financial support for those individuals and families with insufficient resources to meet the full cost of facilities and services. The primary aim is to assist the homeless, unemployed, sick, disabled, handicapped, destitute, and the aged. The concept goes much further than this, however, for the welfare state provides social and health insurance of assistance to middle-income families who are insulated against the shock of unexpected medical bills, while a system of public education relieves all parents who use the system of much of the cost of education. The welfare state, therefore, is not simply a relief system for the poor and unlucky; it provides a steady stream of services to all sections of the community who pay through taxes for buffers against situations when much is needed all at once. The welfare system is a social insurance and health system towards which most pay and all draw in varying degrees throughout life, some very much more than others.

Wellington convention See CONVENTION ON THE PROHIBITION OF FISHING WITH LONG DRIFT NETS IN THE SOUTH PACIFIC.

wet collectors See SCRUBBERS.

wetlands Areas largely inundated with water, yet offering elevated land as a HABITAT for

wet washer

WILDLIFE, notably waterfowl. These areas include swamps, both seasonal and permanent, marsh, open fresh water, shallow saline lagoons, the estuaries of rivers, floodplains, and coastal sand dunes. (See CONVENTION ON WETLANDS OF INTERNATIONAL IMPORTANCE, ESPECIALLY AS WATERFOWL HABITAT.)

wet washer (See SCRUBBERS.)

WFC WORLD FOOD CONFERENCE; WORLD FOOD COUNCIL.

WFP WORLD FOOD PROGRAMME.

whale oil Oil obtained from whales, primarily from their blubber; this oil was important at one time for soap making and as fuel for lamps, later being converted into margarine and cooking fats. It has also provided lubricants, and found use in many industries. The end of commercial whaling in 1986 marked the adoption of substitutes for many whale products.

whales See CONVENTION FOR THE REGULATION OF WHALING; INTERNATIONAL WHALING COMMISSION.

WHO WORLD HEALTH ORGANIZATION.

Who Speaks for Earth A volume of distinguished lectures sponsored by the International Institute for Environmental Affairs in cooperation with the Population Institute presented concurrently with the UN CONFERENCE ON THE HUMAN ENVIRONMENT June 1972. The speakers were Barbara Ward (Lady Jackson), René Dubos, Thor Heyerdahl, Gunnar Myrdal, Carmen Miró, Lord Zuckerman, and Aurelio Peccei.

WICE WORLD INDUSTRY COUNCIL ON THE ENVIRONMENT.

wild and scenic rivers Those rivers which, together with their corridors, represent select parts of the ENVIRONMENT having significant value for recreational, aesthetic, conservation, historical, or other purposes. The US Wild and Scenic Rivers Act 1968 created a national system of designated rivers, restored as necessary to a condition suitable for fishing and swimming.

wilderness A large tract of primitive country with its land and waters, native plant and animal communities, substantially unmodified by humans and their works. Wilderness provides opportunities for primitive recreation, self-education, and observation by scientists. It enables visitors to experience solitude in areas where survival depends on personal resources. Access should be restricted to people travelling only on foot, on skis, or in hand- or sail-powered boats.

wildlife A collective term embracing FLORA and FAUNA that are not under direct domestic care or ownership, including feral, native, and introduced plants and animals. (See CONVENTION ON INTERNATIONAL TRADE IN ENDANGERED SPECIES OF WILD FAUNA AND FLORA; ENDANGERED SPECIES.)

Williamsburg Restoration The restoration of Williamsburg, the capital of colonial Virginia, to much of its 18th-century appearance. Acquisition and renovation has continued since 1926. The first major edifice to be restored was the Wren Building at the college of William and Mary; other restored buildings have been the Governor's Palace, the Capital building, the Bruton Parish Church, the Courthouse, the public gaol, and the magazine and guardhouse.

willingness to pay A principle underlying the valuation of benefits from a project or activity. Benefits are usually of two kinds: services to consumers for which a price is paid; and services to consumers and benefits for the broader community for which a price is not paid. A valuation needs not only a summation of the prices actually paid but an assessment of what people might be expected to willingly pay for the free services and benefits, rather than do without them. A variety of techniques have developed for the assessment of willingness-to-pay values. Some non-traded benefits such as travel-time savings in the case of road construction have well-established methods of valuation.

wind energy The conservation of the kinetic energy of the wind into electricity, by means of wind turbines or windmills. Wind is generated by pressure differences in the ATMOSPHERE, driven by solar power. Studies reveal that the wind could supply an amount of electrical energy equal to the world's electricity demand. Turbines have been supplied for the Californian wind-farm market and in relation to programmes being developed in the EUROPEAN UNION, and in China, India, Russia, and Israel.

wind rose A diagram indicating the frequency and strengths of winds in a definite locality for a given period of years. It is conventional to consider the wind direction as the direction from which the wind blows; for example, a northeast wind will carry pollutants to the southwest of the source.

Windscale nuclear reactor incident A major accident which occurred at the British Atomic Energy Authority plant at Windscale, Cumbria, England on 10 October 1957. A fire occurred within the No. 1 reactor and volatile products were released to the ATMOSPHERE. The accident occurred during a routine Wigner release when local overheating of the uranium elements occurred. A filter in the ventilation stack arrested most of the strontium and caesium, but radioactive fission products escaped to atmosphere including iodine-131 and tellurium-132. (See Figure 25.)

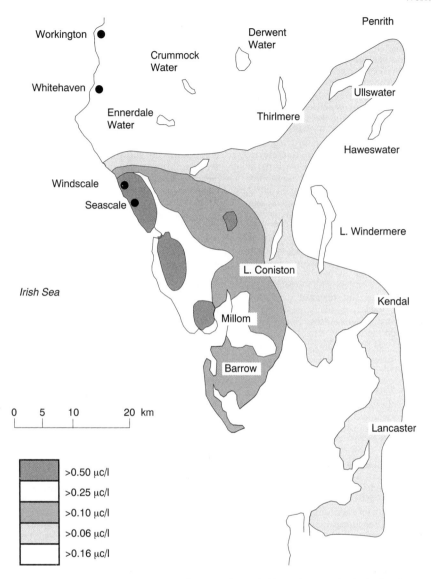

WINDSCALE NUCLEAR REACTOR INCIDENT: Figure 25. Map of Windscale area, England, showing contours of radio-iodine contamination of milk on 13 October, 1957
Source: Medical Research Council, *The Hazards to Man of Nuclear and Allied Radiation*, Cmnd. 1225, HMSO, London, 1960.

within-generational equity See INTRAGENERATIONAL EQUITY.

WMO WORLD METEOROLOGICAL ORGANIZATION.

woodchip industry An industry involving the procurement of timber and its mechanical reduction to chips; woodchips are usually destined for the pulp and paper industry. Timber has been obtained by CLEAR-FELLING nearly all the trees over a wide area, causing soil erosion, water pollution, and loss of native vegetation and wildlife, together with the loss of wilderness and ecological and other values. Environmentalists seek to phase out the woodchipping industry.

World Bank An agency of the United Nations, the World Bank comprises the International Bank for Reconstruction and

Development (IBRD) and its affiliates, the International Development Association (IDA), the International Finance Corporation (IFC), and the Multilateral Investment Guarantee Agency (MIGA). The common objective of these bodies, known collectively as the World Bank, has been to help to raise the standards of living in developing countries, channelling resources from the richer to the poorer nations. The bank was established in 1945. In 1970, the bank created an Office of Environmental and Health Affairs, but with limited resources. Since 1979, investment and assistance began to extend to afforestation and reforestation, soil conservation, flood mitigation and control, range management, wildlife protection, and abatement of air and water pollution. However, in more recent years, the bank has been accused of failing in its original role of reducing poverty and of disregarding in whole or part the environmental effects of major development projects. It has, it is claimed, financed roads for settlers who devastated the Amazon rainforest and financed dams that forced many thousands to be resettled in less satisfactory circumstances. Since 1991, there has been much more attention to these problems; several projects have been modified as a result of environmental assessment. In 1992, the bank published its three-volume sourcebook on EIA. It supports a basic premise that SUSTAINABLE DEVELOPMENT is achieved most efficiently when negative environmental effects are identified and addressed at the earliest possible planning stage. (See PAK MUN HYDROPOWER PROJECT; SARDAR SAROVA DAM PROJECTS.)

World Bank Inspection Panel Created in 1993 by the WORLD BANK, an inspection panel to investigate complaints about its own operations.

World Commission on Environment and Development (the Brundtland Commission) A commission created by the UN General Assembly in 1983 for the purpose of examining potential conflicts between environment protection and economic growth. The 21 members of the commission were drawn from a range of nations with widely different economic and political backgrounds. The commission was chaired by Gro Harlem Brundtland, Prime Minister of Norway. The commission held a large number of meetings and travelled widely. Its report *Our Common Future* to the Governing Council of UNEP was presented in April 1987. The commission envisaged a new era of economic growth, one based on policies of SUSTAINABLE DEVELOPMENT, coupled with an expansion of the environmental resource base. Such growth was seen by the commission as essential to the relief of poverty in much of the developing world, and to sustain coming generations. In respect of the GREENHOUSE EFFECT, the commission recognized greater efficiency in the use of energy as one weapon to combat the problem.

World Congresses on Climate and Development International conferences organized by the WORLD METEOROLOGICAL ORGANIZATION (WMO). The 1988 World Congress was held in Hamburg, Germany. The result was the Hamburg Manifesto on Climatic Change, urging a 30% reduction in the emission of CARBON DIOXIDE by the year 2000, and 50% by the year 2015, largely through improved efficiency in the use of energy. The Second World Congress on Climate and Development was held in Geneva, Switzerland, in 1990 to review progress in the implementation of the Hamburg Manifesto.

World Conservation Monitoring Centre (WCMC) An international centre for the management and dissemination of information on BIODIVERSITY and CONSERVATION, aiming at enhancing the capacity of countries worldwide to manage complex biodiversity information. Founded by its three sponsoring partners, the WORLD CONSERVATION UNION, the WORLD WIDE FUND FOR NATURE, and the UN ENVIRONMENT PROGRAM, the centre is a non-profit-making organization with an international mandate. The centre is based in Cambridge, England.

World Conservation Strategy (WCS Launched) in 1980, a strategy prepared by the WORLD CONSERVATION UNION, the UN ENVIRONMENT PROGRAM, and the WORLD WIDE FUND FOR NATURE. It demonstrated that DEVELOPMENT can only be sustained by conserving the living resources on which that development depends, and by the integration of development and conservation policies. It urged every country to prepare its own national conservation strategy, reflecting the principles in the WCS. The WCS stressed three main objectives for living resource conservation: (1) to maintain essential ecological processes and life-supporting systems (such as soil regeneration, the recycling of nutrients, and the safeguarding of waters); (2) to preserve genetic diversity; and (3) to ensure the sustainable utilization of species and ecosystems (notably fish and other wildlife, forests, and grazing lands). Over 50 national conservation strategies have been adopted since the release of the WCS. The chief successor to the WCS has been the document *Caring for the Earth: a Strategy for Sustainable Living* produced by the World Conservation Union, UNEP and WWF in 1991. It includes a wide range of recommendations for legal, institutional, and administrative reform.

World Conservation Union (WCU) Known previously as the International Union for the Conservation of Nature and Natural Resources (IUCN), the WCU is a voluntary international body whose main objective is trying to ensure the perpetuation of wild nature and natural resources throughout the world. Formed in 1948, the body has its headquarters in Gland, Switzerland. Its membership comprises nations, government agencies, private bodies, and international organizations. With UNEP and WWF, the IUCN participated in the preparation of a WORLD CONSERVATION STRATEGY as a framework for SUSTAINABLE DEVELOPMENT.

World Consumer Rights Day The 15 March each year, marked by consumer rights organizations worldwide. (See CONSUMER ACTIVISM; CONSUMER PROTECTION.)

World Energy Council (WEC) A non-commercial and non-governmental organization involved in the strategic research and analysis of all forms of energy. Its membership is drawn from over 100 countries. The WEC undertakes triennial energy-studies programmes, which form the basis for Congresses, which attract some 5000 to 6000 delegates from all over the world. It is based in London, England.

World Environment Day The 5 June each year, the day adopted by the UN CONFERENCE ON THE HUMAN ENVIRONMENT 1972, as an annual means of focusing attention on national and world environmental problems.

World Food Conference (WFC) A UN Conference held in 1974 that resulted in action to deal with crises in food supplies, including the establishment of a WORLD FOOD COUNCIL and an International Fund for Agricultural Development, and an information system for food stocks and requirements, fertilizer availability, and crop conditions. An international agricultural research and training programme now covers most major food crops and animal production. Environmental considerations are taken into account in all these activities. (See UN FOOD AND AGRICULTURE ORGANIZATION.)

World Food Council (WFC) A UN organization created in 1974 upon the recommendation of the WORLD FOOD CONFERENCE. Based in Rome, Italy, the WFC coordinates information and suggests strategies for food policies, reporting to the UN General Assembly. Attention is focused on the needs of developing countries and on the reduction of food-trade barriers between developing and developed countries. (See WORLD FOOD PROGRAM.)

World Food Program (WFP) A UN organization established in 1961, as a joint project of the UN General Assembly and the UN FOOD AND AGRICULTURE ORGANIZATION (FAO). It seeks to stimulate economic development through food aid and emergency relief. Its chief organs are the Committee of Food Aid Policies and Programs and the joint UN/FAO Administrative Unit. WFPs objectives are specifically geared to eliminating emergency situations that arise from insufficient food supplies. It seeks also to further land reclamation and irrigation.

World Health Organization (WHO) An agency of the United Nations, WHO came into being in 1948. Based in Geneva, WHO absorbed the health activities of the UN Relief and Rehabilitation Administration which had assisted the health departments of many countries with both advice and practical aid, the Paris office of epidemic intelligence, and the health organizations of the former League of Nations. WHO became the sole international health organization; it operates with a high degree of autonomy, though directed and guided by the World Health Assembly which meets annually. The work of WHO encompasses the control of communicable diseases; water supply and waste disposal; air and water pollution; standards for biological and chemical substances; nutrition, food hygiene, and food standards; occupational health; the effects of radiation; psychosocial influences; carcinogenic risks; and environmental health impact assessment.

World Heritage List A list created by the CONVENTION FOR THE PROTECTION OF THE WORLD CULTURAL AND NATURAL HERITAGE (the Paris convention) and administered by the UN EDUCATIONAL, SCIENTIFIC AND CULTURAL ORGANIZATION (UNESCO). The convention came into force in 1975. The list includes properties of great cultural significance and geographic areas of 'outstanding universal value'. To be included, a nation must nominate the property or location to a World Heritage Committee consisting of 21 nations, elected from those nations which are parties to the Convention. The executive body of the World Heritage Committee is the World Heritage Bureau. Well over 300 places now appear on the World Heritage List. They include the Pyramids of Egypt, the Grand Canyon of the United States, the Taj Mahal of India, Westminster Abbey in London, Sagarmatha National Park (containing Mount Everest) in Nepal, the Great Wall of China, and the GREAT BARRIER REEF of Australia.

World Industry Council on the Environment (WICE) A body established in 1993 by leading companies from 23 countries, on the initiative of the International Chamber of Commerce. Its 90 members are committed to high standards of environmental management and the principles of economic growth; its purpose is to provide business leadership in the field of the environment and SUSTAINABLE DEVELOPMENT. WICE is based in Paris, France.

World Meteorological Organization (WMO) A UN agency since 1951, the WMO was created by the World Meteorological Convention;

World population

it evolved from its predecessor, the International Meteorological Organization established initially in 1873. The WMO's primary role is to maintain a worldwide meteorological system, and to assist individual nations with the development of national systems. The WMO has a distinct role in grappling with environmental problems of a global nature. Its contribution has found expression in the Global Atmospheric Research Program (GARP) and in an operational monitoring system called WORLD WEATHER WATCH. In 1979, the WMO undertook a World Climate Programme in cooperation with the UN ENVIRONMENT PROGRAM. In addition, the WMO Background Air Pollution Monitoring Network has been developed as part of the Global Environment Monitoring System (GEMS). (See WORLD CONGRESSES ON CLIMATE AND DEVELOPMENT.)

World population An issue of crucial importance in relation to the possibilities of SUSTAINABLE DEVELOPMENT. According to the UN, the world's population reached one billion at the beginning of the 19th century. By 1950, there were about 2.5 billion people on the planet. Since then the population has more than doubled to 5.6 billion by 1994. Long-term projections suggest that the numbers will continue to grow until after 2200, when the world's population may stabilize at around 11.5 billion, or twice as many again. Since 1950, the rate of growth has been 1.5% a year in rich countries, and more than 2% in poor countries. However, a demographic transition from high birth rates and high death rates to low birth rates and low death rates is expected to characterize future trends. Much of Europe and North America has passed through this demographic transition, with Japan and Asia following.

World Resources Institute (WRI) An international policy research centre, created in 1982; its research is aimed at providing accurate information about global resources and population, identifying emerging issues, and developing politically and economically workable proposals. In conjunction with the International Institute for Environment and Development, the WRI publishes the annual World Resources series. The WRI is supported by a number of private and public organizations.

World Standards Day An annual event celebrated in October by many countries. (See INTERNATIONAL ORGANIZATION FOR STANDARDIZATION.)

World Summit on Global Warming See BERLIN CLIMATE CONFERENCE.

World Trade Organization (WTO) A body created by the Uruguay Round of the GENERAL AGREEMENT ON TARIFFS AND TRADE (GATT), to give expression to and coordinate the application of that agreement. The WTO came into operation in 1995.

Worldwatch Institute A US-based voluntary organization that studies global and regional environmental issues. It has produced a *State of the World* report on the progress towards a sustainable society every year since 1984; the reports have been translated into many languages. Worldwatch owes its continued existence to the generosity of a number of organizations, both private and public, and a whole range of individuals.

World Weather Watch (WWW) A programme established in 1961 by the WORLD METEOROLOGICAL ORGANIZATION to improve the global system of meterological observation and prediction. WWW has four essential elements: a global observing system; a global data-processing system; a complex telecommunications system circling the globe; and a global network of special stations to monitor atmospheric background pollution.

World Wide Fund for Nature (WWF) Formerly the World Wildlife Fund, a worldwide voluntary conservation organization with headquarters in Gland, Switzerland. While covering the major environmental issues, WWF also runs its own practical, scientific research, and conservation projects in the field. WWF seeks to preserve genetic species and ecosystem diversity; promote SUSTAINABLE DEVELOPMENT; minimize pollution and the over-use of resources and energy; and to reverse the accelerating degradation of the planet's natural environment. It has 28 affiliate offices around the world.

WRI WORLD RESOURCES INSTITUTE.
WTO WORLD TRADE ORGANIZATION.
WWF WORLD WIDE FUND FOR NATURE.
WWW WORLD WEATHER WATCH.

X

Xanthophyta Yellow-green ALGAE, a group that contains carotenoid pigments as well as CHLOROPHYLL.

xeromorphic Describes plants which appear to have the ability to restrict water loss during adverse conditions; a plant displaying this characteristic is described as a xerophyte.

xerophyte See XEROMORPHIC.

X-ray ELECTROMAGNETIC RADIATION of an extremely short wavelength produced by the deceleration of charged particles or the transitions of electrons in atoms. X-rays have been applied to a wide range of medical, industrial, and scientific problems. One of the earliest applications of X-rays was in medicine, being used in both diagnosis and therapy. In industry, X-rays have been used to detect flaws in metal.

Y

yellowcake The mixture of uranium oxides and impurities produced at a uranium mill. (See NUCLEAR FUEL CYCLE; RADIOACTIVITY.)

yellow fever Essentially, an acute infectious tropical and subtropical disease caused by a VIRUS, which is transmitted between susceptible hosts by several species of mosquitoes. Immunization is today an essential measure for travellers.

yellow-green algae See XANTHOPHYTA.

Yellowhead Pass A route through the Rocky Mountains at the Alberta-British Columbia border, Canada, leading from Jasper National Park into Mount Robson Provincial Park.

Yellowstone National Park Established in 1872, the oldest national park in the USA, situated in northwest Wyoming, southern Montana and eastern Idaho. It consists mostly of broad volcanic plateaus, with mountain ranges and unusual geological features including fossil forests and eroded basalt lava flows. It has over 10 000 hot springs which find surface expression in a variety of forms, including 200 geysers. The Grand Canyon runs for 31 km through the park, with two majestic waterfalls. Most of the park is forested mainly with lodgepole pine. Animal life includes buffalo, elk, bighorn sheep, deer, moose, black bear, grizzly bear, and coyotes. A scenic roadway connects Yellowstone with Grand Teton National Park. Yellowstone has been included in the WORLD HERITAGE LIST.

Yersinia A genus of BACTERIA, some strains of which can cause gastroenteritis if ingested.

Yoho National Park Established in 1886, in British Columbia, Canada, adjacent to two other national parks - Banff on the east and Kootenay to the south. A mountaineering centre, the park has several high peaks. There are fossil beds thought to be 500 000 000 years old. The fauna includes a wide variety of birdlife, black and grizzly bears, deer, elk, and mountain goats.

Yokkaichi asthma Severe asthma caused by acute levels of sulphur oxides in the atmosphere of Yokkaichi, the site of the largest petrochemical complex in Japan.

Yosemite National Park Established in 1890, a scenic mountainous national park in central California, USA. It embraces areas of the Sierra Nevada Range, the Yosemite Valley, and giant sequoia groves with trees thousands of years old. The flora of the park changes rapidly with elevation, trees ranging from conifers to lodgepole pine. Fauna includes mule deer, squirrels, chipmunks, and black bears. Yosemite has been included in the WORLD HERITAGE LIST.

Yukka Mountain About 160 km northwest of Las Vegas, Nevada, USA, a proposed nuclear waste repository, with the ground-water level about 300 m below the surface; the site has been the subject of much controversy for many years.

Z

zero economic growth An argument advanced by some conservationists that ECONOMIC GROWTH will exhaust or overstress the world's resources and should be restrained, the problem of poverty being addressed by a redistribution of existing resources. (See INTRAGENERATIONAL EQUITY; SUSTAINABLE DEVELOPMENT.)

zero energy growth An argument advanced by some conservationists that energy consumption in the more affluent countries should be restrained and used more efficiently, while demand in developing countries should be allowed to grow with improving living standards.

zero population growth A view by some conservationists that the world's population should be stabilized as soon as possible, without further growth. This is considered central to the target of SUSTAINABLE DEVELOPMENT. (See WORLD POPULATION.)

Zion National Park Established in 1919, in southwest Utah, USA, a national park of deep canyons and high cliffs. It was created to protect the Zion Canyon which contains much evidence of fossils and prehistoric cave dwellers. Wildlife includes deer, lions, and more than 150 species of birds.

zoning The most common control of development, reserving certain areas for specific land uses and separating incompatible uses. Zoning ordinances or local plans generally divide the community into districts, specifying for each district the permitted uses, height, and bulk of structures, minimum lot sizes, and density of development. Open space and environmental characteristics are protected. Basically, housing can be separated from industry and commerce, at least as far as the future is concerned. A successful zoning plan should improve the QUALITY OF LIFE for the community.

zoology A branch of BIOLOGY concerned with members of the animal kingdom and their characteristics, including the relationships between individuals and groups and the ENVIRONMENT.

zooplankton Or protozoa, minute unicellular animals often in the larval stage, swimming or floating in water.

SELECTED BIBLIOGRAPHY

GLOBAL ISSUES

Brown, L. R., Durning, A., Flavin, C., French, H. et al. (1995) *State of the World 1994*, a Worldwatch Institute Report on progress toward a sustainable society, W. W. Norton & Company, New York.

Carson, R. (1962) *Silent Spring*, Houghton Mifflin Company, Boston.

Clark, C. (1977) *Population, Growth and Land Use*, Macmillan, London.

Commoner, B. (1972) *The Closing Circle: Nature, Man and Technology*, Alred A. Knopf, New York.

Darling, F. F. (1970) *Wilderness and Plenty: the Reith Lectures 1969*, Houghton Mifflin Company, Boston.

Dasman, R. F. (1984) *Environmental Conservation*, Collier, New York.

Department of the Environment (1990) *Monitoring Environmental Assessment and Planning*, HMSO, London.

Dubos, R. B. (1982) *Man, Medicine and Environment*, The Pall Mall Press, London.

Economic and Social Commission for Asia and the Pacific (1990) *State of the Environment in Asia and the Pacific 1990*, ESCAP, New York.

Elton, C. S. (1927) *Animal Ecology*, and (1930) *Animal Ecology and Education*.

Erlich, P. R. and Erlich, A. H. (1970) *Population, Resources, Environment: Issues in Human Ecology*, W. H. Freeman and Company, San Francisco.

Federal Republic of Germany (1992) *Global Change: Our World in Transition*, Federal Ministry for Research and Technology, Bonn.

Friday, L. and Laskey, R. (1989) *The Fragile Environment: the Darling College Lectures*, Cambridge University Press, Cambridge.

Hardy, A. (1968) Charles Elton's Influence in Ecology, *Journal of Animal Ecology*, **37**: 1–8 (1968).

International Atomic Energy Agency (1991) *Electricity and the Environment*, IAEA-Tecdoc-624, background papers for a senior expert symposium held in Helsinki 13–17 May 1991, IAEA, Vienna.

Laszio, E. (1977) *Goals for Mankind*, A Report to the Club of Rome, Hutchinson, London.

Lave, L. B. and Seskin, E. P. (1977) *Air Pollution and Human Health*, Johns Hopkins Press, Baltimore.

Mesarovic, M. and Pestel, E. (1975) *Mankind at the Turning Point*, the Second Report of the Club of Rome, Hutchinson, London.

Meadows, D. H., Meadows, D. L., Randers, J. and Behrens, W. W. (1972) *The Limits to Growth*, a report for the Club of Rome's project on the predicament of mankind, Pan Books, London.

Ministry of Health (1954) *Mortality and Morbidity during the London Fog of December 1952*, HMSO, London.

Steffen, W. and Walker, B. H. (1995) *Global Change and Terrestrial Ecosystems*, Cambridge University Press, Cambridge.

Tansley, A. G. (1935) The Use and Abuse of Vegetational Concepts and Terms, in *Ecology*, New York.

United Nations (1995) *Report of the Conference of the Parties to the Framework Convention on Climate Change held in Berlin 28 March to 7 April*, United Nations, New York.

Selected Bibliography

United Nations Children Fund (UNICEF) (1995) *The State of the World's Children*, UNICEF, New York.
United Nations Environment Program (1987) *Drylands Dilemma*, UNEP, New York.
Vos, J. B., Feenstra, J. F., de Boer, J., Braat, L. C. *et al.* (1985) *Indicators for the State of the Environment*, Institute for Environmental Studies, Free University, Amsterdam.
Ward, B. (1966) *Space Ship Earth*, Hamish Hamilton, London.
Ward, B. and Dubos, R. (1972) *Only One Earth: The Care and Maintenance of a Small Planet*, prepared for the UN conference on the Human Environment 1972, Penguin Books, Harmondsworth.
Ward, B., Dubos, R., Heyerdahl, T., Myrdal, G. *et al.* (1973) *Who Speaks for Earth?*, W. W. Norton & Company Inc., New York.
World Commission on Environment and Development (1987) *Our Common Future*, Oxford University Press, Oxford.

ENVIRONMENTAL ECONOMICS

Andrews, R. N. L. and Waits, M. (1978) *Environmental Values in Public Decisions*, School of Natural Resources, University of Michigan, Ann Arbor.
Asian Development Bank (1988) *Guidelines for Integrated Regional Economic-cum-Environmental Development Planning: A Review of Regional Environmental Development Planning Studies in Asia*, ADB, Manila, vol. **1**, 125 pp; vol. **2**, 240 pp.
Barnett, H. J. and Morse, C. (1963) *Scarcity and Growth: the Economics of Natural Resources Availability*, Johns Hopkins Press, Baltimore.
Baumol, W. J. (1968) 'On the social rate of discount', *American Economic Review*, **58**.
Baumol, W. J. and Oates, W. E. (1979) *Economics, Environmental Policy, and the Quality of Life*, Prentice-Hall, New Jersey.
Clayre, A. (ed.) (1977) *Nature and Industrialization*, Oxford University Press in association with the Open University Press, Oxford.
Department of Science and the Environment (1979) *Environmental Economics*, National Environmental Economics Conference, Australian Government Publishing Service, Canberra.
Dixon, J. A. and Hufschmidt, M. M. (1986) *Economic Valuation Techniques for the Environment: a Case Study Workbook*, Johns Hopkins Press, Baltimore.
Dixon, J. A., Carpenter, R. A., Fallon, L. A., Sherman, P. B. *et al.* (1988) *Economic Analysis of the Environmental Impacts of Development Projects*, Earthscan Publications, London.
Fisher, A. C. (1981) *Economic Efficiency and Air Pollution Control*, research paper 8, Environment and Policy Institute, US East-West Center, Honolulu.
Fisher, A. C. and Peterson, F. M. (1976) 'The environment in economics: a survey', *Journal of Economic Literature*, vol. **14**, no. 1, March, pp. 1-33.
Gittinger, J. P. (1982) *Economic Analysis of Agricultural Projects*, Johns Hopkins Press, Baltimore.
Herfindahl, O. C. and Kneese, A. V. (1965) *Quality of the Environment: an Economic Approach to Some Problems in using Land, Water, and Air*, Resources for the Future, Washington DC.
Herfindahl, O. C. and Kneese, A. V. (1974) *Economic Theory of Natural Resources*, Charles, E. Merrill Publishing, Columbus.
Hufschmidt, M. M. and Hyman, E. L. (eds) (1982) *Economic Approaches to Natural Resource and Environmental Quality Analysis*, Bray Co. Tycooly International, Wicklow.
Hufschmidt, M. M., Bower, B. T., James, D., Meister, A. D. *et al.* (1981) *The Role of Benefit Cost Analysis in Environmental Quality and Natural Resource Management*, Environment and Policy Institute, East-West Center, Hawaii.

Selected Bibliography

Hufschmidt, M. M., James, D. E., Meister, A. D., Bower, B. T. *et al.* (1983) *Environment, Natural Systems, and Development: an Economic Valuation Guide*, Johns Hopkins Press, Baltimore.

James, D. (1993) *Economic Instruments for Meeting Environmental Objectives: Australian Experience*, Ecoservices Pty Ltd, Canberra.

Krutilla, J. V. and Eckstein, O. (1958) *Multiple Purpose River Development: Studies in Applied Economic Analysis*, Johns Hopkins Press, Baltimore.

Krutilla, J. V. and Fisher, A. C. (1975) *The Economics of Natural Environments*, Johns Hopkins Press, Baltimore.

Norton, G. A. (1984) *Resource Economics*, Edward Arnold, London.

Organization for Economic Cooperation and Development (1974) *Environmental Damage Costs*, OECD, Paris.

Organization for Economic Cooperation and Development (1976) *Economic Measurement of Environmental Damage*, OECD, Paris.

Organization for Economic Cooperation and Development (1977) *Pollution Control Costs in the Primary Aluminium Industry*, OECD, Paris.

Organization for Economic Cooperation and Development (1981) *The Costs and Benefits of Sulphur Oxide Control*, OECD, Paris.

Pearce, D. W. (1978) *The Valuation of Social Cost*, Macmillan, London.

Pearce, D. W. (1983) *Cost-Benefit Analysis*, Macmillan, London.

Pearce, D. W. and Nash, C. A. (1981) *The Social Appraisal of Projects: a Text in Cost-Benefit Analysis*, Macmillan, London.

Sinden, J. A. (1991) *An Assessment of our Environmental Valuations*, Twentieth Annual Conference of Economists, Hobart, Tasmania.

Sinden, J. A. and Worrell, A. (1979) *Unpriced Values: Decisions without Market Prices*, Wiley Interscience, New York.

Squire, L. and Van der Tak, H. G. (1976) *Economic Analysis of Projects, World Bank*, Johns Hopkins Press, Baltimore.

Yang, E. J., Dower, R. C. and Menefee, M. (1984) *The Use of Economic Analysis in Valuing Natural Resource Damage*, Environmental Law Institute, Washington.

ENVIRONMENTAL IMPACT ASSESSMENT

Aguilo, M., Alonso, S., Blair, W. *et al.* (1986) *Foundations for Visual Project Analysis*, John Wiley & Sons, London.

Asian Development Bank (1991a) *Guidelines for Social Analysis of Development Projects*, ADB, Manila.

Asian Development Bank (1991b) *Environmental Risk Assessment: Dealing with Uncertainty in Environmental Impact Assessment*, ADB, Manila.

Atelier Central de l'Environnement (1990) *Environmental Impact Assessment: the French Experience*, ACE, Neuilly, Paris.

Australian and New Zealand Environment and Conservation Council (1991) *A National Approach to Environmental Impact Assessment in Australia*, ANZECC Secretariat, Canberra.

Australian International Development Assistance Bureau (1991) *Annual Audit of the Environment in the Australian International Development Cooperation Program*, Australian Government Publishing Service, Canberra.

Ball, S. (1991) 'Implementation of the environmental assessment directive in Britain', *Integrated Environmental Management* **5**: 9-11.

Bamber, R. N. (1990) 'Environmental impact assessment: the example of marine biology and the UK power industry', *Marine Pollution Bulletin* **21** (6): 270-274.

Selected Bibliography

Barrett, B. F. D. and Therivel, R. (1991) *Environmental Policy and Impact Assessment in Japan*, Routledge, London.
Beanlands, G. E. and Duinker, P. N. (1983) *An Ecological Framework for Environmental Impact Assessment in Canada*, Federal Environmental Assessment Review Office, Quebec.
Bingham, G. (1986) *Resolving Environmental Disputes: a Decade of Experience*, The Conservation Foundation, Washington DC.
Biswas, A. K. and Agarwala, S. B. C. (eds) (1992) *Environmental Impact Assessment for Developing Countries*, Butterworth–Heinemann, Oxford.
Boer, B., Craig, D., Handmer, J. and Ross, H. (1991) *The Potential Role of Mediation in the Resource Assessment Commission Inquiry Process*, discussion paper 1, January, Resource Assessment Commission, Canberra.
Boes, M. (1990) 'Environmental impact statements in Belgium', *Northwestern Journal of International Law and Business* **10** (3): 522–540.
Buchan, D. and Rivers, M.-J. (1990) 'Social impact assessment: development and application in New Zealand', *Impact Assessment Bulletin* **8** (4): 97–105.
Canadian Environmental Assessment Research Council (1988) *The Assessment of Cumulative Effects: a Research Prospective*, CEARC, Quebec.
Coulson, R. (1989) *Business Mediation: What You Need to Know*, American Arbitration Association, New York.
Council of Europe (1989) *Model Outline Environmental Impact Statement from the Standpoint of Integrated Management or Planning of the Natural Environment*, Nature and Environment Series 17, Council of Europe, Strasbourg.
Department of the Arts, Sport, the Environment, Tourism and Territories (1987) *Environmental Impact Assessment as a Management Tool*, Workshop Proceedings, November, Canberra.
Department of the Environment/Welsh Office (1989) *Environmental Assessment: a Guide to the Procedures*, HMSO, London.
Department of Environment (1990a) *Environmental Impact Assessment (EIA): Procedure and Requirements in Malaysia*, Ministry of Science, Technology and the Environment, Kuala Lumpur.
Department of Environment (1990b) *A Handbook of Environmental Impact Assessment Guidelines*, Ministry of Science, Technology and the Environment, Kuala Lumpur.
Economic and Social Commission for Asia and the Pacific, Environment and Development Series (1985–90) *Environmental Impact Assessment* (1) *Guidelines for Planners and Decision-Makers*, (2) *Guidelines for Agricultural Development*, (3) *Guidelines for Industrial Development*, (4) *Guidelines for Transport Development*, (5) *Guidelines for Water Resources Development*, United Nations, New York.
Economic Commission for Europe (1991) *Policies and Systems of Environmental Impact Assessment*, UN Economic Commission for Europe, Geneva; United Nations, New York.
Environment Agency (1991) *Quality of the Environment in Japan*, EA, Tokyo.
Environmental Impact Management Agency (Bapedal) (1991) *EIA: a Guide to Environmental Assessment in Indonesia*, Jakarta.
Environmental Protection Administration (1985) *A National EIA Program*, EPA, Taipei.
Environmental Protection Agency (1990) *Bibliography and Abstracts of Environmental Impact Assessment Methodologies*, EPA Office of Federal Activities, Washington DC.
Farmer, A. (1980) *Habitat Evaluation Procedures*, US Fish and Wildlife Services, Washington DC.
Federal Environmental Assessment Review Office (1988) *Manual on Public Involvement in Environmental Assessment: Planning and Implementing Public Involvement Programs*, FEARO, Calgary.
Federal Republic of Germany (1975) *Resolution on Adopting Environmental Assessment Principles*, 22 August, Bonn.

Selected Bibliography

Fisher, R. and Ury, W. (1981) *Getting to Yes*, Hutchinson, London.

Geddes, P. (1915) *Cities in Evolution*, London. New Edition, London 1949.

Gilpin, A. (1995) *Environmental Impact Assessment: Cutting Edge for the Twenty-first Century*, Cambridge University Press, Cambridge.

Gilpin, A. and Lin, S. J. (1990) *Environmental Impact Assessment in Taiwan and Australia: a Comparative Study*, Environment and Policy Institute, US East-West Center, Hawaii.

Goudie, A. (1986) *The Human Impact on the Natural Environment*, Basil Blackwell, Oxford.

Govorushko, S. M. (1989) *Environmental Impact Assessment in the USSR: the Current Situation*, Pacific Institute of Geography, Far East Branch, The USSR Academy of Sciences, Vladivostok.

Japanese Ministry of Foreign Affairs (1991) *Environment and Development: Japan's Experience and Achievement*, Tokyo (the Japanese National Report to the United Nations Conference on Environment and Development, 1992).

Khanna, P. (1988) *Environmental Impact Assessment in Land Use Planning and Urban Settlement Projects in India*, Asian Development Bank, Manila.

Krutilla, J. V. and Fisher, A. C. (1975) *The Economics of Natural Environments: Studies in the Valuation of Commodity and Amenity Resources*, Johns Hopkins Press, Baltimore.

Matarrese, G. (1991) 'Environmental impact assessment in Italy', *EIA Newsletter* **6**, EIA Centre, Department of Planning and Landscape, University of Manchester, Manchester.

Mathews, W. H. and Carpenter, R. A. (1981) 'The growing international implications of the US requirements for environmental impact assessments', *Environmental Law and Policy in the Pacific Basin Area*, University of Tokyo Press, Tokyo, pp. 159-169.

Maudgal, S. (1988a) *Environmental Impact Assessment in India: an Overview*, Asian Development Bank, Manila.

Maudgal, S. (1988b) *Environmental Impact of a Thermal Power Station in Bombay, India*, Asian Development Bank, Manila.

McHarg, I. L. (1969) *Design with Nature*, Natural History Press, New York.

Ministry for Population and Environment (1991) *Documents Relating to the Environmental Impact Analysis Process in Indonesia*, 2nd edn, Jakarta.

Ministry of the Environment (1991) *Environmental Impact Assessment in Denmark*, ME, Copenhagen.

National Environmental Protection Council (1977) *The Environmental Impact Statement System*, NEPC, Quezon City, Philippines.

Netherlands (1992) *Prediction in EIA*, being 10 volumes on the prediction of effects in: air; surface water; soil and ground water; flora, fauna and ecosystems; landscape; the acoustic environment; radiation; risks; and health; plus an introductory volume, Government Printing Office, Amsterdam.

Office of the Commissioners of Inquiry (1988) *Commissions of Inquiry for Environment and Planning: How they Work*, OCI, Sydney.

Office of the National Environment Board (1988) *Environmental Impact Assessment in Thailand*, ONEB, Bangkok.

Overseas Development Administration (1992) *Manual of Environmental Appraisal*, ODA, London.

Prieur, M. and Lambrechts, Mrs C. (1980) *Model Outline Environmental Impact Statement from the Standpoint of Integrated Management or Planning of the Natural Environment*, Nature and Environment Series, No. 17, Council of Europe, Strasbourg.

Richardson, M. (ed.) (1995) *Environmental Toxicology Assessment*, Taylor and Francis, Basingstoke.

Salim, E. (1988) *Social Impact Assessment: the Indonesian Experience*, presentation to the seventh meeting of the International Association of Impact Assessment, Griffith University, Queensland, 5-9 July.

Selected Bibliography

Street, L. (1990) 'The Court system and alternative dispute resolution procedures', *Australian Dispute Resolution Journal*, 1 (1): 5.

Tarrant, J., Barbier, E., Greenburg, R. J., Higgins, M. L. et al. (1987) *Natural Resources and Environmental Management in Indonesia: an Overview*, United States Agency for International Development, Jakarta.

United Nations Economic Commission for Europe (1987-1991) Environment series: (1) *Application of Environmental Impact Assessment: Highways and Dams* (1987), (2) *National Strategies for Protection of Flora, Fauna and their Habitats* (1988), (3) *Post-project Analysis in Environmental Impact Assessment* (1990), (4) *Policies and Systems of Environmental Impact Assessment* (1991), (5) *Application of Environmental Impact Assessment Principles to Policies, Plans and Programs* (1992) Economic Commission for Europe, Geneva; United Nations, New York.

United Nations Economic Commission for Europe (1990) *Post-project Analyses in Environmental Impact Assessment*, environmental series 3, UN, New York.

United Nations Economic Commission for Europe (1992) *Application of Environmental Impact Assessment Principles to Policies, Plans and Programs*, environmental series 5, UN, New York.

United Nations Environment Program (1987) *Goals and Principles of Environmental Impact Assessment*, UNEP, Nairobi.

United Nations Environment Program (1990) *An Approach to Environmental Impact Assessment for Projects Affecting the Coastal and Marine Environment*, UNEP regional sea report No. 122, Nairobi.

Ward, D. V. (1978) *Biological Environmental Impact Studies: Theory and Methods*, Academic Press, New York.

Westman, W. E. (1985) *Ecology Impact Assessment and Environmental Planning*, John Wiley, London.

World Health Organization (1987) *Health and Safety Components of Environmental Impact Assessment*, WHO, Regional Office for Europe, Copenhagen.

World Health Organization (1989) *Environmental Impact Assessment: an Assessment of Methodological and Substantive Issues Affecting Human Health Considerations*, WHO report 41, University of London, London.

ENVIRONMENTAL POLICY AND MANAGEMENT

Asian Development Bank (1986) *Manual of Environmental Guidelines for Development Projects*, series of 12 manuals, ADB, Manila.

Asian Development Bank (1987) *Environmental Guidelines for Selected Agricultural and Natural Resources Development Projects*, ADB, Manila.

Asian Development Bank (1990) *Environmental Guidelines for Selected Infrastructure Projects*, ADB, Manila.

Asian Development Bank (1990b) *Environmental Guidelines for Selected Industrial and Power Development Projects*, ADB, Manila.

Bass, R. E. and Herson, A. I. (1992) *Successful California Environmental Quality Act Compliance: a Step-by-step Approach*, Solano Press, Point Area, California.

Bolton, K. F. and Curtis, F. A. (1990) 'An environmental procedure for siting solid waste disposal sites', *Environmental Impact Assessment Review*, **10**: 285-296.

Bowonder, B. and Arvind, S. S. (1989) 'Environmental regulations and litigation in India', *Project Appraisal* **4** (4): 182-196.

Brandt Commission Report (1980) *North-South: a Program for Survival*, The report of the Independent Commission on International Development Issues, Pan Books, London.

Selected Bibliography

British Standards Institute (1992) *BS7750 British Standard for Environmental Management*, BSI, London.
Buckley, R. (1988) 'Critical problems in environmental planning and management', *Environmental Planning and Law Journal (Australia)*, September, pp. 206-225.
Buckley, R. (1989) *Precision in Environmental Impact Predictions: First National Environmental Audit, Australia*, Centre for resource and environmental studies, Australian National University, Canberra, Australia.
Commission of the European Communities (1986) *EEC Fourth Environmental Action Program 1987-1992*, CEC, Brussels.
Committee on Air Pollution (1954) *Report*, Cmnd. 9322, HMSO, London.
Council on Environmental Quality (1986) *Regulations for Implementing the Procedural Provisions of the National Environmental Policy Act*, CEQ, Washington DC.
Council on Environmental Quality (1987) *Memorandum to Agencies: Forty Most Asked Questions Concerning CEQ's National Environmental Policy Act Regulations*, CEQ, Washington DC.
Council on Environmental Quality (1995) *Environmental Quality: Twenty-fifth Annual Report*, executive office of the President, CEQ, Washington DC.
Department of the Environment (1990) *This Common Inheritance: Britain's Environmental Strategy*, HMSO, London.
Department of the Environment (1991) *Policy Appraisal and the Environment*, HMSO, London.
Environment Canada (1983) *Environmental Planning for Large-scale Development Projects: Recommendations and Actions for Implementation*, proceedings of an international workshop, October 2-5, EC, Vancouver.
Environmental Resources Ltd (1985a) *Environmental Health Impact Assessment of Urban Development Projects: Guidelines and Recommendations*, WHO, Regional Office for Europe, Copenhagen.
Environmental Resources Ltd (1985b) *Assessment of Irrigated Agricultural Development Projects: Guidelines and Recommendations*, WHO, Regional Office for Europe, Copenhagen.
European Union (1982) *The Major Hazards of Certain Industrial Activities*, Directive 82/501/EEC bd 24, June, Brussels.
Gilpin, A. (1986) *Environmental Planning: a Condensed Encyclopedia*, Noyes Publications, New Jersey, p. 103.
Henderson, J. E. (1982) *Handbook of Environmental Quality Measurement and Assessment: Methods and Techniques*, Instruction Report E-82-2, US Army Engineer Waterways Experiment Station, CE, Vicksburg, Mississippi.
Interim Committee for Coordination of Investigations of the Lower Mekong Basin (1982) *Nampong Environmental Management Research Project: Final Report for Phase 3*, Mekong River Commission, Bangkok.
Jänicke, M., Mönch, H., Ranneberg, T., and Simonus, U. E. (1987) *Improving Environmental Quality through Structural Change: a Survey of Thirty-one Countries*, Research Unit Environmental Policy, International Institute for Environment and Society, Berlin.
Kormondy, Edward J. (1989) *International Handbook of Pollution Control*, Greenwood Press, Westport, Connecticut.
Leopold, L. B., Clarke, F. E., Hornshaw, B. R., and Balsley, J. R. (1971) *A Procedure for Evaluating Environmental Impact*, US Geological Survey, circular 645, Washington DC.
Lohani, B. N. (1988) *Environmental Assessment and Management in the Bank's Developing Member Countries*, Asian Development Bank, Manila.
Ministry for Planning and Environment (1990) *Planning and Environment Act 1987: Social, Economic and Environmental Effects*, MPE, Victoria, Australia.
Ministry of the Environment (1990) *Swedish Environmental Legislation*, MOE, Stockholm.

Ministry for the Environment (1990) *Report on the Regional Conference at Ministerial Level being the Follow-up to the Report of the World Commission on Environment and Development (the Brundtland Report) in the ECE Region*. ME, Bergen.
Ministry of the Environment (1992) *Environmental Management in Singapore*, MOE, Singapore.
Republic of Indonesia (1978) *Guidelines for State Environmental Policy: Decree 4 of the General Session of the Consultative Peoples' Assembly*, Jakarta.
Sunkel, O., Gligo, N., Koolen, R., Ballesteros, R. B. *et al.* (1990) *The Environmental Dimension in Development Planning*, HMSO, London.
Tarrant, J., Barbier, E., Greenburg, R. L., Higgins, M. L., *et al.* (1987) *Natural Resources and Environmental Management in Indonesia*, US Agency for International Development, Jakarta.
Turnbull, R. G. H. (ed.) (1992) *Environmental and Health Impact Assessment of Development Projects: a Handbook for Practitioners*, Elsevier Applied Science, London.
United Nations Environment Program (1982) *Guidelines on Risk Management and Accident Prevention in the Chemical Industry*, UNEP, Paris.
United Nations Environment Program (1987) *Environmental Guidelines for Settlements Planning and Management*, UN Centre for Human Settlements (Habitat), Nairobi, vols **1**, **2**, and **3**.
Williams, D. C. (1988) *National Environmental Policy Act Handbook*, Bureau of Land Management, US Department of the Interior, Washington DC.

NATIONAL AND INTERNATIONAL PROGRESS

Australian Trade Commission (1991) *Protection and Conservation of the Environment in Singapore*, Austrade.
Austrian Federal Government (1992) *Austria: National Report to the United Nations Conference on Environment and Development 1992*, Austrian Federal Ministry of Environment, Youth and Family, Vienna.
Canada (1991) *Canada's National Report to the United Nations Conference on Environment and Development 1992*, Ministry of Supply and Services, Ottawa.
Chien, E. (1990) *Working towards Environmental Quality in the 21st Century*, Environmental Protection Administration, Taipei.
Commission of the European Communities (1992) *Report from the Commission of the European Communities to the United Nations Conference on Environment and Development 1992*, Brussels.
Department of Environment (1989) *Environmental Quality Report*, DoE, Kuala Lumpur.
Department of the Environment (1992) *Ireland: National Report to the United Nations Conference on Environment and Development*, DoE, Dublin.
Economic and Social Commission for Asia and the Pacific (1992) *The Asian and Pacific Input to the United Nations Conference on Environment and Development, Brazil, 1992*, ESCAP, New York.
Environmental Protection Administration (1988) *Environmental Protection in the Republic of China*, EPA, Taipei.
Federal Republic of Germany (1991) *National Report to the United Nations Conference on Environment and Development 1992*, FRG, Bonn.
Ministry for Foreign Affairs (1991) *Finland National Report to the United Nations Conference on Environment and Development 1992*, MFA, Helsinki.
Ministry for the Environment/Ministry of External Relations and Trade (1991) *New Zealand's National Report to the United Nations Conference on Environment and Development 1992*, ME, Wellington.
Ministry for the Environment (1992) *Norway's National Report to the United Nations Conference on Environment and Development, 1992*, ME, Oslo.

Selected Bibliography

Ministry of Foreign Affairs (1992) *Malaysia's Report to the United Nations Conference on Environment and Development 1992*, Ministry of Foreign Affairs, Kuala Lumpur.

Ministry of the Environment (1991b) *The State of the Environment in Denmark*, Danish State Information Office, Copenhagen.

Ministry of the Environment (1991a) *National Report to the United Nations Conference on Environment and Development 1992*, Danish State Information Office, Copenhagen.

Ministry of the Environment (1991) *National Report of the Republic of Korea to the United Nations Conference on Environment and Development 1992*, Ministry of the Environment, Seoul.

Ministry of the Environment (1991) *Sweden: National Report to the United Nations Conference on Environment and Development 1992*, MOE, Stockholm.

Ministry of the Environment, Physical Planning and Public Works, et al. (1991) *National Report of Greece to the United Nations Conference on Environment and Development*, MOE, Athens.

Ministry of the Interior (1972) *Report of the Federal Republic of Germany on the Human Environment*, Ministry of the Interior, Bonn.

Office of Environment, Bureau of Planning and Coordination, Republic of Korea (1983) *Han River Basin Environment Master Plan: Final Report*, Office of Environment, Seoul.

Singapore Interministry Committee (1991) *Singapore's National Report to the United Nations Conference on Environment and Development 1992*, SIMC, Singapore.

Thailand (1992) *National Report to the United Nations Conference on Environment and Development 1992*, Bangkok.

United States of America National Report (1992) *National Report to the United Nations Conference on Environment and Development*, Council on Environmental Quality, Washington DC.

Walton, T. (1989) *The World Bank Operational Manual: Operational Directive 4.00*, WB, Washington DC.

World Bank (1985) *Manual of Industrial Hazard Assessment Techniques*, WB Office of Environmental and Scientific Affairs, Washington DC.

World Bank (1991) *The World Bank and the Environment: a Progress Report Fiscal 1991*, WB, Washington DC.

World Bank (1995) *World Development Report*, WB, Washington DC.

SUSTAINABLE DEVELOPMENT

Brown, L. R. et al. (1995) *State of the World: a Worldwatch Institute Report on Progress towards a Sustainable Society*, W. W. Norton & Company, New York/London.

Commission of the European Communities (1992) *Towards Sustainability: a European Community Program of Policy and Action in Relation to the Environment and Sustainable Development*, COM (92) 23 final: vol. **2**, CEC, Brussels.

Department of Environment and Natural Resources (1990) *Philippine Strategy for Sustainable Development: A Conceptual Framework*, DENR, Quezon City, Philippines.

Dovers, S. (1994) *Sustainable Energy Systems*, Cambridge University Press, Cambridge.

Economic and Social Commission for Asia and the Pacific (1990) *Ministerial Declaration on Environmentally Sound and Sustainable Development in Asia and the Pacific*, ESCAP, New York.

Economic and Social Commission for Asia and the Pacific (1991) *Regional Strategy on Environmentally Sound and Sustainable Development in Asia and the Pacific*, ESCAP, New York.

Ehrlich, P. R. and Ehrlich, A. H. (1987) *The Population Bomb*, Ballantine, New York.

Htun, N. (1988) *EIA and Sustainable Development*, presentation to the seventh annual meeting of the International Association of Impact Assessment, Griffith University, Queensland, Australia, 5-9 July.

Selected Bibliography

International Institute of Environment and Development/World Resources Institute (1995) *World Resources 1995*, Basic Books, New York.

Jacobs, P. and Sadler, B. (eds) (1990) *Sustainable Development and Environmental Assessment: Perspectives on Planning for a Common Future*, Canadian Environmental Assessment Research Council, Quebec.

Organization for Economic Cooperation and Development (1992) *Environment and Development: the OECD Approach*, a report prepared by OECD for the United Nations Conference on Environment and Development, Rio de Janeiro, Brazil 3-14 June, OECD, Paris.

Salim, E. (1990) 'Alleviating poverty through sustainable development, Jakarta, Republic of Indonesia', paper presented in Perth, Australia.

United Nations Conference on Environment and Development (1992) *Agenda 21*, as adopted by the plenary session in Rio de Janeiro on 14 June 1992, UN New York.

World Conservation Union, UN Environment Program, World Wide Fund for Nature (1991), *Caring for the Earth: A Strategy for Sustainable Living*, WCU, Gland.